Carbon Management for Promoting Local Livelihood in the Hindu Kush Himalayan (HKH) Region

Zhanhuan Shang • A. Allan Degen
Muhammad Khalid Rafiq • Victor R. Squires
Editors

Carbon Management for Promoting Local Livelihood in the Hindu Kush Himalayan (HKH) Region

Springer

Editors
Zhanhuan Shang
School of Life Sciences
Lanzhou University
Lanzhou, Gansu, China

A. Allan Degen
Ben-Gurion University of the Negev
Beer Sheva, Israel

Muhammad Khalid Rafiq
Pakistan Agricultural Research Council
Islamabad, Pakistan

Victor R. Squires
University of Adelaide
Adelaide, SA, Australia

ISBN 978-3-030-20590-4   ISBN 978-3-030-20591-1 (eBook)
https://doi.org/10.1007/978-3-030-20591-1

© Springer Nature Switzerland AG 2020
This work is subject to copyright. All rights are reserved by the Publisher, whether the whole or part of the material is concerned, specifically the rights of translation, reprinting, reuse of illustrations, recitation, broadcasting, reproduction on microfilms or in any other physical way, and transmission or information storage and retrieval, electronic adaptation, computer software, or by similar or dissimilar methodology now known or hereafter developed.
The use of general descriptive names, registered names, trademarks, service marks, etc. in this publication does not imply, even in the absence of a specific statement, that such names are exempt from the relevant protective laws and regulations and therefore free for general use.
The publisher, the authors, and the editors are safe to assume that the advice and information in this book are believed to be true and accurate at the date of publication. Neither the publisher nor the authors or the editors give a warranty, express or implied, with respect to the material contained herein or for any errors or omissions that may have been made. The publisher remains neutral with regard to jurisdictional claims in published maps and institutional affiliations.

This Springer imprint is published by the registered company Springer Nature Switzerland AG
The registered company address is: Gewerbestrasse 11, 6330 Cham, Switzerland

Preface

In the context of globalization, 'carbon' has emerged as a 'global currency' and is taking on increasingly important functions in global politics, ecological projects and economy. One of the most important achievements of the 2018 COP24 conference in Katowice, Poland, was the commitment of all contracting states to 'nationally report greenhouse gas emissions and emission reduction efforts to establish a unified and transparent set of guidelines'. This means that a set of standards for carbon emission reduction will be adopted worldwide that could link actions by global organizations, countries and regions. It also means that carbon will be the most effective tool to measure regional differences in the world.

In view of this, carbon management has become crucial for regional and local livelihood development. This book applies these concepts to the HKH region and proposes a carbon management assessment framework at national and regional levels to provide options for further development strategies.

The HKH region, known as the world's third pole, includes a number of countries and is one of the world's largest mountain regions. The area is characterized by extreme harsh environmental conditions and by inhabitants who are among the poorest in the world. For there to be a progress in the region, livelihood development must be linked closely to the climate governance of the region.

In many areas of the HKH, due to travel constraints, the exchange within and outside the system is limited. Establishing basic communication with the outside world may depend largely on investments from the local governments, even though the governments lack finances in much of the HKH. Implementation of the suggestions proposed at the COP24 conference that 'developed countries must provide assistance to developing countries in climate governance actions' via water and carbon management, and, therefore, livelihood development is closely related to the natural resources of the area. The first task is to solve the livelihood development of local residents in the HKH, such as by the application of the 'community forest' and 'community pasture' systems' models. The 'global carbon trading market' could assist in developing these poorer countries as substantial global carbon assets could make it possible to strengthen the HKH region and make it more accessible to the rest of the world.

In the HKH region, zoning and classification is a problem as it involves many countries and regions, different ecosystems and geographic units and a number of ethnic groups and cultures among the local residents. For example, although it has been argued that the livelihood approach of directly using local materials by the residents is detrimental, the carbon and ecological balances and the forest ecosystem have not been damaged in the long term by forest management. In addition to the spatial scale and management level, the time scale is also vital in establishing accurate and operational evaluation schemes. From current results, the implementation of the REDD+ project has achieved a long-term improvement in the HKH. However, it is crucial for the local governments and residents to be more aware of the potential of the program and improve the capacity building for project implementation. For this to occur successfully, there must be an active cooperation between the local residents and the government.

In local policies and livelihoods, more and more activities are directly involving carbon management issues, especially in the restoration, improvement and sustainable management of ecological functions of natural ecosystems. For example, in alpine grassland pastoral areas, livelihoods have always been a key consideration in ecological function enhancing projects, such as sustainable grazing, artificial grasslands and integrated pasture utilization. Optimal utilization of water resources is related to the balances of ecological functions and production, especially in cultivated lands where effective water use is essential for sustained carbon sequestration and livelihood acquisition. As an important linkage of the ecosystem to maintain carbon output and carbon cycle, animal and their products, as well as their management and marketing, affect the livelihood of the stakeholders and the sustainable exchange of material and technical sources from external systems.

Specific methods for improving ecosystem carbon management and ecological functions are emphasized in this book. Forests, grasslands, wetlands, deserts and farmlands are important ecosystems in the HKH region. In this regard, this book discusses the overall carbon pool assessment, the management of each ecosystem for enhancing the carbon sink and the strategies to improve livelihood activities. The ecosystem service functions are important in the national and regional policy decisions and have gained long-term attention in the HKH region. It will be important in the future to develop strategies to effectively combine carbon management with livelihood development. For this to occur, it is necessary to clearly understand the complexity of the geographic, social, economic, ecological and ethnic aspects of the HKH region. This is the dilemma that must be faced in this region and presents a barrier to the sustainable development in the HKH region.

In conclusion, through this book, we hope to better understand the livelihood needs of local residents and to attract more attention to the development of the HKH region by effective carbon management, policies and improved technologies and to work more closely with global climate governance initiatives. We hope that this book succeeds in providing a base for promoting HKH as a model for sustainable development in an extremely harsh environment.

Lanzhou, Gansu, China	Zhanhuan Shang
Beer Sheva, Israel	A. Allan Degen
Islamabad, Pakistan	Muhammad Khalid Rafiq
Adelaide, SA, Australia	Victor R. Squires

Acknowledgements

The publication of this book was made possible as a result of the efforts and involvement of many people. We thank the authors for submitting chapters in their fields of expertise. Dr. Sherestha Saini, Ms. Kiruthika Kumar, Ms. Margaret Deignan and Ms. Agnes Felema, the editorial team from Springer, provided excellent advice on the presentation and organization of the book. Prof. Shikui Dong (Beijing Normal University) made many helpful suggestions on the subject matter and content of the book. Comments and suggestions from many reviewers improved the chapters greatly. We are very grateful to them all for their help in preparing this book.

Financial support for many of the projects, field research, meetings and preparation of the book was provided by the Belt and Road Special Project of Lanzhou University (2018ldbryb023), the Second Tibetan Plateau Scientific Expedition and Research (STEP) program (Grant No. 2019QZKK0302), the National Key Research and Development Project (2016YFC0501906), the Natural Science Foundation of China (41671508; 31870433) and the Qinghai Innovation Platform Construction Project (2017-ZJ-Y20).

Contents

Part I Assessment and Management of Carbon Dynamics

1 **Managing Carbon Cycle Linkage to Livelihood in HKH Region** ... 3
   Zhanhuan Shang, A. Allan Degen, Devendra Gauchan,
   Bhaskar Singh Karky, and Victor R. Squires

2 **Climate Change Mitigation and Pastoral Livelihood
   in the Hindu Kush Himalaya Region: Research Focuses,
   Opportunities and Challenges** 25
   Wenyin Wang, Devendra Gauchan, A. Allan Degen,
   and Zhanhuan Shang

3 **Tracking of Vegetation Carbon Dynamics
   from 2001 to 2016 by MODIS GPP in HKH Region** 45
   Zhenhua Chao, Mingliang Che, Zhanhuan Shang,
   and A. Allan Degen

4 **Livelihood and Carbon Management by Indigenous
   People in Southern Himalayas** 63
   Dil Kumar Limbu, Basanta Kumar Rai, and Krishna Kumar Rai

**Part II Livestock, Grasslands, Wetlands, and Environment
on Carbon Dynamics**

5 **Effects of Different Grassland Management Patterns
   on Soil Properties on the Qinghai-Tibetan Plateau** 91
   Jianjun Cao, Xueyun Xu, Shurong Yang, Mengtian Li,
   and Yifan Gong

6 **Carbon Management of the Livestock Industry
   in the HKH Region** 109
   Yu Li, A. Allan Degen, and Zhanhuan Shang

7   Wetlands as a Carbon Sink: Insight into the Himalayan Region.... 125
    Awais Iqbal and Zhanhuan Shang

8   Milk and Dung Production by Yaks (*Poephagus grunniens*):
    Important Products for the Livelihood of the Herders
    and for Carbon Recycling on the Qinghai-Tibetan Plateau 145
    A. Allan Degen, Shaher El-Meccawi, and Michael Kam

9   The Effect of Ecology, Production and Livelihood on
    the Alpine Grassland Ecosystem of the Tibetan Plateau........... 163
    Xingyuan Liu

Part III Impacts of Restoration, REDD
          and Biochar on Carbon Balance

10 Prospects of Biochar for Carbon Sequestration
    and Livelihood Improvement in the Tibetan Grasslands 185
    Muhammad Khalid Rafiq, Jamila Sharif, Zhanhuan Shang,
    Yanfu Bai, Fei Li, Ruijun Long, and Ondřej Mašek

11 Optimizing the Alpine Grazing System to Improve
    Carbon Management and Livelihood for Tibetan Herders 197
    Xiaoxia Yang, Quanmin Dong, and Chunping Zhang

12 Promoting Artificial Grasslands to Improve Carbon
    Sequestration and Livelihood of Herders...................... 211
    Huakun Zhou, Dangjun Wang, Meiling Guo, Buqing Yao,
    and Zhanhuan Shang

13 Prospects for REDD+ Financing in Promoting
    Forest Sustainable Management in HKH...................... 229
    Shambhavi Basnet, Jagriti Chand, Shuvani Thapa,
    and Bhaskar Singh Karky

Part IV Policies and Strategies for Livelihood Improvements
         and Carbon Management

14 Designing Water Resource Use for Poverty Reduction
    in the HKH Region: Institutional and Policy Perspectives......... 245
    Madan Koirala, Udhab Raj Khadka, Sudeep Thakuri,
    and Rashila Deshar

15 Indigenous Practice in Agro-Pastoralism and Carbon
    Management from a Gender Perspective: A Case from Nepal...... 267
    Rashila Deshar and Madan Koirala

16 Adaptation by Herders on the Qinghai-Tibetan Plateau
    in Response to Climate Change and Policy Reforms:
    The Implications for Carbon Sequestration and Livelihoods....... 281
    Haiying Feng and Melissa Nursey-Bray

17 **Developing Linkages for Carbon Sequestration, Livelihoods and Ecosystem Service Provision in Mountain Landscapes: Challenges and Opportunities in the Himalaya Hindu Kush (HKH) Region** 299
Victor R. Squires

18 **Experience for Future Good Practice and Policy of Combined Carbon Management and Livelihood in HKH Region** 315
Zhanhuan Shang, A. Allan Degen, Devendra Gauchan, and Victor R. Squires

Index 337

Contributors

Yanfu Bai School of Life Sciences, State Key Laboratory of Grassland Agro-Ecosystems, Lanzhou University, Lanzhou, China

Shambhavi Basnet International Centre for Integrated Mountain Development (ICIMOD), Lalitpur, Kathmandu, Nepal

Jianjun Cao College of Geography and Environmental Science, Northwest Normal University, Lanzhou, China

Jagriti Chand International Centre for Integrated Mountain Development (ICIMOD), Lalitpur, Kathmandu, Nepal

Zhenhua Chao School of Geographic Science, Nantong University, Nantong, China

Mingliang Che School of Geographic Science, Nantong University, Nantong, China

A. Allan Degen Desert Animal Adaptations and Husbandry, Wyler Department of Dryland Agriculture, Blaustein Institutes for Desert Research, Ben-Gurion University of Negev, Beer Sheva, Israel

Rashila Deshar Central Department of Environmental Science, Tribhuvan University, Kirtipur, Kathmandu, Nepal

Quanmin Dong State Key Laboratory of Plateau Ecology and Agriculture, Qinghai University, Xining, China

Qinghai Academy of Animal and Veterinary Sciences, Key Laboratory of Alpine Grassland Ecosystem in the Three River Head Waters Region Jointly Funded by Qinghai Province and Ministry of Education, Xining, China

Shaher El-Meccawi Achva Academic College, Arugot, Israel

Haiying Feng Research Institute for Qinzhou Development, Beibu Gulf University, Qinzhou, China

Devendra Gauchan Biodiversity International, Lalitpur, Nepal

Yifan Gong College of Geography and Environmental Science, Northwest Normal University, Lanzhou, China

Meiling Guo Key Laboratory of Restoration Ecology of Cold Area in Qinghai Province, Northwest Institute of Plateau Biology, Chinese Academy of Sciences, Xining, China

Awais Iqbal School of Life Sciences, State Key Laboratory of Grassland Agro-Ecosystems, Lanzhou University, Lanzhou, Gansu, China

Michael Kam Desert Animal Adaptations and Husbandry, Wyler Department of Dryland Agriculture, Blaustein Institutes for Desert Research, Ben-Gurion University of the Negev, Beer Sheva, Israel

Bhaskar Singh Karky International Centre for Integrated Mountain Development (ICIMOD), Lalitpur, Kathmandu, Nepal

Udhab Raj Khadka Central Department of Environmental Science, Tribhuvan University, Kirtipur, Kathmandu, Nepal

Madan Koirala Central Department of Environmental Science, Tribhuvan University, Kirtipur, Kathmandu, Nepal

Fei Li School of Life Sciences, State Key Laboratory of Grassland Agro-Ecosystems, Lanzhou University, Lanzhou, China

Dil Kumar Limbu Central Campus of Technology, Tribhuvan University, Dharan, Nepal

Mengtian Li College of Geography and Environmental Science, Northwest Normal University, Lanzhou, China

Xingyuan Liu State Key Laboratory of Grassland Agro-Ecosystems, College of Pastoral Agricultural Science and Technology, Lanzhou University, Lanzhou, Gansu, China

Yu Li School of Life Sciences, State Key Laboratory of Grassland Agro-Ecosystems, Lanzhou University, Lanzhou, China

Ruijun Long School of Life Sciences, State Key Laboratory of Grassland Agro-Ecosystems, Lanzhou University, Lanzhou, China

Ondřej Mašek UK Biochar Research Centre, School of GeoSciences, University of Edinburgh, Edinburgh, UK

Melissa Nursey-Bray Department of Geography, Environment and Population, University of Adelaide, Adelaide, SA, Australia

Muhammad Khalid Rafiq UK Biochar Research Centre, School of GeoSciences, University of Edinburgh, Edinburgh, UK

Rangeland Research Institute, National Agricultural Research Center, Islamabad, Pakistan

Basanta Kumar Rai Central Campus of Technology, Tribhuvan University, Dharan, Nepal

Krishna Kumar Rai Mahendra Multiple Campus, Tribhuvan University, Dharan, Nepal

Zhanhuan Shang School of Life Sciences, State Key Laboratory of Grassland Agro-Ecosystems, Lanzhou University, Lanzhou, China

Jamila Sharif School of Sociology and Philosophy, Lanzhou University, Lanzhou, China

Victor R. Squires Institute of Desertification Studies, Beijing, China

University of Adelaide, Adelaide, SA, Australia

Sudeep Thakuri Central Department of Environmental Science, Tribhuvan University, Kirtipur, Kathmandu, Nepal

Shuvani Thapa International Centre for Integrated Mountain Development (ICIMOD), Lalitpur, Kathmandu, Nepal

Dangjun Wang Key Laboratory of Restoration Ecology of Cold Area in Qinghai Province, Northwest Institute of Plateau Biology, Chinese Academy of Sciences, Xining, China

Wenyin Wang School of Life Sciences, State Key Laboratory of Grassland Agro-Ecosystems, Lanzhou University, Lanzhou, China

Xueyun Xu College of Geography and Environmental Science, Northwest Normal University, Lanzhou, China

Shurong Yang College of Geography and Environmental Science, Northwest Normal University, Lanzhou, China

Xiaoxia Yang State Key Laboratory of Plateau Ecology and Agriculture, Qinghai University, Xining, China

Qinghai Academy of Animal and Veterinary Sciences, Key Laboratory of Alpine Grassland Ecosystem in the Three River Head Waters Region Jointly Funded by Qinghai Province and Ministry of Education, Xining, China

Buqing Yao Key Laboratory of Restoration Ecology of Cold Area in Qinghai Province, Northwest Institute of Plateau Biology, Chinese Academy of Sciences, Xining, China

Chunping Zhang State Key Laboratory of Plateau Ecology and Agriculture, Qinghai University, Xining, China

Qinghai Academy of Animal and Veterinary Sciences, Key Laboratory of Alpine Grassland Ecosystem in the Three River Head Waters Region Jointly Funded by Qinghai Province and Ministry of Education, Xining, China

Huakun Zhou Key Laboratory of Restoration Ecology of Cold Area in Qinghai Province, Northwest Institute of Plateau Biology, Chinese Academy of Sciences, Xining, China

About the Editors

Zhanhuan Shang is a full professor at School of Life Sciences, Lanzhou University. He graduated from Ningxia University (2003) and Gansu Agricultural University (2006) for his MSc and PhD in Rangeland Science and Ecology. From 2006, he started his research career in Lanzhou University. From 2008 to 2012, he did an on-the-job post-doc at the Institute of Tibetan Plateau Research, Chinese Academy of Sciences. From 2011 to 2012, he was visiting scholar and joined in the Terrestrial Ecosystem Research Network (TERN) at the University of Adelaide, Australia. During the past 15 years, he conducted and participated in more than 20 projects funded by the NSFC, MOST, MOE, CAS, FAO, GEF/WB, EC, IAEA, ICIMOD, DFG, etc. His research interests include ecological restoration, grassland management, rural sustainable development, carbon management and biodiversity on the Tibetan plateau. He has more than 150 publications in international and Chinese journals and books.

A. Allan Degen is an emeritus professor at Ben-Gurion University, Be'er Sheva, Israel, since 2014. He received his MSc (1972) and PhD (1976) in Animal Science and in Zoology at the University of McGill and University of Tel Aviv, respectively, and did a post-doctorate programme (1978–1980) at the Department of Animal Science, University of Alberta, where he was granted a Natural Sciences and Engineering Research Council of Canada (NSERC) fellowship. He joined Ben-Gurion University of the Negev in 1980 where he was head of the Desert Animal Adaptations and Husbandry Unit from 1985 to 2015 and chairman of the Wyler Department of Dryland Agriculture from 2001 to 2004 and from 2013 to 2015. In 1986, he received the Chaim Sheba Award for outstanding contribution by a young scientist in Israel, and from 2004 to 2015, he was the incumbent of the Bennie Slome Chair for Applied Research in Desert Animal Production. He served as the coeditor in chief of the *Israel Journal of Zoology* from 1999 to 2007 and has supervised numerous students for their MSc and PhD degrees. He does research on livestock production under harsh conditions and has co-operated in projects in Kazakhstan, Uzbekistan, Nepal, Ethiopia, Afghanistan, Kenya, South Africa and China. In Israel, he is doing research on rural and urban agriculture among the Bedouin. He has authored or coauthored 16 books or chapters and over 220 publications in peer-reviewed journals. His research has been presented at numerous conferences worldwide.

Muhammad Khalid Rafiq is working as Commonwealth Rutherford fellow at UK Biochar Research Center, School of Geosciences, the University of Edinburgh (UoE), United Kingdom. He has been awarded a highly competitive and globally prestigious Rutherford fellowship for his research at the UoE. He also holds a position of Senior Scientific officer at Rangeland Research Institute (RRI), National Agricultural Research Centre (NARC) Islamabad, Pakistan. He obtained his PhD from the Lanzhou University, China, in Grassland Science. He has working experience with the national and international organizations, including the Pakistan Agricultural Research Council (PARC), Islamabad, the International Union for Conservation of Nature (IUCN) and the World Wide Fund for Nature (WWF). His research interests include grassland ecology, biochar production and applications for carbon sequestration. He has also attended conferences at the national and international level in the United Kingdom, China, Nepal, Pakistan, etc. He has published many research articles in the journals of international repute, including *Journal of Hazardous Materials*, *Science of the Total Environment*, etc., and book chapters.

About the Editors

Victor R. Squires is a guest professor in the Institute of Desertification Studies, Beijing. He is an Australian who when still a young man studied animal husbandry and rangeland ecology. He has a PhD in Rangeland Science from Utah State University, USA. He is a former foundation dean of the Faculty of Natural Resource Management at the University of Adelaide, where he worked for 15 years in Australia after a 22-year career in Australia's CSIRO. He is author/editor of 15 books, including 2012 *Rangeland Stewardship in Central Asia: Balancing Livelihoods, Biodiversity Conservation and Land Protection*, Springer, Dordrecht, p. 458; 2013 *Combating Desertification in Africa, Asia and the Middle East: Proven Practices*, Springer, Dordrecht, p. 476; 2014 *River Basin Management in the Twenty-First Century: Understanding People and Place*, CRC Press, USA, p. 476; 2015 *Rangeland Ecology, Management and Conservation Benefits*, Nova Science Publishers, N.Y. 353 p.; and 2018 *Desertification: Past, Current and Future Trends*, Nova Science Publishers, N.Y., and *Grasslands of the World: Diversity, Management and Conservation*, CRC Press, USA, p. 411; and numerous research papers and invited book chapters. Since his retirement from the University of Adelaide, he was a visiting fellow in the East-West Center, Hawaii, and an adjunct professor in the University of Arizona, Tucson, and at the Gansu Agricultural University, Lanzhou, China. He has been a consultant to the World Bank, Asian Development and various UN agencies in Africa, China, Central Asia and the Middle East.

He was awarded the 2008 *International Award and Gold Medal for International Science and Technology Cooperation* and, in 2011, the *Friendship Award* by the Government of China. The Gold Medal is the highest award for foreigners. In 2015, he was honoured by the Society for Range Management (USA) with an *Outstanding Achievement Award*. In December 2018, he was admitted to the degree of DSc (*honoris causa*) in the Gansu Agricultural University, Lanzhou – the first person to be honored in this way at that university.

Part I
Assessment and Management of Carbon Dynamics

Chapter 1
Managing Carbon Cycle Linkage to Livelihood in HKH Region

Zhanhuan Shang, A. Allan Degen, Devendra Gauchan, Bhaskar Singh Karky, and Victor R. Squires

Abstract The ratchet effects of global climate change make all countries and regions vulnerable. It is believed that if countries/regions are not involved in climate change mitigation initiatives, they may be overwhelmed by 'climate flood', and their citizens may be victims of climate change. The HKH, as the biggest, poorest mountain area in the world, faces a big challenge, and efforts should be made to understand the status of the HKH and to develop a blueprint for mitigating climate change. This chapter integrates several components including: (1) the carbon management status of the HKH region and the urgent need for livelihood improvement and research and development linked with climate issues; (2) the framework for evaluating the level and mode of carbon compensation; (3) the strategy of sharing benefits from carbon management with indigenous people in the HKH region; and (4) the options for carbon management in HKH over future decades. Finally, this chapter provides a short summary of the contents and purpose of this book.

Z. Shang (✉)
School of Life Sciences, State Key Laboratory of Grassland Agro-Ecosystems, Lanzhou University, Lanzhou, China
e-mail: shangzhh@lzu.edu.cn

A. A. Degen
Desert Animal Adaptations and Husbandry, Wyler Department of Dryland Agriculture, Blaustein Institutes for Desert Research, Ben-Gurion University of Negev, Beer Sheva, Israel
e-mail: degen@bgu.ac.il

D. Gauchan
Biodiversity International, Lalitpur, Nepal

B. S. Karky
International Centre for Integrated Mountain Development (ICIMOD), Lalitpur, Kathmandu, Nepal
e-mail: Bhaskar.karky@icimod.org

V. R. Squires
Institute of Desertification Studies (formerly University of Adelaide, Australia), Beijing, China

Keywords Carbon cycle · Livelihood improvement · Ecological compensation · Indigenous people · Potential framework · HKH region

1.1 Introduction

The Paris Agreement, signed by 200 contracting parties at the Paris Global Conference in December 2015, was the first-ever global climate deal, and was an important milestone in tackling global change. Under the challenge of climate change, the motto of 'only advance, not retreat' was initiated in the Agreement. The Agreement was under the climate change framework of UN, 'Kyoto Protocol' and 'Bali roadmap' and was based on the principle of 'common but differentiated responsibilities' and equity and individual competency. Its aim was to strengthen 'The United Nations convention on climate change'. In the Agreement the relation between the control of climate change (including action, effect, obtaining sustainable development equally) and eliminating poverty was emphasized. Priority subjects included food security and the eradication of hunger, in particular, the production of more food despite the adverse impact of climate change (Paris Agreement 2016). In addition, the Paris Agreement encouraged the development of sustainable livelihoods in backward and developing areas within the common action of global climate governance.

Of all ecosystems, the mountain ecosystem, in particular, in developing countries or areas, is the most fragile and the most vulnerable to the pressure of global climate change (IPCC 2013). The major reason for this is the low level of management, infrastructure, and ease of adopting techniques in the mountain ecosystem of poor areas. The climate governance should assist these impoverished areas and encourage and facilitate their participation in the global strategy. The Hindu Kush-Himalayan region (HKH), which includes eight countries (Afghanistan, Bangladesh, Bhutan, China, India, Myanmar, Nepal, and Pakistan) lies across the 3500 km from east to west and requires such assistance (Fig. 1.1). This region supports about 20% of the world population, either directly or indirectly and is characterized by a fragile ecosystem, low livelihood security and vulnerable agricultural ecosystem (Sandhu and Sandhu 2015; Shukla et al. 2016; Pandey et al. 2017). Linking the HKH's carbon management ability to livelihood benefits and recognizing the HKH's contribution to global carbon management could prompt the HKH to participate in the global climate governance campaign initiative (Pacala and Socolow 2004; Banskota et al. 2007; Devi et al. 2012; Sharma et al. 2015).

In the history of the global industrialization process, the HKH region was always a huge carbon sink, and one of the poorest areas in the world. Poverty problems led to residents deforesting much of the area in the past 30 years, which resulted in a large increase in carbon emission in non-industrial regions (Singh et al. 1985). In addition, over-grazing, forest burning and harvesting trees for fuelwood and construction (Naudiyal and Schmerbeck 2018) are having detrimental effects on the ecosystem. The HKH should be converting the non-industrial carbon emission areas

Fig. 1.1 Photography of the Hindu Kush-Himalayan mountain (Source from ICIMOD)

to carbon sinks. To encourage such a conversion, a program of carbon compensation should be incorporated in the economy to tackle climate change in the HKH region.

1.2 The Common Problem of Carbon Pool Changes, Environment and Livelihood

1.2.1 The Basic Status of the Carbon Pool in the HKH

Accurate evaluation of the carbon budget (sequestration and emission) is the basis for carbon management, compensation and trade. Developed countries have established an accurate evaluation system, but that is not the case in the HKH where the system is still very weak (Quéré et al. 2018). Only parts of the carbon budget have been determined in areas of the HKH but a more complete, comprehensive carbon budget is required (Ward et al. 2014).

The forest ecosystem is the main vegetation type in HKH, and the forest has most of the carbon storage (Fig. 1.2). In India, forest carbon storage increased from 4327 million tons in 1995 to 4680 million tons in 2007 (Dasgupta et al. 2015). In Nepal, carbon fixation by forests in alpine area amounted to 2.4 Mt from 1990 to 2010, but in Terai, forests were a net carbon emitter of about 1.64 Mt., so that in these 20 years (1990–2010) the total forest carbon fixation was about 2.07 Mt(Shrestha et al. 2015). In the Wang Chhu watershed of western Bhutan, the total soil carbon pool to a depth of 1 m was about 27.1 Mt (Dorji et al. 2014a). In the forested area of the southern part of India Kashmir-Himalaya, carbon storage decreased in the past

Fig. 1.2 Alpine forest landscape in Nepal Himalaya region (Photography by Yu Li 2017)

Table 1.1 National estimates of carbon stored in mountain shrubland and grassland ecoregions of HHK region (Ward et al. 2014)

Country	Land area (ha)	Carbon stock range (Pg C)
Afghanistan	7,068,260	0.33–0.44
Bhutan	682,494	0.03–0.05
India	10,244,045	0.48–0.64
Myanmar	666,758	0.03–0.04
Nepal	2,946,719	0.14–0.19
Pakistan	10,945,241	0.52–0.68
China	122,400,000	7.17–9.66

20 years compared with other India-Himalayan temperate coniferous forest (Wani et al. 2014). In the Garhwal mountain region of Himalaya, the highest above-ground biomass density and carbon density was in the *Cedrus deodara* forests woodland with 464.2 Mg ha^{-1} and 208.9 Mg C ha^{-1} (TOC), which is not the situation in *Abies spectabilis* woodland although it has the highest number of tree species and greatest individual density. The lowest biomass and carbon densities is in the *Quercus semecarpifolia* forests with 283.4 Mg ha^{-1} (TBD) and 127.5 Mg C ha^{-1} (TCD) respectively, *Quercus semecarpifolia* woodland has the lowest number of tree species and lower individual density, than the woodland area with dominant tree species of *Abies spectabilis* should enhance the biodiversity for increasing its carbon sinking ability (Sharma et al. 2016). Besides forests, the brushwood and grassland are important carbon storage pools the in HKH region (Ward et al. 2014) (Table 1.1).

The topographical change is very dramatic in the HKH, and the carbon storage pattern is uneven. In the eastern Himalayas of southwestern Bhutan, from the altitude of 317 to 3300 m, the soil carbon of surface layer (0–30 cm) increased about 4.3 Mg C ha^{-1} for each 100 m of altitude (Tashi et al. 2016). Bhattacharyya et al. (2000) estimated a carbon storage in the surface soil (0–30 cm) of about 7.89 Pg in the northern mountain land of India. The soil layer of 0–150 cm has very high carbon storage of about 18.31 Pg. In the region of Garhwal Himalaya, where the forest grows from an altitude of 350 to 3100 m, its carbon storage increased from 59 to 245 Mg C ha^{-1}, and among the different forest types in this area, carbon storage of coniferous forest was more than in broad-leaf forest (Sharma et al. 2010). At an altitude 6350 m in north-eastern Himalayan region of India, the soil carbon storage of forest and grassland was about 35.2–42.1 Mg ha^{-1}, which was higher than soil carbon storage of 27.4–28.4 Mg ha^{-1} in farmland (Choudhury et al. 2016).

Afforestation activity is an important pathway to enhance carbon storage in the HKH region. In the temperate mountains of Kashmir Himalayas, the *Juglans regia* woodland (1800–2000 m) and *Betula utilis* (2800–3200 m) has an average soil carbon storage of 39.07–91.39 Mg C ha^{-1} at a depth of 0–30 cm. The soil organic carbon storage of *Juglans regia* woodland (18.55, 11.31, and 8.91 Mg C ha^{-1}, respectively in soil layers of 0–10, 10–20, 20–30 cm) is lower than that of *Betula utilis* (54.10, 21.68, and 15.60 Mg C ha^{-1}, respectively) (Dar and Somaiah 2015). The planting of *Populus deltoides* could enhance the carbon storage in Terai area of middle Himalaya, for example the planting age of *P. deltoides* from 1 to 11 years, carbon storage increased 64.4–173.9 Mg ha^{-1} (Arora et al. 2014). In the farmland area, the afforestation also proved to be of high carbon benefit. After 4 and 23 years of planting *Grewia optiva*, the major tree species for planting in the farmland, soil carbon storage was 1.99 Mg ha^{-1} and 15.27 Mg ha^{-1}, respectively. The carbon sequestration rate in plantations of *G. optiva* is about 0.63–0.81 Mg ha^{-1} year^{-1}, and at a soil depth of 0–30 cm the total carbon storage is about 25.4–33.6 Mg ha^{-1} (Verma et al. 2014).

Land utilization pattern is the major driving factor for global carbon pool change and is a major factor in HKH. Much of the area has been degraded due to human activity. Enhancing vegetation coverage is the basic factor to increase carbon storage (Debasish-Saha et al. 2014). In the foothill areas of the Himalaya, the tropical sal forest (*Shorea ia robusta*) in the terai of Nepal is common in protected reserves, and community and government forests, and have increased carbon storage significantly. Consequently, the conversion of forest land ownership could increase the carbon sink capacity (Gurung et al. 2015). In the Lesser Himalayan Foothills of Kashmir (900–2500 m), soil carbon storage of four land types (closing forest land, open woodland, disturbed woodland and farmland) has decreased with altitude, and with land cultivation, disturbance and over-grazing. Protection of vegetation on these sites is important to enhance carbon storage (Shaheen et al. 2017). In the Kullu district of Himachal Pradesh of northwest Indian Himalaya, the major land uses include agriculture, agro-horticulture, horticulture, silvi-pasture, and forest, of which the forest has the most carbon storage (404.3 Mg C ha^{-1}) (Rajput et al. 2017).

1.2.2 Carbon Emission of the HKH

The big challenge in the HKH is curbing carbon emission, which has been increasing in recent years due to activity of the residents such as over-cropping, land conversion to cropland, deforestation, over-grazing and burning of fuelwood (Fig. 1.3). In Nepal,

Fig. 1.3 Forest deforestation in southeastern Tibet of China Himalayas (Photography by Zhanhuan Shang 2009)

watersheds, forests, grasslands, farmlands and shrublands all decreased in the last century, which has resulted in large amounts of carbon emission. For example, since 1978, 29% of the carbon was lost in the region of Mardi and 7% in the region of Fewas (Sitaula et al. 2004). On the carbon budget of whole Nepal, the land use change resulted in a carbon emission of $6.9–42.1 \times 10^6$ Mg year^{-1}, of which only the cutting of fuelwood contributed about 1.47×10^6 Mg year^{-1} (Upadhyay et al. 2005).

Protection of forest ecosystems in developing and poor counties or areas is low-cost and the best option for tackling climate change. In 30 years (1980–2009) in the western Himalayas, the total carbon pool declined about 135–145 Mt. The social economic aspect should be encouraged to enhance the carbon pool level for tackling the climate change in developing countries or regions (Wani et al. 2017). In the Mamlay watershed of India Sikkim Himalaya, the total area of forest declined by 28%, and the area of cropland doubled, which led to a carbon loss of 55% in 2007 compared with that in 1988. However, in the less disturbed cropping areas, the land use change had only a small effect on carbon emission, which demonstrated the large impact of land use change on carbon influx in the HKH (Sharma and Rai 2007).

The grassland ecosystem of the Tibetan plateau, a large terrestrial biome in the HKH region, occupies 1.02% of the total global land area and 16.9% of Chinese national territorial area, and is an important carbon sink (Wang et al. 2002). However, because of grassland degradation and land use change, the Tibetan plateau's grassland has lost about 3.02 Pg of its carbon pool in the past 30 years, when the whole grassland carbon emission rate was about 1.27 Pg C year^{-1} (Wang et al. 2002). The carbon emission status of the HKH region demonstrated that the basic livelihood activities of residents influences carbon fluxes. Enhancing the livelihood ability of residents and minimizing livelihood activities less dependent on carbon emissions are the keys to improving the carbon storage capacity of the HKH.

1.2.3 Climate Change Problem in HKH Region

The Paris Agreement presented the vision of 'long term of low greenhouse gas emission's strategy' (Ross and Fransen 2017), a strategy that included a general plan for the whole earth's climate governance. For the HKH region, climate change governance's tactics are very important for sustainable development. A study was made in the mid-Himalayan for six areas using the IPCC model of CENTURY (Gupta and Kumar 2017). Two scenarios were generated: in one, the carbon pool would decline by 11.6–19.2% and in the other the pool would decline by 9.6–17.0% by 2099. Policy decision makers must take these projected carbon losses into account when establishing climate governance regulations (Kumar 2005; Gupta and Kumar 2017).

Air temperature change has different impacts in different areas of the HKH. For example, in the Himalayas of northern India, soil carbon change has been more sensitive to atmospheric moisture than air temperature. In the past 7000 years, C3 plants have dominated the vegetation, and soil carbon accumulation and conversion rate were influenced by climate, vegetation and topography (Longbottom et al. 2014). In the past 50 years, the soil carbon accumulation rate (1.9 g m^{-2} year^{-1}) was lower than in the past 3300 years (47.3 g m^{-2} year^{-1}). Increased precipitation could improve the carbon storage in this area (Longbottom et al. 2014). Global warming has increased soil carbon emission, a potentially large problem of uncertain magnitude. For example, it has been predicted that increases in air temperature of 1.88 and 3.19 °C in the Tibetan plateau, would increase CO_2 release from the frozen soil of alpine grassland by about 18% and 29%, respectively (Feng et al. 2016).

1.2.4 Livelihood and Ecological Problem in HKH Region

Climate governance faces the big challenge of balancing the tradeoff between environment and livelihood. The long-term goal is critical for the poor and developing areas and must differ substantially from the current condition. Free trade and appropriate agricultural strategies are important factors to promote international collaboration and enhance sustainable livelihoods to mitigate climate change (Schmidhuher and Tubiello 2007). Data analysis on 80 community forests of Asia, Africa and Latin America showed that community forests with larger areas and higher self-government management ability had more carbon storage and a higher standard of living. The ownership of large areas of public forests should be controlled by local communities, and compensation payment should be offered for the carbon storage benefit to residents, which will contribute towards mitigating climate change (Chhatre and Agrawal 2009). On the Tibetan plateau, the Chinese government has invested about one billion dollars in the past 10 years to mitigate grassland ecosystem degradation and restore degraded land to protect the Tibetan plateau (Zhao et al. 2018). However, this investment was faced with the difficulty of balancing ecological protection, livestock production and livelihood development (Zhao et al. 2018). Models incorporating these variables have proven to be effective in recent years (Zhao et al. 2018), but its sustainability needs long term assessments. The poor infrastructure, weak science and development, lack of credit institutions and difficulty in transferring techniques in the HKH impede the implementation of strategies to mitigate climate change (Adenle et al. 2015).

Forest community institutions have been established in the HKH region, and they have enhanced carbon storage and the livelihoods of residents. The community institutions must also consider the fragile ecosystem and the acquisition of benefits by stakeholders and avoid conflicts from border communities asking for forest ownership (Kant 2011). Appropriate afforestation methods must be initiated, and pure wood logging activity must be prohibited (Fig. 1.4). In the India-Himalaya region, the logging areas of forests contain just 1/3 of the carbon pool of above- and underground when compared to the non-logging areas (Yadav et al. 2016).

Carbon-fixation-livelihood is the sustainable development strategy in the HKH. For example, Pecan nut (*Carya illinoinensis*) forests in the Himalayan region of India are very important for the livelihood of the residents and for the ecosystem. A study indicated that two mixed agro-forest systems of (1) pecan nut-wheat, and (2) lentil-pecan nut provided both production and ecological function, yielding about 56.5 t ha^{-1} and 53.2 t ha^{-1} respectively, and, storing 25.3 t ha^{-1} and 23.9 t ha^{-1} carbon, respectively. A single crop of lentils produced 2.75 t ha^{-1} of grain and stored 1.17 t ha^{-1} carbon (Kaul et al. 2010). An agro-forest system can contribute about 1.67 t C ha^{-1} year^{-1} to the carbon pool; consequently, developing well-managed agro-forests, particularly with economic trees, can produce fuel wood, timber, fruits, and other agricultural products, and concomitantly, decrease the emission of carbon by agricultural activities (Kaul et al. 2010). By lengthening the logging interval, forests can maximize wood production and carbon fixation that can mitigate carbon emission by deforestation (Kaul et al. 2010). In Nepal, management of governmental forest land has been strengthened to enhance the carbon sink and to protect biodiversity. In the HKH region, the residents rely heavily on forest fuel wood for their livelihood, so alternative materials must be found for fuel to reduce the pressure on the forests (Suwal et al. 2014).

Sustainable agriculture linked with carbon balance is important in adapting to climate change in the HKH region for carbon-fixation. In this vein, the proper use of fertilizers must be considered in the context of the social and economic conditions, not only for crop yield but also for soil quality (Paul et al. 2016). In the Indian Himalayas, the restoration of degraded and cultivated land to grassland and forest reduced land use intensity and increased carbon storage (Meena et al. 2018). In general, the clean energy and REDD+ projects planted trees and crops with high carbon fixation ability, such as *Eucalyptus* species and sal trees (*Shorea robusta*) (Pandey et al. 2016a, b). The grassland ecosystem management should pay special attention to increasing the carbon pool's stabilization of soil and curbing land degradation. This could lead to two-fold benefits: ecological building and livelihood development (Wen et al. 2013) (Fig. 1.5).

Fig. 1.4 Timber products in Himalayas (Photography by Zhanhuan Shang 2009)

Fig. 1.5 Fencing enclosure for restoring degraded grasslands in North Tibet (Photography by Xiaopeng Chen 2011)

1.3 Proposing a Framework for Accurately Evaluating the Level and Mode of Carbon Compensation

1.3.1 The Requirement of Accurate and Coordinative-Evaluation Systems Supporting Carbon Compensation in the HKH

The accurate evaluation of the carbon pool is important in the campaign of global climate governance in the HKH region. For example, the main purpose of the REDD+ program is enhancing livelihood efficiency through environmental improvement, but the lack of policy on a regional scale is one of the barriers in the HKH region (Sharma et al. 2015; Pandey et al. 2016a). The mechanism for carbon compensation and a roadmap for assistance in the HKH region are important outcomes that are presented in this book.

As there is no accurate evaluation of carbon balance of vegetation on the whole HKH, estimates rely on modeling deduction (Sitaula et al. 2004; Shrestha et al. 2009). The carbon data and model of HKH region was generated from several studies (Dorji et al. 2014a, b; Lekhendra et al. 2015), and not based on an overall research project in the region (Upadhyay et al. 2005, 2006). Although the REDD+ project has been practiced widely in the HKH, it still requires a more quantitative overall carbon evaluation for the global carbon management campaign (Pandey et al. 2016a, b). A base of vegetation zoning, vegetation succession, land use, and the ability of providing a reference system must be established with the help of consultants and field experts (Upadhyay et al. 2006).

1.3.2 Innovative Action of Carbon and Livelihood Evaluation in HKH

A stratification-model should be generated for the carbon assessment project and for the livelihood benefits from carbon compensation program in the HKH (Fig. 1.6). The main content of the proposed model in the HKH region should include: the vegetation partition framework based on vegetation classification, vegetation carbon pool variation, carbon sink potential capacity, effect of carbon pool change on agricultural and pastoral productive/ecological functions, carbon benefits, ecological compensation pathway and strategy, and strategy for tackling climate change based on carbon management. The HKH is the typical global poor region, with a very high diversity of societies and nationalities, and with a simple economy and slow social development. However, the HKH region consumes natural resources directly for social, economic and livelihood purposes at a high rate. This has caused much of the area to become a large non-industrial carbon emission source. The whole region of HKH should strengthen the carbon benefit compensation program which would be important in the alleviation of poverty in the HKH region. The model should provide a carbon benefit compensation scheme based on vegetation and land use, with an aim of improving livelihoods.

This stratification-model concept of carbon evaluation in the HKH demonstrates the research program on a natural scale and work level (Fig. 1.6). In the model, the horizontal axis is the logic stratification with a sequence of three task levels, including field, data and management. The vertical axis is the research content scale stratification within the three tasks. In the field work level, five sampling stratifications are needed for the baseline of the whole evaluation including from small scale to large scale to identify local sites with records derived from previous historic monitoring, best available system, land system classification, and bioregion type. In the data task, there are three levels including carbon measurement, ecological surveys

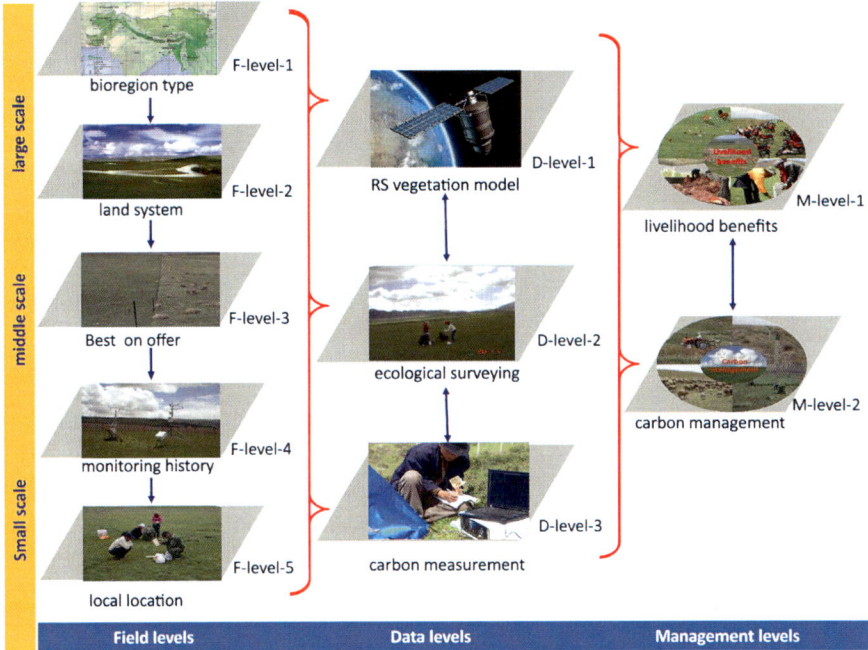

Fig. 1.6 The diagram of stratification-model concept for carbon evaluation in the HKH

and details on vegetation. The management level includes carbon and livelihood for the strategy of social-carbon benefits. In the overall model, the first important step is identifying the sites for data collection. Data at all levels must be collected from the past and present to benefit the future. In the management levels, the database of field work survey and laboratory measurements will be useful for carbon management and livelihood benefits by the social-carbon-metrology model.

1.3.3 The Field Work Design of the Potential Concept Model

The field evaluation work should cover all the HKH regions in all countries (Pakistan, India, Nepal, Afghanistan, Bhutan, Bangladesh, Myanmar and China). Three types of field sampling sites will be selected to collect accurate data on plants and soil (Fig. 1.7). The first sites (marked with red circles), which were determined according to vegetation types, land use types, remote sensing, and field investigation, are for soil and plant sampling. The second sites (marked with pink stars), which were determined by intact vegetation area or undisturbed status (best on offer), are used to provide reference for determining the carbon data variation with climate change and human activity. The third sites (marked with blue triangles) were drawn from study documents and publications and from the field station's monitoring document (monitoring history plots).

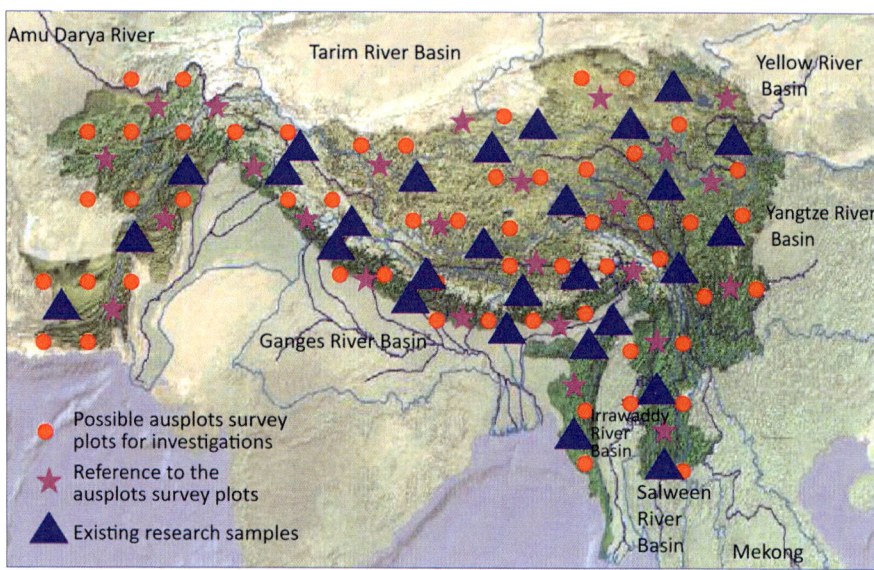

Fig. 1.7 The field work design of the potential concept model in the HKH region

1.3.4 Strategy of Ecological Service and Carbon Benefit Accounting

Including ecological service in climate change governance through the carbon accounting pathway requires more compensation in the cost of ecosystem management. The carbon benefit analysis should refer to the Millennium Ecosystem Assessment, in which the United Nations established the methodology and technical guidelines of global ecosystem assessment (MA 2003, 2005). The HKH region has special social and natural needs in this aspect. For example, Bhutan proposed the concept of a happiness index or GHP. The question arises as how to include the GHP in the carbon benefit solution. This example stresses the need to maintain livelihood as one core variable in coordinating the balance between ecological function and productivity.

The carbon benefit model of ecological service should be based on field investigations, data analysis, and historical records. The stable model of ecological service and carbon sink benefit could be based on long term vegetation succession that is linked with biodiversity change over the years. The HKH region offers one of the highest ecological services in the world, providing extensive biodiversity and water resources. Consequently, the model with biodiversity, ecological service and population as variables could be generated as a base for ecological compensation, in which the carbon sink value is linked with ecological service. The carbon sink function should receive maximum compensation in the HKH region, and the output of carbon products such as forest lumber and animal by-products could be beneficial to the residents.

The fifth report of IPCC recommended the transformative adaption strategy to tackle global climate change by using regional strategies and approaches. In this strategy, the global technological, economic, and social assistance in the global climate governance's system chain is directed through carbon trade by accounting for carbon emission and carbon footprint. However, in different regions, there are different carbon emissions quota per capita and different livelihoods, making it difficult to reach common criteria for the carbon trade.

Here we propose a method of social-carbon accounting for carbon storage accumulation in the HKH region, based on the increased rate of livelihood and economy. The carbon accounting would differ for different livelihood systems, but the accounting model would be unified within the system. For example, livestock husbandry, land cultivation, forestry, tourism, mining, and nature reserves would consider the livelihood cost for carbon storage and emission within the local system and within the global system. For livestock husbandry, the potential carbon accounting model can use the GLEAM-i model and technological system by FAO (2016). For forestry, the CBM-CFS3 model could be used to correct the carbon accounting (Kurz et al. 2009). The agricultural industry can use the IPCC emission coefficient as the criterion. Road building, mining and other industries can use the area involved in accounting for the carbon pool. The potential carbon model (Figs. 1.6 and 1.7) should account for the livelihood pathway and the carbon balance associated with the labor to complete the carbon accounting inventory for the household and enterprise.

1.3.5 The Scheme and Approach for Determining the Livelihood Benefits in Carbon Management

According to the social-carbon model of carbon sinking and emission in the past and predicted in the future, livelihood compensation should be determined by the different livelihood types such as forestry, livestock industry, agriculture, and tourism in the HKH region. For example, within this context, we should calculate the effect on livelihood due to bans on lumbering, selective tree cutting and under-use of forest resources. The method of calculating carbon-livelihood equivalent is based on the livelihood effect value ($\triangle L$) and carbon income value ($\triangle C$) per household. That is the method used for carbon-livelihood equivalent of local residents living activities in forestry, pastoral industries, agriculture, and fishery industries, and for setting out the inventory of carbon-livelihood. The carbon compensation method can be applied to calculate cash compensation in services such as construction, education, medical treatment, food supply and water treatment; by deducting the cost of assistance, the surplus of compensation is the compensation from carbon benefit. Poverty is an important issue in the HKH region that should be alleviated through carbon compensation. Multi-dimension poverty in HKH should be determined by different indices for receiving carbon benefits through the compensation model. The Data Envelopment Analysis Model—(carbon-livelihood equivalent model) is useful to calculate the multi-dimension poverty compensation from the carbon benefit base computed from input and output (WRI 2006).

1.4 Obtaining and Utilizing Carbon Compensation in the HKH

1.4.1 The Requirement of Sustainable Development in the HKH Region

The threat of global climate change on livelihoods has become severe in recent years. Consequently, all countries should act in the campaign for the common aim of ameliorating climate change. Because of the different levels of development, historic background, and technological and governance levels, each country should form its own regional strategy and roadmap in the campaign. The theme of regional carbon pool variation and livelihood is of priority in the global climate change, with a goal of global carbon emission reduction and neutralization. Because the HKH region has the biggest fresh water resource area in world but is also one of poorest regions, tackling climate change and livelihood development are always common topics in the region. Consequently, the proposed social-carbon-management framework is important for development of the HKH during climate change, particularly in the future.

1.4.2 Obtaining Livelihood Compensation from Carbon Sinking Benefit

There are two approaches to account for carbon benefits in determining ecological compensation. The first should be based on the past status of carbon balance due to vegetation and land change (incoming and outgoing), that can be used in modelling the future role of carbon sinking and emission. If there is always carbon capture in the survey area, then this should be maintained by improving livelihoods through carbon compensation. Secondly, if the survey area exhibits carbon emission, then investments and improvements are needed to convert the emission to sinks. Reasons should be detailed for the carbon emission in the area, for example land use, over-cropping, over-utilization of woodland and over-grazing. Based on the findings, changes should be initiated for converting carbon emission to sinks. How to assist, how much assistance is required and for how long the assistance is needed should be calculated? Carbon compensation and the methods, timetable and policies for promoting the carbon neutralization campaign for tackling climate change should all be included in the strategy.

In the HKH region, carbon compensation in government decision-making programs should include more livelihood benefits to alleviate poverty in the regional context. The benefit of the carbon pool towards livelihood of the HKH region should be divided into multi-poverty and multi-pathway livelihood compensations as follows: (1) carbon sink benefit for different livelihoods, and the general cost of social economic growth; (2) livelihood change and improvement cost following the conversion of carbon emission to sinks; (3) cost of food subsidy or assistance due to compensation from carbon sinking benefit; and (4) livelihood benefit compensation for increasing carbon sinking, including sustainable agricultural improvement cost. In addition, the gender issue needs special support in social carbon compensation from the global carbon trade. Attention should be paid to the women's contribution in carbon management. Adolescent education should also receive compensation from the carbon sinking benefit.

1.4.3 Applications in Climate Governance in the HKH

There are four approaches that can be used for the social-carbon compensation model, including carbon management and livelihood improvement based on vegetation, land use and integrated livelihood-carbon system. The first is a partitioned management model, especially in degraded ecosystem areas, applying the 'classification-zoning-grading-staging' system which accounts for carbon and livelihood benefits. The second is compensation for livelihood transformation, which should be used especially in carbon emission areas that rely heavily and directly on natural resources. At the beginning of the transformation, the compensation from government or carbon trade should offer strong support for any lost livelihood. The third approach is the carbon management roadmap of 2020 in HKH (HKH-CM2020), which provides government aid in education, science development, and infrastructure building. The fourth is the promoting of industries with carbon sinks, which would benefit the climate governance's strategy of ecological and livelihood development.

1.4.4 The Key Technique Requiring Solution

In the proposed social-carbon model (Figs. 1.6 and 1.7), there are three key points that are needed to implement the social-economy strategy. The first is the classification and zoning of carbon management in the HKH region. The HKH covers different countries and administrative regions, each having social management and economic systems, which should be linked with the carbon management proposal. The second is the social-carbon-metrology link, which is used to analyze carbon benefits and compensation based on gender, nationality and livelihoods. The third is the transformation of livelihood pathway, which differs among regions and livelihoods.

1.5 Tracking the Trajectory for Carbon Management in the HKH in the Future

1.5.1 Building Up Repeatable Evaluation Techniques

Carbon management linked with livelihood is a local and global concern today. Methodology should establish repeatable techniques for the sustainable application of carbon evaluation in the HKH. This should include four aspects: (1) benchmark plots in all regions and countries of the HKH that are based on vegetation type, land use, and social economy; (2) systematic database, including field measurement data and historical records, with information, such as elevation, soil types, plants, carbon and soil properties; (3) a model relating the carbon pool and vegetation should be established by remote sensing, land use data (at least 30 years), field data, historical data and climate scenarios; and (4) mechanism and policy of ecological compensation for effective livelihood transformations in the future.

1.5.2 The Long-Time Monitoring and Evaluating Plots System

A long-term monitoring plot system can provide accurate data for vegetation and soil dynamics under climate change and human activity. The 'best on offer' system (White et al. 2012), which selects an undisturbed or top-level succession stage area, can be used as a long-term station.

The baseline plots system was generated by the Australian TERN program (White et al. 2012). However, the HKH region must develop its own long-term baseline plots system and methodology because of the great difference in landscapes between the HKH region and Australia. Data describing the ecology, biology and environment, and social aspects should be collected. Plots should be selected to include typical vegetation, land use and topography and hydrology.

Plants, vegetation, soil properties and water should be sampled. Remote sensing data should be included in the model (Hou 2015). Plant specimens should be collected, and high-resolution photograph should be taken (preferably from fixed photo points) for visualization of change. Vegetation biomass should be determined for biomass carbon. In addition, soil carbon content should be determined to a depth of 30 cm.

1.5.3 Collaborative Network and Feedback Mechanism

In general, the international and regional coordinating mechanisms for climate change management in the HKH are weak and must be strengthened. 'The International Centre for Integrated Mountain Development (ICIMOD)' is a very active organization for regional scientific development and capacity building in HKH region. The center can coordinate climate change management in the HKH, especially livelihood and sustainable transformations. ICIMOD has very close collaborative relations with local governments, organizations and even residents in eight countries in the HKH region and has built a valuable data base. Through its information center, ICIMOD can promote the sharing of programs and systems in the HKH region for climate change governance and livelihood development. This is especially essential for carbon compensation and social and livelihood benefits from carbon evaluation.

1.6 Prospects and Conclusion

The urgent activity is to devise transparent methods for presenting the status and potential action on carbon management and livelihood benefits in the context of global climate governance in HKH region as other part of the world. It has been reported that restricting the rise in global air temperature to 1.5 °C as agreed in the Paris Climate Accord could be translated into a benefit of 20 trillion US dollars (Burke et al. 2018). If the target is met it would allow a substantial source of capital for carbon compensation for developing areas such as the HKH region. The biggest recipients of compensation benefits should be poor areas in the world, as this would provide them with the impetus for participating in the global climate governance campaign (Burke et al. 2018). The HKH region supports the world's largest population of pastoralists and hundreds of millions of livestock. Carbon stored in the soil makes up approximately 3% of the world's stocks but the area is fragile and is now a net emitter of carbon into the atmosphere that contributes to reduced carbon stocks. This book is unique in that all aspects of the HKH region will be discussed. The HKH is considered to be a laboratory and museum for scientific research due to its multi-dimensional ecological, economic, cultural and environmental significance, not only for Asia but also on a global scale. Managing the carbon cycle, with respect to livelihood perspectives, is urgently needed and requires immediate attention in the HKH region. We cover social, economic, policy perspectives and propose innovative approaches to solve current and future problems.

References

Kumar, P. 2005. Ecosystems and Human Well Being: Synthesis. *Future Survey* 34(9): 534–534.

Adenle, A.A., H. Azadi, and J. Arbiol. 2015. Global assessment of technological innovation for climate change adaptation and mitigation in developing world. *Journal of Environmental Management* 161: 261–275.

Arora, G., S. Chaturvedi, R. Kaushal, et al. 2014. Growth, biomass, carbon stocks, and sequestration in an age series of *Populus deltoides* plantations in Tarai region of central Himalaya. *Turkish Journal of Agriculture and Forestry* 38: 550–560.

Banskota, K., B.S. Karky, and M. Skutsch. 2007. *Reducing carbon emission through community-managed forests in the Himalaya*. Kathmandu, Nepal: International Centre for Integrated Mountain Development.

Bhattacharyya, T., D.K. Pal, C. Mandal, et al. 2000. Organic carbon stock in Indian soils and their geographical distribution. *Current Science* 79 (5): 655–660.

Burke, M., W.M. Davis, and N.S. Diffenbaugh. 2018. Large potential reduction in economic damages under UN mitigation targets. *Nature* 557: 549–553. https://doi.org/10.1038/s41586-018-0071-9.

Chhatre, A., and A. Agrawal. 2009. Trade-offs and synergies between carbon storage and livelihood benefits from forest commons. *PNAS* 106 (42): 17667–17670.

Choudhury, B.U., A.R. Fiyaz, K.P. Mohapatra, et al. 2016. Impact of land uses, agrophysical variables and altitudinal gradient on soil organic carbon concentration of north-Eastern Himalayan region of India. *Land Degradation & Development* 27 (4): 1163–1174.

Dar, J.A., and S. Somaiah. 2015. Altitudinal variation of soil organic carbon stocks in temperate forests of Kashmir Himalayas, India. *Environmental Monitoring and Assessment* 187: 11. https://doi.org/10.1007/s10661-014-4204-9.

Dasgupta, S., T.P. Singh, R.S. Rawat, et al. 2015. Assessment of above ground biomass and soil organic carbon stocks in the forests of India. In *Geospatial information systems for multiscale forest biomass assessment and monitoring in the Hindu Kush Himalayan region*, 20–29. Kathmandu: ICIMOD.

Debasish-Saha, S.S. Kukal, and S.S. Bawa. 2014. Soil organic carbon stock and fractions in relation to land use and soil depth in the degraded Shiwaliks hills of lower Himalayas. *Land Degradation & Development* 25: 407–416.

Devi, B., D.R. Bhardwaj, P. Panwar, et al. 2012. Carbon allocation, sequestration and carbon dioxide mitigation under plantation forests of north western Himalaya, India. *Annual of Forest Research* 56 (1): 123–135.

Dorji, T., I.O.A. Odeh, D.J. Field, et al. 2014a. Digital soil mapping of soil organic carbon stocks under different land use and land cover types in montane ecosystems, Eastern Himalayas. *Forest Ecology and Management* 318: 91–102.

Dorji, T., J.O.A. Odeh, and D.J. Field. 2014b. Vertical distribution of soil organic carbon density in relation to land use/cover, altitude and slope aspect in the Eastern Himalayas. *Land* 3: 1232–1250.

FAO. 2016. *Global livestock environmental assessment model-interactive (GLEAM-i), a tool for estimating greenhouse gas emissions in livestock production and assessing intervention scenarios (Revision 1)*. FAO. http://www.fao.org/gleam/resources/en/.

Feng, J.L., H.P. Hu, and F. Chen. 2016. An eolian deposit–buried soil sequence in an alpine soil on the northern Tibetan Plateau: Implications for climate change and carbon sequestration. *Geoderma* 266: 14–24.

Gupta, S., and S. Kumar. 2017. Simulating climate change impact on soil carbon sequestration in agro-ecosystem of mid-Himalayan landscape using CENTURY model. *Environmental Earth Science* 76: 394. https://doi.org/10.1007/s12665-017-6720-8.

Gurung, M.B., H. Bigsby, and R. Cullen. 2015. Estimation of carbon stock under different management regimes of tropical forest in the Terai Arc Landscape, Nepal. *Forest Ecology and Management* 365: 144–152.

Hou, Y.C. 2015. *Evaluation method of alpine meadow based on remote sensing and forage feeding value*. Master's Degree Thesis in Lanzhou University.

IPCC. 2013. *Chapter 6 Working group I contribution to the IPCC fifth assessment report climate change 2013: The physical science basis*. Cambridge University Press, Cambridge, United Kingdom and New York, NY, USA.

Kant, P. 2011. The critical importance of forest carbon sink in the green economy of the Hindu Kush-Himalayan Mountain Systems. In *IGREC Working Paper IGREC-26: 2011*. New Delhi: Institute of Green Economy.

Kaul, M., G.M.J. Mohren, and V.K. Dadhwal. 2010. Carbon storage and sequestration potential of selected tree species in India. *Mitigation and Adaptation Strategies for Global Change* 15: 489–510.

Kurz, W.A., C.C. Dymond, T.M. White, et al. 2009. CBM-CFS3: A model of carbon-dynamics in forestry and land-use change implementing IPCC standards. *Ecological Modelling* 220 (4): 480–504.

Lekhendra, T., R. Dipesh, S. Prakriti, et al. 2015. Carbon sequestration potential and chemical characteristics of soil along an elevation transect in southern Himalayas, Nepal. *International Research Journal of Environment Sciences* 4 (3): 28–34.

Longbottom, T.L., A. Townsend-small, L.A. Owen, et al. 2014. Climatic and topographic controls on soil organic matter storage and dynamics in the Indian Himalaya: Potential carbon cycle-climate change feedbacks. *Catena* 119: 125–135.

Meena, V.S., T. Mondal, B.M. Pandey, et al. 2018. Land use changes: Strategies to improve soil carbon and nitrogen storage pattern in the mid-Himalaya ecosystem, India. *Geoderma* 921: 69–78.

Millennium Ecosystem Assessment. 2003. *Ecosystems and human wellbeing: A framework for assessment (MA)*. Washington, DC: Island Press.

Naudiyal, N., and J. Schmerbeck. 2018. Impacts of anthropogenic disturbances on forest succession in the mid-montane forests of Central Himalaya. *Plant Ecology* 219: 169–183.

Pacala, S., and R. Socolow. 2004. Stabilization wedges: Solving the climate problem for the next 50 years with current technologies. *Science* 305: 968–972.

Pandey, R., S.K. Hom, S. Harrison, et al. 2016a. Mitigation potential of important farm and forest trees: A potentiality for clean development mechanism afforestation reforestation (CDMA R) project and reducing emissions from deforestation and degradation, along with conservation and enhancement of carbon stocks (REDD+). *Mitigation and Adaptation Strategies Global Change* 21: 225–232.

Pandey, R., S.K. Jha, J.M. Alatalo, et al. 2017. Sustainable livelihood framework-based indicator for assessing climate change vulnerability and adaptation for Himalayan communities. *Ecological Indicators* 79: 338–346.

Pandey, S.S., G. Cockfield, and T.N. Maraseni. 2016b. Assessing the roles of community forestry in climate mitigation and adaptation: A case study from Nepal. *Forest Ecology and Management* 360: 400–407.

Paris Agreement. 2016. *United Nations treaty collection*, July 8, 2016.

Paul, J., A.K. Choudhary, S. Sharma, et al. 2016. Potato production through bio-resources: Long-term effects on tuber productivity, quality, carbon sequestration and soil health in temperate Himalayas. *Scientia Horticulturae* 213: 152–163.

Quéré, C.L., R.M. Andrew, P. Friedingstein, et al. 2018. Global carbon budget 2017. *Earth System Science Data* 10: 405–448.

Rajput, B.S., D.R. Bhardwaj, and N.A. Pala. 2017. Factors influencing biomass and carbon storage potential of different land use systems along an elevational gradient in temperate northwestern Himalaya. *Agroforestry Systems* 91: 479–486.

Ross, K., and T. Fransen. 2017. Early insights on long-term climate strategies. In *Working paper*. Washington, DC: World Resources Institute.

Sandhu, H., and S. Sandhu. 2015. Poverty, development, and Himalayan ecosystems. *AMBIO* 44: 297–307.

Schmidhuher, J., and F.N. Tubiello. 2007. Global food security under climate change. *PNAS* 104 (50): 19703–19708.

Shaheen, H., Y. Saeed, M.K. Abbasi, et al. 2017. Soil carbon stocks along an altitudinal gradient in different land-use categories in Lesser Himalayan Foothills of Kashmir. *Eurasian Soil Science* 50 (4): 432–437.

Sharma, B.P., M. Nepal, B.S. Karky, et al. 2015. *Baseline considerations in designing REDD + pilot projects: Evidence from Nepal*. Kathmandu, Nepal: South Asian Network for Development and Environmental Economics (SANDEE).

Sharma, C.M., N.P. Baduni, S. Gairola, et al. 2010. Tree diversity and carbon stocks of some major forest types of Garhwal Himalaya, India. *Forest Ecology and Management* 260: 2170–2179.

Sharma, C.M., A.K. Mishra, R. Krishan, et al. 2016. Variation in vegetation composition, biomass production, and carbon storage in ridge top forests of high mountains of Garhwal Himalaya. *Journal of Sustainable Forestry* 35 (2):119–132.

Sharma, P., and S.C. Rai. 2007. Carbon sequestration with land-use cover change in a Himalayan watershed. *Geoderma* 139: 371–378.

Shrestha, B.M., S. Williams, M. Easter, et al. 2009. Modelling soil organic carbon stocks and changes in a Nepalese watershed. *Agriculture, Ecosystems and Environment* 132: 91–97.

Shrestha, H.L., K. Uddin, H. Gilani, et al. 2015. Forest carbon flux assessment in Nepal using the gain-loss method. In *Geospatial information systems for multi-scale forest biomass assessment and monitoring in the Hindu Kush Himalayan region*, ed. A.B. Murray, A. Sellmyer, D.R. Maharjan, et al., 178–190. Kathmandu, Nepal: ICIMOD.

Shukla, R., K. Sachdeva, and P.K. Joshi. 2016. Inherent vulnerability of agricultural communities in Himalaya: A village-level hotspot analysis in the Uttarakhand state of India. *Applied Geography* 74: 182–198.

Singh, J.S., A.K. Tiwari, and A.K. Saxena. 1985. Himalayan forests: A net source of carbon for the atmosphere. *Environmental Conservation* 12 (1): 67–69.

Sitaula, B.K., R.M. Bajracharya, B.R. Singh, et al. 2004. Factors affecting organic carbon dynamics: In soils of Nepal/Himalayan region-a review and analysis. *Nutrient Cycling in Agroecosystems* 70: 215–229.

Suwal, A.L., D.R. Bhuju, and I.E. Maren. 2014. Assessment of forest carbon stocks in the Himalayas: Does legal protection matters? *Small-Scale Forestry* 14: 103–120.

Tashi, S., B. Singh, C. Keithel, et al. 2016. Soil carbon and nitrogen stocks in forests along an altitudinal gradient in the eastern Himalayas and a meta-analysis of global data. *Global Change Biology* 22: 2255–2268.

Upadhyay, T.P., P.L. Sankhayan, and B. Soleberg. 2005. A review of carbon sequestration dynamics in the Himalayan region as a function of land-use change and forest/soil degradation with special reference to Nepal. *Agriculture, Ecosystems and Environment* 105: 449–465.568.

Upadhyay, T.P., B. Solberg, and P.L. Sankhayan. 2006. Use of models to analyse land use changes, forest/soil degradation and carbon sequestration with special reference to Himalayan region: A review and analysis. *Forest Policy and Economics* 9: 349–371.

Verma, A., R. Kaushal, N.M. Alam, et al. 2014. Predictive models for biomass and carbon stocks estimation in Grewiaoptiva on degraded lands in western Himalaya. *Agroforestry Systems* 88: 895–905.

Wang, G., J. Qian, G. Cheng, et al. 2002. Soil organic carbon pool of grassland soils on the Qinghai-Tibetan Plateau and its global implication. *The Science of Total Environment* 291: 207–217.

Wani, A.A., P.K. Joshi, O. Singh, et al. 2017. Forest biomass carbon dynamics (1980-2009) in western Himalaya in the context of REDD+ policy. *Environmental Earth Sciences* 76: 673. https://doi.org/10.1007/s12665-017-6903-3.

Wani, A.A., W.P.K. Joshi, O. Singh, et al. 2014. Estimating soil carbon storage and mitigation under temperate coniferous forests in the southern region of Kashmir Himalayas. *Mitigation and Adaptation Strategies for Global Change* 19: 1179–1194.

Ward, A., P. Dargusch, S. Thomas, Y. Liu, et al. 2014. A global estimate of carbon stored in the world's mountain grasslands and shrublands, and the implications for climate policy. *Global Environmental Change* 28: 14–24.

Wen, L., S. Dong, Y. Li, et al. 2013. The impact of land degradation on the C pools in alpine grasslands of the Qinghai-Tibet Plateau. *Plant Soil* 368: 329–340.

White, A., B. Sparrow, E. Leitch, et al. 2012. *Ausplots rangelands surveys protocols manual (Version1.2.9)*. Adelaide, SA: The University of Adelaide Press.

WRI. 2006. *The greenhouse gas protocol: The land use, land-use change, and forestry guidance for GHG project accounting*. Washington, DC: World Resources Institute.

Yadav, R.P., J.K. Bisht, and B.M. Pandey. 2016. Cutting management versus biomass and carbon stock of oak under high density plantation in Central Himalaya, India. *Applied Ecology and Environmental Research* 14 (3): 207–214.

Zhao, X., L. Zhao, Q. Li, et al. 2018. Using balance of seasonal herbage supply and demand to inform sustainable grassland management on the Qinghai-Tibetan plateau. *Frontiers of Agricultural Science and Engineering* 59 (1): 1–8.

Chapter 2
Climate Change Mitigation and Pastoral Livelihood in the Hindu Kush Himalaya Region: Research Focuses, Opportunities and Challenges

Wenyin Wang, Devendra Gauchan, A. Allan Degen, and Zhanhuan Shang

Abstract The authoritative global climate report of the Intergovernmental Panel on Climate Change (IPCC) urged each country to take action on climate change mitigation. The vast HKH region should make a more consistent and concise evaluation that can augment the global level evaluation. In this chapter, the whole HKH's carbon emission were described and some advice about increasing carbon sink were given. This chapter sets out to: (1) clarify our understanding about the HKH region's natural resources, ecological functions, social economy and livelihood of its peoples; (2) promote understanding and cognition of carbon dynamics and its driving forces that could better equip HKH's governments and people to mitigate and adapt to future climate change; and (3) encourage anti-poverty assistance be involved in the carbon trade in the world platform for HKH sustainable development.

Keywords Hindu Kush Himalaya · Ecosystem services · Natural source · Carbon sink · Anti-poverty

---

W. Wang · Z. Shang (✉)
School of Life Sciences, State Key Laboratory of Grassland Agro-Ecosystems, Lanzhou University, Lanzhou, China
e-mail: wangwy2015@lzu.edu.cn; shangzhh@lzu.edu.cn

D. Gauchan
Bioversity International, Lalitpur, Nepal

A. A. Degen
Desert Animal Adaptations and Husbandry, Wyler Department of Dryland Agriculture, Blaustein Institutes for Desert Research, Ben-Gurion University of Negev, Beer Sheva, Israel
e-mail: degen@bgu.ac.il

2.1 Introduction

2.1.1 The Severity of Climate Change

Due to climate change, some areas of the HKH have been seriously affected. For example, high-intensity monsoon rains in northern India and western Nepal caused landslides in 2013, which were the worst flood disasters in history. Since 2000, western Nepal has been affected by drought due to natural and human factors, and in 2008–2009, the degree of drought was extreme. In the Tamakoshi basin of central Nepal, between 2000 and 2009, 18% of the runoff per year was melted snow and ice (Khadka et al. 2014), and climate warming has caused a large number of glaciers to melt, especially during the fast warming season. Warming and melting glaciers have led to rapid expansion of some glacial lakes in the HKH region.

In the last few decades, the entire Himalayan region has become warmer and drier, causing inconsistent precipitation in the HKH region. Glaciers in the eastern Himalayas (Nepal-Bhutan) are more susceptible to climate change due to reduced snowfall and increased snow melt (Gurung and Sherpa 2014). Extreme weather events disrupted the supply of food production and freshwater resources, damaged infrastructure and residents' settlements, and forced some residents to migrate. At the same time, morbidity and mortality have increased, which endangered human health. The hazards associated with climate change will reduce crop yields and destroy homes. It has a major negative impact on livelihood of people, especially for the poor.

2.1.2 The Importance of Actively Participating in Climate Governance

The high-altitude region of the HKH ranges is susceptible to natural disasters such as heavy rain, sudden floods, blizzards, squalls and landslides due to the topographic features (Molden et al. 2014). The hydropower industry is sensitive to changes in seasonal climate and extreme weather events. It is often occurred in the Yarlung Zangbo river, the Ganges river and the Indus river basin, which are complex and climate change may have a major impact on hydropower generation (Molden et al. 2014). Therefore, climate change not only affects hydropower, biodiversity systems, forestry, etc., but also has a profound impact on the residents of the HKH region. To reduce the rate of natural resource reduction caused by climate change in HKH, actively participating in climate governance, strengthening the construction of network observation points, obtaining more basic meteorological data are needed to improve the climate environment and promote the sustainable development of local resources.

2.1.3 Paris Agreement Opportunity

The IPCC fourth assessment report predicted that climate change would be the main driver of global change in the twenty-first century and pointed to the lack of long-term monitoring in the HKH region and called for national, regional and global efforts to fill this gap. The entry into force of the "Paris Agreement" provided an institutional framework for global cooperation to address climate change after 2020, indicated the direction and objectives, provided a new impetus to global climate change cooperation to reduce greenhouse gas emissions and transform to low carbon. It formed a new pattern of harmonious development between people and nature. In the process, many different opportunities should be created.

2.1.4 The Way to Improve Livelihoods in HKH

The grassland ecosystem forms the source of the main river system in the HKH region, and the water flow will play an important role in the development of future for hydropower generation and agricultural irrigation in low-altitude areas. The vast grazing land provides forage for livestock, and the herd can convert plant biomass into animal products for the herdsmen to consume and improve income. Due to the unique characteristics of the terrain in the area, farmers and herders living in cultivated grassland could improve tourism infrastructure increase income by developing tourism.

Food security issues have been affected in the HKH due to poor development and climate change and self-sufficiency has not been reached by local residents. With the support of national and local governments, food production could increase by introducing new technologies and promoting their application to improve efficiency. In addition, it is necessary for more young people to stay in the region to continue and develop agriculture. It has a great potential and a broad market in developing a low carbon economy in HKH. Thus, it is possible to reduce greenhouse gas emissions into the atmosphere through afforestation, sustainable forest management and reduced deforestation. It is also possible to promote economic development through carbon trading with developed countries, would increase the sustainability of local resources and improve livelihoods.

2.2 Natural Resources and Ecosystem Services in HKH

2.2.1 Basic Situation of Natural Resources

The HKH region, which has many key ecological regions of global importance, has different landscapes and soil formations, and is known for its unique flora and fauna (Mittermeier et al. 2017). It includes many south and central Asian countries:

Afghanistan, Pakistan, China, India, Nepal, Bhutan, Bangladesh and Myanmar and the world's largest high-altitude area and largest high-latitude permafrost region in the world, from the Tibetan Plateau in northern China to the Ganges Basin in the south, covering an area of 4.3×10^6 km^2. More than half of the land in the HKH is covered by grassland (50.5%), 22.6% by forests, 9% by agricultural land, less than 0.1% by urban development, and 17.8% by bare land or by water, snow and ice. There are ten main river basins, namely, Indus, Ganges, Amu Darya, Yarlung Zangbo, Brahmaputra (Yarlungtsanpo), Irrawaddy, Salween (Nu), Mekong (Lancang), Tarim, Yangtse (Jinsha) and Yellow (Huanghe) covering about 8.6×10^6 km^2 (Table 2.1).

The biodiversity and environmental services in the HKH have an impact on the lives of local people, as well as many people living in downstream regions. And the ecosystems of the HKH are vital in terms of the many important ecosystem services they provide for native and domestic animals. The HKH is a critical source of wood, medicine, plants, wild food, fiber, and freshwater for humans, and provide essential habitats for many endangered wildlife species (Dong et al. 2010). It is known as the "Asian Water Tower" and contain a large amount of water sources of plateau lakes and wetlands. It is another freshwater reservoir that provides lifeline for millions of people and increases groundwater recharge through watershed functions in the form of snow, glaciers, permafrost, wetlands and rivers (Rasul 2014; Chettri et al. 2012).

Table 2.1 The ten major river basins of the Himalayan region (IUCN/IWMI 2003)

Rivers	Basin area (×1000 km^2)	Countries	Population (million)	Population density (/km^2)
Amu Darya	534.7	Afghanistan, Tajikistan, Turkmenistan, Uzbekistan	20.9	39
Brahmaputra	651.3	China, India, Bhutan, Bangladesh	118.5	182
Ganges	1016.1	India, Nepal, China, Bangladesh	407.5	401
Indus	1081.7	China, India, Pakistan	178.5	165
Irrawaddy	413.7	Myanmar	32.7	79
Mekong	805.6	China, Myanmar, Laos, Thailand, Cambodia, Vietnam	57.2	71
Salween	271.9	China, Myanmar, Thailand	6	22
Tarim	1152.4	Kyrgyzstan, China	8.1	7
Yangtze	1722.2	China	368.5	214
Yellow	945	China	147.4	156
Total	8594.8		1345.2	

2.2.2 Biodiversity and Its Value

The rich natural and cultural heritage in the HKH is the home to many ethnic groups and has the richest ecosystem and natural resources in the world. Due to the complex and diverse climate, altitude and geological conditions, the HKH region is rich in biodiversity, including 4 global biodiversity hotspots, 8 world natural heritage sites, 30 Ramsar Sites, 488 protected areas and 330 important bird areas, and is ranked tenth among the 34 most bio-diverse regions in the world, and 53 Important Plant Areas (IPAs) for medicinal plants (Sharma et al. 2009). Many of the world's endangered species live in the HKH region, which provides a breeding ground for a rich variety of medicinal plants. An estimated 25,000 fauna species and 30% of flora are endemic throughout the Himalayas (Lópezpujol et al. 2006).

The HKH provides a habitat for many wild animals and is the home to all four big cats in Asia: the snow leopard (*Uncia uncia*), tiger (*Panthera tigris*), common leopard (*Panthera pardus*) and clouded leopard (*Neofelis nebulosa*). The Tibetan wild ass (*Equus kiang*), wild yak (*Bos grunniens*), Chiru (*Pantholops hodgsoni*) and Tibetan gazelle (*Procapra picticaudata*) inhabit the area and are all endangered animals (Schaller and Kang 2008). There are also about 75,000 species of insects (10% of the world) and 1200 species of birds (13% of the world) (Sharma et al. 2009). The HKH also plays an important role for many long-distance migratory birds. The raptor migration occurs near the Indus river in Pakistan and the Tsangpo-Brahmaputra river in Tibet and eastern India. Birds living in wetlands, such as black-necked cranes (*Grus nigricollis*), barheaded goose (*Anser indicus*), uddy shelduck (*Tadorna ferruginea*) use a variety of mountain passes and wetlands to feed, inhabit or migrate as a temporary habitat (Chettri et al. 2012). These biological resources are undoubtedly of great value.

2.2.3 Ecosystem Services

The HKH has the largest snow and ice protection area, except for the Antarctica and Arctic. In the 60,000 km^2 range, there are 6000 km^3 of ice and 760,000 km^2 of snow. The HKH area represents a unique source of freshwater for agriculture, industry and household use, and an important source of economics for tourism and hydropower (Mats et al. 2009). The HKH provides water for the top ten major river basins. There are more than 16,000 glaciers in the Himalayas, that has a huge natural reservoir that stores more than 12,000 km^3 of fresh water. The HKH mountain system plays a critical role in agriculture of South Asia through the water supply, climate and wind regulation, groundwater recharge and the maintenance of wetland ecosystems (Eriksson et al. 2009).

The ten major river basins in the HKH region provide water, food, energy and ecosystem services for more than 1.3 billion people in south Asia. Glacier and snow contributes significantly to river flows, from a minimum of 1.3% in the Yellow river to 40.2% in the Tarim river, and the highest contribution from the Indus valley (44.8%). According to statistics, about 30% of the water in the east Himalayas is derived directly from the melting of snow and ice. The proportion is about 50% in the central and western Himalayas, and 80% in the Karakoram. The HKH is the source of freshwater resources and plays an important role in water storage and regulation in wetlands (Xu et al. 2010).

2.2.4 Carbon Sink Resources

The HKH region has a vast territory and many natural ecosystems, including carbon sink capacity such as forests, grasslands, farmlands and wetland ecosystems. In terms of the potential for carbon sink resources, forest ecosystems have the carbon sink function. 22.6% of the area is forest in HKH. Research demonstrated that for every 1 m^3 of forest accumulation, there is an average absorption of 1183 tons of CO_2 and the release of 1162 tons of O_2. Forest ecosystems can control soil erosion, restore degraded land, and improve soil organic matter content (Jackson et al. 2017). The forest (42%) fixes 779 million tons of carbon in Nepal. Grassland ecosystem, one of the widely distributed vegetation types, is of great significance in the global carbon cycle. The plants on the grassland absorb the CO_2 in the atmosphere through photosynthesis and fix it in the soil vegetation, which can reduce the concentration of CO_2 in the atmosphere and effectively suppressing the greenhouse effect. The grassland ecosystem carbon sequestration accounts for 50.5% of the total in HKH (Sharma et al. 2009).

2.3 Relationship Between Social Economy, Population and Natural Resources in the HKH Region

2.3.1 Relationship Between Social Economy and Resource

Water and energy are interrelated, and the water needed for industry, agriculture, and family life requires pumps to process and transport. Hydropower plays a critical role in global energy and supplies approximately 16% of the world's electricity. The HKH has a great potential for power generation. A total of 162 hydropower projects are in the process of development in India and will generate more than 50,000 MW (Choudhury 2014). The water resources owned in Nepal can generate 83,290 MW (Surendra et al. 2011). There is a potential of about 30,000 MW of power generation in Bhutan (Uddin et al. 2007), and about 42,000 MW in Pakistan (Asif 2009).

In South Asia, agricultural land accounts for a large proportion of the area in most countries: 70% in Bangladesh, 60% in India and 35% in Nepal. India,

Pakistan, Nepal, Bangladesh and Bhutan are the main food producers in South Asia, and the Himalayas are the main water source of farmland irrigation. With the significant increase in food production, India, Pakistan and Nepal transformed themselves from countries with long-term food shortages to almost being self-sufficient in the early 1990s. In the late 1990s, except for Afghanistan, all countries in HKH exported grain (Kumar et al. 2008). The snow in the Himalayas, the huge and complex watershed networks, monsoon rainfall and abundant underground aquifers provide water in South Asia. It could increase the income of residents and the diversification of agricultural products in high mountains (Chand et al. 2008).

2.3.2 People and Resources

Ten major rivers in the HKH provide goods and services directly or indirectly to 3.2×10^9 people. Most of the population in HKH depends on subsistence agriculture and natural resources for their livelihood. China and India account for 44% and 41% of the population in the HKH region, respectively. Based on growth trends, the populations of countries in HKH region in 2030 and 2050 were predicted (Table 2.2). It is well documented that traditional agriculture alone is insufficient to meet the food requirements of the growing population. The number of people who are still undernourished remains high in the HKH. In 2013, the under-nourished people in China, India, Nepal and Pakistan numbered about 406 million people or 48% of the world's under-nourished people (FAO, IFAD, WFP 2013).

Table 2.2 Population of the countries of 2017 in the HKH and predicted populations in 2030 and 2050 (United Nations, Department of Economic and Social Affairs (UNDESA), Population Division 2017)

Country	Population (million)		
	2017	2030	2050
Afghanistan	35.5	46.7	61.9
Bangladesh	164.7	185.6	201.9
Bhutan	0.808	0.914	0.994
China	1409.5	1441.2	1364.5
India	1339.2	1513	1659
Myanmar	53.4	58.9	62.4
Nepal	29.3	33.2	36.1
Pakistan	197	244.2	306.29

2.3.3 Carbon Pool and Economy

The carbon pool plays a critical role in mitigating climate change in HKH region. For example, the peat land of the Qinghai-Tibet Plateau is one of the main carbon pools and is an essential carbon source for maintaining carbon cycle balance. It plays an important role in determining the physical and chemical properties and the fertility of the soil. The forest stores most of the carbon on land, in biomass (trunks, branches, leaves, roots, etc.) and soil organic matter. The organic carbon in forests and soils is the main body of terrestrial carbon pools. It contains about 2400–2500 Gt of organic carbon in soils up to 2 m deep, which is three times that of living biomass (Kirschbaum 2000). The carbon stocks in the HKH provides tangible and intangible goods and services, offers conditions for fuel wood, feed, pharmaceuticals and aromatic products. Forests are the major source of fuel, in addition to be the main carbon pool in Nepal, especially in rural areas, where the country's energy comes from traditional sources, including firewood, agricultural residues and animal waste.

2.4 Carbon Stock Changes

2.4.1 Carbon Sink Changes

The HKH is particularly vulnerable to soil erosion due to steep slopes, glaciers, heavy rainfall, and sparse vegetation cover. It is estimated that the average annual soil erosion rate of farmland in the hilly areas of central Nepal, the Himalayas in eastern India, the highlands of northeastern Myanmar and the highlands of southern China is 35, 54, 55 and 57 Mg/ha per year, respectively (Partap and Watson 1994). The natural CO_2 emissions are much higher compared with the annual CO_2 emissions from fossil fuel combustion (Singh et al. 2010). The soil is considered to be a huge carbon pool, small changes in soil carbon sinks may have a serious impact on the CO_2 content in the atmosphere (Lal 2016; Schlesinger and Bernhardt 2013). Satellite observations in the HKH indicate that some pastures may be degraded due to climate warming, over-grazing and human activities, and the degraded pastures account for more than 40% of the dry lands of the Qinghai-Tibet Plateau (Gao et al. 2005). Grassland and forests have changed from carbon sinks to carbon sources due to over-grazing and deforestation.

2.4.2 Carbon Source Changes

Figure 2.1 shows that the carbon emissions have increased in all countries of the HKH. Humans are accelerating the CO_2 concentrations in the atmosphere by burning fossil fuels, land use, land-use change, and forestry activities. Lal (2004) reported that the reduction of the soil organic carbon pool contributed 78 Pg of carbon to the atmosphere, and some cultivated soils have lost half to two-thirds of the original soil organic carbon pool, with cumulative emissions of 30–40 Mg carbon/ha. The emissions of fossil fuels in the HKH region since 1978 are shown in Fig. 2.1. It is obvious that the emission of China and India are much higher than other countries. The emissions in China decreased slightly in 2014. Cumulative CO_2 emissions from land in China and India in 2005–2016 were 104.3 and 21.8 Gt CO_2 respectively, and the cumulative carbon emissions per capita were 74.02 and 16.30 t CO_2, respectively. The carbon emissions from other countries in HKH are shown in Table 2.3.

2.4.3 Driving Force for Change

2.4.3.1 Land Use

In most parts of the HKH, rapid population and economic growth has increased the demand for natural resources, resulting in large changes in land use and land cover, over-exploitation, and habitat fragmentation. Land use change can be both a carbon source and carbon sink, but most land use changes increased CO_2 emissions in the atmosphere. During 1995–2015, land use changes led to emissions of 87 Pg of CO_2 into the atmosphere (Houghton et al. 2012). However, Lal (2002) believes that scientific and rational land use and management can restore about 60–70% of the carbon that has been lost, and contributes to the reduction of carbon emissions. Conversion of land to construction land has led to a significant increase in carbon emissions. In the process of converting forests to farmland, not only do the greenhouse gases in the atmosphere increase, but there is also a reduction in soil organic carbon (Watson et al. 2000). When the grassland is converted into artificial forest,

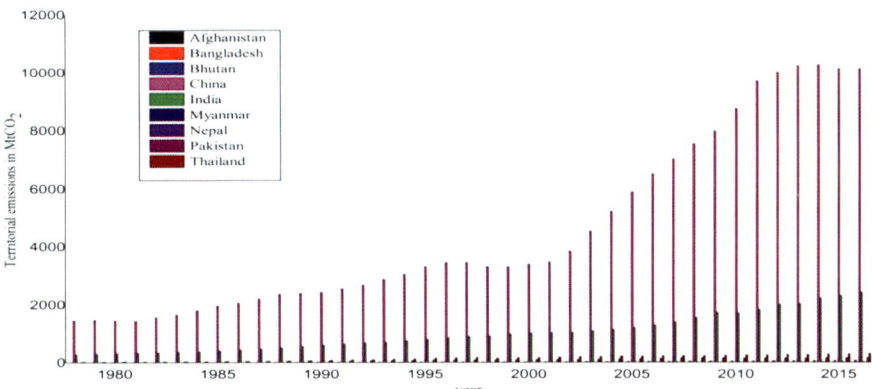

Fig. 2.1 Territorial emissions in Mt CO_2 in countries of the HKH in 1978–2016 (data from Boden et al. 2016)

Table 2.3 Comparison of emission indicators in countries of HKH region (data from Boden et al. 2016)

Country	Territorial emissions in 2005 (MtCO$_2$)	Total territorial emissions in 2005–2016 (MtCO$_2$)	Territorial emissions in 2016 (tCO$_2$ per person)	Territorial emissions in 2005–2016 (tCO$_2$ per person)
Afghanistan	1.3264	89.594	11.7063	2.52
Bangladesh	39.4466	725.916	82.402	4.41
Bhutan	0.39571	8.120	1.1262	10.05
China	5892.1334	104331.692	10150.8163	74.02
India	1221.5629	21831.236	2430.8016	16.30
Myanmar	11.5892	177.083	24.3515	3.32
Nepal	3.0814	64.675	9.1212	2.21
Pakistan	136.5243	1939.289	188.9061	9.84

Note: Territorial emissions in 2005–2016 per person was calculated with total territorial emissions in 2005–2016 divided by the population in 2016

with an increase of surface vegetation and defoliation, soil carbon storage increases. It can be seen that land use and vegetation cover changes can directly or indirectly affect the process of greenhouse gas exchange and carbon cycle between terrestrial ecosystems and the atmosphere (Watson et al. 2000).

2.4.3.2 Industry

The most direct manifestation of industrial development is the large consumption of fossil fuels. In terms of CO_2 emissions from coal combustion and cement production, the emissions in China and India are ranked as the top two countries in the past 30 years in the HKH. In 2016, the amounts of CO_2 emitted by coal combustion in China and India were 7171 and 1586 Mt, respectively, and the emissions from cement production were 1201 and 145 Mt CO_2 (Boden et al. 2016). In China, cement and coal combustion per capita CO_2 emissions are 5.1 and 0.9 t, and India is 1.2 and 0.1 t respectively. China's total emissions of CO_2 rank among the highest in the world, and the emissions from burning fossil fuels and cement production increased by nearly 75% between 2010 and 2012 (Boden et al. 2016). Liu et al. (2015) reassessed the carbon emissions from fossil fuel and cement production in China as 249 million tons of carbon in 2013. Compared with other sectors such as electricity and steel, the cement industry not only emits CO_2 directly, but also emits it from the decomposition of calcium carbonate, which is the main component of limestone in raw materials. Therefore, it has attracted special attention from the IPCC.

2.4.3.3 Agriculture

Agricultural production is an important foundation of the national economy. With the development of society, production and efficiency have greatly improved, and the scale of agricultural production has greatly developed in HKH. However, chemical fertilizers, pesticides, and straw burning have led to an increase in CO_2 emissions while agriculture is rapidly developing. In summary, AFOLU (agriculture, forestry, and other land use) emissions from high-income countries are primarily agricultural activities, and from low-income countries are mainly deforestation and land degradation (IPCC 2014). Bennetzen et al. (2016) divided the world into nine regions and used the Kaya-Porter Identity method to decompose the carbon emissions from agricultural production and land use in each region from 1970 to 2007. The results showed that underdeveloped areas increased agricultural land by 447 million hectares. The agricultural output in underdeveloped areas increased from 17.9 to 37.71 EJ/year, the livestock production increased from 1.18 to 3.36 EJ/year, and total carbon emissions increased by 34% (Bennetzen et al. 2016).

2.5 Reduce Emissions and Improve Carbon Sinks

2.5.1 *Gap with Future Paris Agreement Requirements*

According to the Paris Agreement, countries will ensure that the rise in global mean temperature is limited to 2.0 °C until the end of the twenty-first century, achieving a balance between carbon emissions and absorption in the second half of twenty-first century (net zero emissions). From the beginning of industrialization (1875) to the present, global temperatures have risen by nearly 1 °C, and it is expected to rise another 1 °C in the next 82 years. Based on CMIP5 simulation in the 32 models of the RCP 2.6 scenario, if the greenhouse gas emissions peak in 2010–2020 and then drop significantly, the net emissions will be negative. China and India were in the world's top 10 greenhouse gas emitters in 2016. China is both a developing country and the largest carbon emitter, but has committed to peak carbon emissions by 2030. In this process, it would be necessary to develop its own industrial economy and accelerate the construction of industrialization and urbanization with India.

In the "Intended Nationally Determined Contribution (INDC)", China proposed four major operational goals: (1) increase the proportion of non-fossil energy to about 20%; (2) lower CO_2 emissions per unit of GDP in 2030 by 60–65% compared to 2005; (3) reach peak CO_2 emissions at 2030; and (4) increase forest reserves by 4.5 billion m^3 in 2030 when compared to 2005. Since the reform and opening up, China's fossil energy consumption and greenhouse gas emissions have risen rapidly. In 2007, CO_2 emissions in China has exceeded the United States for the first time, making it the largest emitter in the world. It put more pressure on China to achieve its goals. In the process of achieving the INDC goals, there were difficulties and challenges in the supply of natural resources, population growth, industrial restructuring, low carbon and technology development with huge investment and construction. There was a large gap between the requirements of the Paris Agreement and the need to comply with the policies.

2.5.2 Reduction of Emission and Fixed Numerical Criteria

The Kyoto Protocol stipulated that the amount of six greenhouse gases emitted by all developed countries in 2008–2010 should be 5.2% lower than 1990, developing countries, including China and India, should voluntarily set emission reduction targets. The Chinese government promised that the CO_2 emissions per unit of GDP will fall by 40–50% by 2020 when compared to 2005. Since developing countries did not specify emission reduction targets, it can be adjusted according to national conditions and characteristics of the development stage. As China is still in the stage of rapid industrialization and urbanization, with rapid economic growth, energy consumption will continue to grow for a long period of time. Therefore, the absolute amount of emission reduction targets should not be chosen in China, but relative indicators should be selected, which are related to economic development, energy efficiency and economic improvement of carbon emissions.

2.5.3 Practice of Improving Carbon Sinks and Reducing Emissions and Recommendations

2.5.3.1 Ecological Industry

The materials and energy use would be highly efficient, and the resources and environment would be systematically developed and continuously utilized when extend and broaden the ecological industry. To date, 51 state-level eco-industrial parks have been built in China, and 82 state-level eco-industrial parks are under construction (Wang et al. 2017). In developing the ecological industry, some cooperation is based on economic, while others are based on sharing infrastructures. Such as sugar mill in Pakistan, it not only increased the income of local people, but also promoted the development of other industries in the process of producing sugar (Qureshi and Mastoi 2015). The HKH region has a vast territory and a large variety of natural resources. Developing ecological industry combines local resources with enterprises and products to reduce carbon emissions and contribute to emission reduction. These improvements could increase the income of residents and reduce poverty in the HKH.

2.5.3.2 Public-Private Reasonable Configuration

The lifestyles of residents in the HKH are mostly self-sufficient. In the development of countries in the HKH region, due to the improper management of public and private lands, there is often excessive deforestation of forest and over-grazing on pastures. The economic distribution inequality and economic disputes are caused by unreasonable public-private distributions. Since forests and grasslands are the main carbon sinks, public-private proper management of land could increase terrestrial vegetation and soil carbon storage, reduce deforestation, improve harvesting and grazing practices, improve utilization efficiency, and effectively control natural disasters (fires, pests, etc.). By rationally allocating resources, carbon sinks could be increased to reduce carbon emissions, so as to adapt to climate change, reduce conflicts, and bring economic benefits to local residents.

2.5.3.3 Afforestation Reward

With forest degradation or over-harvesting, organic carbon stored in forests and soils is decomposed and released into the atmosphere. Increasing carbon sinks occurs by reducing emissions, afforestation, strengthening forest protection, increasing forest accumulation, and enhancing the carbon absorption function of forests. Forests have strong capacity for carbon sequestration, so they should be restoring and protected in response to climate change. In 2011, the State Council of China allocated 1.36 billion RMB of financial funds to implement the grassland ecological protection subsidy and reward mechanism policy. The number of farmers and herdsmen who enjoyed the policy reached 10.567 million. The project fenced 450 ha of grassland, replanted 1.459 million ha of servely degraded grassland, and planted 47,000 ha of artificial grassland in 2011, which increased the grassland carbon sink.

2.5.3.4 Infrastructure

In terms of the overall quality of infrastructure services, Bangladesh, India, Nepal and Pakistan in the HKH are substantially lower than the global average (Table 2.4). Floods are the most common water-induced hazard in HKH. Government efforts to protect people from flood waters and mitigate the impacts of flood through the structural measures like embankments (ICIMOD 2013). Chen (2015) studied the carbon balance of 35 major cities in China and found that the green space (the urban green space infrastructure dominated) accounted for 6.38% of the total urban area of the country by the end of 2010, accounted for 51.7% of the total urban green space (657 cities) in China. The urban carbon infrastructure of 35 cities had an estimated carbon storage of 18.7 million tons and an average carbon density of 21.34 tons/ha. In 2010, the total carbon sequestration was 1.9 million tons, and the average carbon sequestration rate was 2.16 tons/ha per year.

Table 2.4 Quality of infrastructure services (World Economic Forum 2014)

Country	Total infrastructure	Roads	Airport	Electricity supply
World	4.23	4.02	4.36	4.50
Bangladesh	2.82	2.88	3.02	2.55
Bhutan	4.63	4.31	3.51	5.85
China	4.36	4.61	4.72	5.22
India	3.75	3.79	4.27	3.43
Nepal	2.93	2.90	2.92	1.83
Pakistan	3.32	3.81	3.92	2.07

Note: Scores on a scale of 1 (low) to 7 (high)

2.5.3.5 Education

Despite the abundance of food, biodiversity and energy, there are still a large number of poor people living in the HKH region. Due to rugged terrain, rapid population growth, poor management and other reasons, there is a degradation of resources and a decline in agricultural productivity. The people in the region are under-educated and have a low awareness of emission reductions. In the HKH region, agricultural production is mainly practiced by old people (Kurvits and Kaltenborn 2014). Therefore, it is necessary to increase education to attract young and skilled people to engage in agriculture and help preserve the cultures, traditions and natural environment of the mountains. Encouraging young people to maintain and develop mountain farming systems for future sustainable food production, could help alleviate poverty in the HKH region, and increase efforts to develop mountain agriculture and livestock production.

In addition, strengthening education and building an effective knowledge sharing network in HKH is essential to stimulate a range of adaptations and innovations through new policies, pilot projects and monitoring to reduce carbon emissions and combat climate change (Kurvits and Kaltenborn 2014). There is a need to make people aware of the importance of appropriate harvesting and grazing, to protect forests and pastures, improve the quality of life, and reduce carbon emissions caused by cutting young leaves, deforestation and over-grazing.

2.5.3.6 Technical Assistance

At present, it is necessary to continuously adapt to climate change in the process of reducing emissions. Due to the uncertainties of the impact of climate change, the development and application of advanced technologies is the means to combat climate change. The technology obviously affects the lifestyles, consumption patterns and social and cultural customs of local residents in the HKH and brings new development opportunities for people in remote areas. Many rural areas in the HKH region are shifting from subsistence to market agriculture farming dominated by cash crops. Increasing the technical assistance in global services, software and data storage makes the HKH region easier to understand. In the process of increasing carbon sinks, not only man-power and financial resources, but also management and technical support were required.

2.6 Anti-Poverty and Carbon Trade, Carbon Compensation

2.6.1 Challenges of Poverty in the HKH Region

The poverty rate in the HKH mountain is 5% higher than the average of the entire HKH (Kurvits and Kaltenborn 2014). The factors that contribute to poverty are also different, such as access to basic conveniences, poor roads and high dependence on natural resources. Although the HKH area could regulate by itself, there are inadequate and insufficient access to resources, technology and funds. The basic natural resources are rapidly degraded and cannot be fully integrated into the value chain and market. It will seriously affects the ability of the HKH to cope with climate change and to take advantage of increasing opportunities to escape poverty. In solving the poverty problem in the HKH region, there is an urgent need to improve the resilience of vulnerable mountain families and communities by providing essential ecosystem services, especially to challenges faced by women and vulnerable groups. An international conference was held in Nepal, with the main purpose of collecting the latest data on the poverty line and promoting a sustainability policy in the HKH and setting the tone for strengthening regional partners to promote sustainable mountain development.

2.6.2 How to Use Carbon Trade to Reduce Poverty

India and China are big carbon emitters, and in implementing the emission reduction tasks, there will inevitably be some enterprises that cannot complete the task. In addition, the cost of industrial emission reduction is much higher than the cost in forests, grasslands and farmland. Therefore, some enterprises will definitely purchase emission reduction quotas in areas with high carbon sequestration potential. In this process, the carbon sink will be gradually commercialized, and the carbon trade will gradually be formed. Except for mutual transactions between countries in the HKH region, it also provides resources for purchase by European countries, which promotes cooperation between countries and increases the income of the HKH region.

At present, China has launched pilot projects for carbon trading. The main purpose is examining own carbon emissions trading system. The seven provinces are mainly based on energy reduction and are economically developed and industrialized. Some enterprises in these regions will be moved to the western provinces, especially the three rivers source region, to purchase emission reduction quotas when they fail to complete their emission reduction tasks. The economically developed provinces will improve the economy in undeveloped regions by buying fixed carbon sinks in the process of ecological construction from countries with less carbon emissions.

2.6.3 How to Use Carbon Compensation to Reduce Poverty

In the HKH region, most of the carbon trading is conducted from the aspects of emission reduction quota. The HKH region does not make mandatory emission reduction credits. Among the carbon standards in China, one is the "Sanjiangyuan Standard" of the Qinghai's environmental energy exchange. In the ecological construction of the "Sanjiangyuan Nature Reserve" implemented by the central government, food production and the standard of living, as well as the conditions of the local farmers and herdsmen have improved. At the same time, the ecological construction funds in Sanjiangyuan are still dominated by government supported programs. The funds are limited, and the ecological compensation standards are low. Translating the more fixed carbon sinks in the ecological recovery process through the relevant trading platforms such as the Qinghai environmental exchange, would certainly broaden the source of ecological compensation funds and ensure the sustainability of the funds and long-term ecological protection. Farmers and herders in the HKH could obtain ecological compensation fees from carbon trading funds, thereby increasing the income of each household, improving living conditions and reducing poverty.

2.6.4 Advantage and Disadvantage

In the HKH region, poverty could be reduced mainly through carbon trade and carbon compensation. Due to the high forest coverage and wide grassland area in the region and the broad terrain, the HKH could maintain water and soil, increase carbon sinks, reduce carbon emissions, provide carbon emission quotas for developed countries. The high vegetation coverage is an advantage for the development of the HKH region, but due to low education, many areas of the HKH are not aware it.

All countries in the HKH are developing country, and the poor are often concentrated in remote areas with low economy, poor transportation, inadequate education and backward concepts. Children in the family must earn money to support the family instead of going to school. Women are expected to raise many children to maintain the livelihood of their members. Due to the poor education of the people in the HKH, the poor can only engage in simple labor. It is difficult for the poor to find employment and they are often unemployed. In addition, the HKH is a multi-ethnic region, and religious beliefs are widespread, which can often lead to social problems and conflicts.

2.7 Conclusion

The HKH region has been recognized globally as an important independent region. The biodiversity and cultural diversity of HKH are very rich, and includes a wide variety of habitats, micro climates and environmental conditions. Climate change has brought new challenges to the HKH, which has been threatened by problems like poverty, environmental degradation, depletion of natural resources, water shrinkage and desertification. In addition, melting glaciers, greenhouse gases, air pollution, biodiversity conservation, and disasters caused by extreme events require collective action and cooperation among nations. One of the main challenges facing the HKH region is the limited availability of spatial information, due to the complex landscape and lack of appropriate regional frameworks and mechanisms. The climate change situation in the HKH region has received worldwide attention. Understanding the dynamic changes of the mountain environment in the HKH region is necessary to formulate appropriate policies and to design rational development interventions.

References

Asif, M. 2009. Sustainable energy options for Pakistan. *Renewable & Sustainable Energy Reviews* 13 (4): 903–909.

Bennetzen, E.H., P. Smith, and J.R. Porter. 2016. Agricultural production and greenhouse gas emissions from world regions-The major trends over 40 years. *Global Environmental Change* 37: 43–55.

Boden, T.A., G. Marland, and R.J. Andres. 2016. Global, regional, and national fossil-fuel CO_2 emissions. Oak Ridge, TN, USA: Carbon Dioxide Information Analysis Center, Oak Ridge National Laboratory, US Department of Energy.

Chand, R., S.S. Raju, and L.M. Pandey. 2008. Progress and potential of horticulture in India. *Indian Journal of Agricultural Economics* 63 (3): 299–309.

Chen, W.Y. 2015. The role of urban green infrastructure in offsetting carbon emissions in 35 major Chinese cities: A nationwide estimate. *Cities* 44: 112–120.

Chettri, N., A.B. Shrestha, Z. Yan, et al. 2012. Real world protection for the "Third Pole" and its people. In *Protection of the three poles*, ed. F. Huettmann, 113–133. Tokyo: Springer.

Choudhury, N. 2014. Environment in an emerging economy: The case of environmental impact assessment follow-up in India. In *Large dams in Asia*, ed. M. Nüsser, 101–124. Dordrecht: Advances in Asian Human-Environmental Research. Springer.

Dong, S.K., L. Wen, L. Zhu, et al. 2010. Implication of coupled natural and human systems in sustainable rangeland ecosystem management in HKH region. *Frontiers of Earth Science in China* 4 (1): 42–50.

Eriksson, M., J.C. Xu, A.B. Shrestha, et al. 2009. The changing Himalayas: impact of climate change on water resources and livelihoods in the greater Himalayas. ICIMOD. Kathmandu.

FAO, IFAD, WFP. 2013. *The state of food insecurity in the world 2013. The multiple dimensions of food security*. Rome: FAO.
Gao, Q., Y. Li, E. Lin, et al. 2005. Temporal and spatial distribution of grassland degradation in northern Tibet. *Acta Geographica Sinica* 60 (6): 965–973.
Gurung, P., and T.Y.C. Sherpa. 2014. Freshwater scarcity and sustainable water management in the Hindu Kush-Himalayan (HKH) Region. *Hydro Nepal Journal of Water Energy & Environment* 15 (15): 42–47.
Houghton, R.A., G.R.V.D. Werf, R.S. DeFries, et al. 2012. Chapter G2 carbon emissions from land use and land-cover change. *Biogeosciences* 9 (1): 5125–5142.
ICIMOD. 2013. Policy and institutions in adaptation to climate change: Case study on flood mitigation infrastructure in India and Nepal. In *ICIMOD working paper 2013/4*. Kathmandu: ICIMOD.
IPCC. 2014. Climate change: Mitigation of climate change. Contribution of working group III to the fifth assessment report of the intergovernmental panel on climate change. Cambridge University Press, Cambridge, United Kingdom and New York, NY, USA.
IUCN/IWMI. 2003. *Ramsar Convention and WRI*. Watersheds of The world. https://www.wri.org/publication/watersheds-world.
Jackson, R.B., K. Lajtha, S.E. Crow, et al. 2017. The ecology of soil carbon: Pools, vulnerabilities, and biotic and abiotic controls. *Annual Review of Ecology Evolution & Systematics* 48 (1): 419–445.
Khadka, D., M.S. Babel, S. Shrestha, et al. 2014. Climate change impact on glacier and snow melt and runoff in Tamakoshi basin in the Hindu Kush Himalayan (HKH) region. *Journal of Hydrology* 511 (4): 49–60.
Kirschbaum, M.U.F. 2000. Will changes in soil organic carbon act as a positive or negative feedback on global warming? *Biogeochemistry* 48 (1): 21–51.
Kumar, P., S. Mittal, and M. Hossain. 2008. Agricultural growth accounting and total factor productivity in South Asia: A review and policy implications. *Agricultural Economics Research Review* 21 (2): 145–172.
Kurvits, T., and B. Kaltenborn. 2014. *The Last Straw: Food security in the Hindu Kush Himalayas and the additional burden of climate change*. Kathmandu: International Centre for Integrated Mountain Development (ICIMOD).
Lal, R. 2002. Soil carbon dynamics in cropland and rangeland. *Environmental Pollution* 116 (3): 353–362.
Lal, R. 2004. Soil carbon sequestration to mitigate climate change. *Geoderma* 123 (1–2): 1–22.
Lal, R. 2016. Soil health and carbon management. *Food & Energy Security* 5 (4): 212–222.
Liu, Z., D. Guan, W. Wei, et al. 2015. Reduced carbon emission estimates from fossil fuel combustion and cement production in China. *Nature* 524 (7565): 335–346.
Lópezpujol, J., F.M. Zhang, and S. Ge. 2006. Plant biodiversity in China: Richly varied, endangered, and in need of conservation. *Biodiversity and Conservation* 15 (12): 3983–4026.
Mats, E., J.C. Xu, S. Arun Bhak, et al. 2009. *The changing Himalayas: Impact of climate change on water resources and livelihoods in the greater Himalayas*. Kathmandu: International Centre for Integrated Mountain Development (ICIMOD).
Mittermeier, R.A, P.R. Gil, M. Hoffmann, et al. 2017. Hotspots: Earth's biologically richest and most endangered terrestrial ecoregions. *Journal of Mammalogy* (2): 237–238.
Molden, D.J., R.A. Vaidya, A.B. Shrestha, et al. 2014. Water infrastructure for the Hindu Kush Himalayas. *International Journal of Water Resources Development* 30 (1): 60–77.
Partap, T., and H.R. Watson. 1994. Sloping agricultural land technology (SALT): A regenerative option for sustainable mountain farming. In *ICIMOD occasional paper no.23*, Kathmandu, Nepal.
Qureshi, M.A., and G.M. Mastoi. 2015. The physiochemistry of sugar mill effluent pollution of coastlines in Pakistan. *Ecological Engineering* 75 (75): 137–144.
Rasul, G. 2014. Food, water, and energy security in South Asia: A nexus perspective from the Hindu Kush Himalayan region. *Environmental Science & Policy* 39 (5): 35–48.

Schaller, G.B., and A.L. Kang. 2008. Status of Marco Polo sheep *Ovis ammon polii* in China and adjacent countries: Conservation of a vulnerable subspecies. *Oryx* 42 (1): 100–106.

Schlesinger, W.H., and E.S. Bernhardt. 2013. *Biogeochemistry: An analysis of global change*, 665–672. New York: Academic.

Sharma, E., I. Khadka, and G. Rana. 2009. *Mountain biodiversity and climate change*. Kathmandu: International Centre for Integrated Mountain Development (ICIMOD).

Singh, B.K., R.D. Bardgett, P. Smith, et al. 2010. Microorganisms and climate change: Terrestrial feedbacks and mitigation options. *Nature Reviews Microbiology* 8 (11): 779–790.

Surendra, K.C., S.K. Khanal, P. Shrestha, et al. 2011. Current status of renewable energy in Nepal: Opportunities and challenges. *Renewable & Sustainable Energy Reviews* 15 (8): 4107–4117.

Uddin, S.N., R. Taplin, and X. Yu. 2007. Energy, environment and development in Bhutan. *Renewable & Sustainable Energy Reviews* 11 (9): 2083–2103.

United Nations, Department of Economic and Social Affairs (UNDESA), Population Division. 2017. World population prospects: The 2017 revision, key findings and advance tables. In *Working paper no. ESA/P/WP/248*.

Wang, Q.S., S.S. Lu, X.L. Yuan, et al. 2017. The index system for project selection in ecological industrial park: A China study. *Ecological Indicators* 77: 267–275.

Watson, R.T., I.R. Noble, B. Bolin, et al. 2000. Land use, land-use change and forestry: a special report of the Intergovernmental Panel on Climate Change. Cambridge University Press.

World Economic Forum. 2014. *Competitiveness dataset*. http://reports.weforum.org/global-competitiveness-report-2014-2015/defining-sustainable-competitiveness/

Xu, J., R.E. Grumbine, A. Shrestha, et al. 2010. The melting Himalayas: Cascading effects of climate change on water, biodiversity, and livelihoods. *Conservation Biology* 23 (3): 520–530.

Chapter 3
Tracking of Vegetation Carbon Dynamics from 2001 to 2016 by MODIS GPP in HKH Region

Zhenhua Chao, Mingliang Che, Zhanhuan Shang, and A. Allan Degen

Abstract Carbon dynamics, a key index to evaluate ecosystems, are very complex in the Hindu Kush Himalayan (HKH) region due to the topography, diverse regional climate, and different land cover types. MODIS GPP was used to evaluate carbon sequestration in the HKH region from 2001 to 2016. In general, the spatio-temporal variation of the average daily gross primary productivity (GPP) was very heterogeneous due to the changing terrain, diverse regional climate, and different land cover types in the region. Many factors should be considered for GPP measurements, including satellite, airplane, ground-based and modelling data. We concluded that it is necessary to determine the driving forces of GPP in the future in order to establish scientific policies and development programs for the HKH region.

Keywords Hindu Kush Himalayan region · Carbon dynamics · MODIS GPP · Spatio-temporal variation · Qinghai-Tibet Plateau

---

Z. Chao (✉) · M. Che
School of Geographic Science, Nantong University, Nantong, China
e-mail: chaozhenhua@lzb.ac.cn

Z. Shang
School of Life Sciences, State Key Laboratory of Grassland Agro-Ecosystems, Lanzhou University, Lanzhou, China
e-mail: shangzhh@lzu.edu.cn

A. A. Degen
Desert Animal Adaptations and Husbandry, Wyler Department of Dryland Agriculture, Blaustein Institutes for Desert Research, Ben-Gurion University of the Negev, Beer Sheva, Israel
e-mail: degen@bgu.ac.il

Abbreviations and Nomenclature

APAR	Absorbed photosynthetically active radiation
a.s.l.	Above sea level
BPLUT	Biome parameter look-up table
DEM	Digital elevation model
EROS	Earth resources observation and science
EOS	Earth observing system
FPAR	The fraction of photosynthetically active radiation
GIS	Geographic information system
GMAO	Global modelling and assimilation office
GPP	Gross primary productivity
HKH	Hindu Kush Himalayan region
km	Kilometre
LAI	Leaf area index
LULC	Land use and land cover
LUE	Light use efficiency
mm	Millimetre
MODIS	Moderate resolution imaging spectrometer
NASA	National aeronautics and space administration
NDVI	Normalized difference vegetation index
NPP	Net primary productivity
PAR	Photosynthetically active radiation
QTP	Qinghai-tibetan plateau
RS	Remote sensing
SIF	Solar-induced chlorophyll fluorescence
TRHR	Three-river headwaters region
USGS	The United States geological survey

3.1 Introduction

The Hindu Kush Himalayan (HKH) region comprises an area of more than 4.3 million km^2 and is characterized by a large number of physiographic landscapes, various regional climate types and bio-systems, the largest cryosphere in the world apart from the Antarctica and the Arctic, the source of many highly important large rivers and large stores of water in the form of snow and glaciers (You et al. 2017). With its vast terrain, the HKH region has a substantial influence on the East Asian monsoon, and even on the global atmospheric circulation (Sharma et al. 2016).

The melting rate of glaciers has been accelerating in the HKH in recent years, especially in the south and the east, due to global warming and anthropogenic impacts such as increased greenhouse gas emissions. In addition, the combined action of rising temperatures and more frequent intense precipitation in the mid to high mountains increases the risk of flooding disasters in the lower courses of the main rivers (Ren and Shrestha 2017). This region, which includes fragile alpine ecosystems, irrigated cropland and snow and glaciers, is one of the most sensitive areas to climate change and human activities (Xu et al. 2009). It is very important for the livelihood of the local people and for ecological functions and, consequently, it is necessary to carry out relevant research to ensure sustainable development and beneficial ecosystem services. However, relevant data are deficient and surveys are difficult to carry out in the HKH region.

Carbon dynamics are influenced by the combined effects of climate change, environmental factors, and land use/land cover (LULC), and can be regarded as a key index to evaluate ecosystem change (Almeida et al. 2018; Chao et al. 2018). Vegetation productivity of terrestrial ecosystems effectively reflects the absorption of CO_2 from the atmosphere by plants through photosynthesis. In this process, light energy is converted into chemical energy and organic dry matter is produced. Therefore, vegetation productivity can be used to estimate the support capacity and to evaluate the carbon dynamics of the ecosystem. Close to half of the vegetation produced is either consumed as food or used as fuel directly or indirectly by people, being the foundation for human survival and sustainable development. Vegetation primary productivity is commonly split into two components: gross primary productivity (GPP) and net primary productivity (NPP). GPP is the overall rate of biomass production at the ecosystem scale, whereas NPP is the biomass produced after accounting for energy lost to cellular respiration and maintenance of the plants. As a key component of the global carbon cycle, terrestrial GPP plays an important role in the global carbon, water, and energy cycles and shows meaningful spatial and temporal variations (Zhang et al. 2017). It is of crucial importance to predict GPP and the total carbon assimilation rate of vegetation for evaluating the role of the global carbon cycle in an ecosystem (Sánchez et al. 2015).

Terrestrial GPP is influenced by environmental factors (e.g. nutrients, water, light, and CO_2), biotic factors (e.g. canopy structure, leaf phenology) and LULC, and provides important information on the ecosystem function in response to local and global environmental changes. Remote sensing (RS) is a popular technique in measuring the effect of continuous temporal and spatial information on biophysical land surface properties (Vitousek et al. 1997). With RS, information related to GPP such as leaf area index (LAI) and the fraction of radiation absorbed by green vegetation can be determined. Recent research questions employing this methodology have centred around: (1) the increase in the atmospheric concentration of greenhouse gases, mainly CO_2, and the resulting climate change; (2) the terrestrial vegetation responses to these changes; and (3) the capacity to sequester and store atmospheric carbon for counterbalancing the anthropogenic CO_2 emissions (Almeida et al. 2018). GPP variation in space and time is controlled by climate and physiological processes of vegetation photosynthesis and autotrophic respiration, and is limited primarily by absorbed photosynthetically active radiation (APAR). These controls vary in importance in daily and seasonal time scales.

Currently, several GPP datasets including ground upscaling FLUXNET observations (MPI-BGC) and moderate resolution imaging spectrometer estimations (MODIS GPP) can provide GPP information over large regions, but they often exhibit different estimation accuracies for different vegetation types (Lin et al. 2018). MODIS is a widely used global monitoring sensor for various environmental parameters of the earth surface onboard the NASA Earth Observing System (EOS) satellites Terra/Aqua. A standard suite of global measurements characterizing vegetation cover, LAI, GPP, and NPP at the 500 m spatial resolution are now being produced operationally on observations from MODIS. Being largely dependent on the photosynthetically active radiation (PAR) conversion efficiency, MODIS GPP is based on the Monteith's theory that NPP is related linearly to the amount of absorbed photosynthetically active radiation. Generally, MODIS GPP can accurately reproduce seasonal and spatial variability (Zhao et al. 2005; Zhao and Running 2010; Frazier et al. 2013; Running et al. 2015). Carbon dynamics is more subjected to climate change and anthropogenic activity in the HKH region than in other regions. In such a complex, fragile region, RS technology can provide timely information to evaluate the carbon cycle. In this chapter, MODIS GPP was used to evaluate carbon sequestration in this region from 2001 to 2016.

3.2 HKH Region and Data Analysis

3.2.1 HKH Region

The HKH region extends from Afghanistan in the northwest to Myanmar in the southeast, covering all or part of eight countries, namely, Afghanistan, Bangladesh, Bhutan, China, India, Myanmar, Nepal and Pakistan (Fig. 3.1; Table 3.1). Elevation varies greatly, ranging from sea level to above 8000 m a.s.l., with many mountain peaks above 6000 m a.s.l. The region is known for its unique flora and fauna, high level of endemism and numerous eco-regions of global importance (Chettri et al. 2008; Elalem and Pal 2015). Elevation zones across the HKH region range from

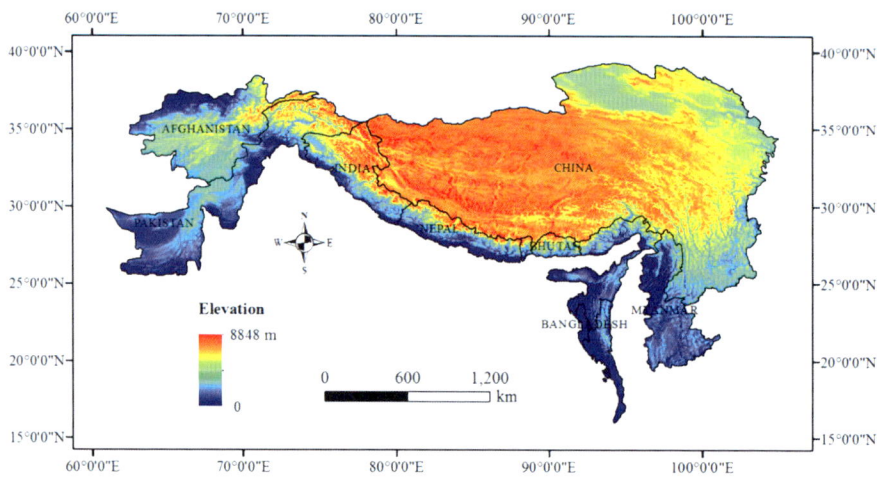

Fig. 3.1 The range of elevations in the Hindu Kush Himalayan region

Table 3.1 The area and population of countries in the Hindu Kush Himalayan region (Chettri et al. 2008; United Nations 2017)

Country	Total area (km²)	Total area within HKH (km²)	Population of each country within HKH (million)
Afghanistan	652,225	390,475	28.28
Bangladesh	143,998	13,295	1.33
Bhutan	46,500	46,500	0.81
China	9,596,960	2,420,266	29.48
India	2,387,590	461,139	72.36
Myanmar	676,577	317,629	11.01
Nepal	147,181	147,181	29.30
Pakistan	796,095	489,988	39.36
Total	**14,447,126**	**4,286,473**	**211.93**

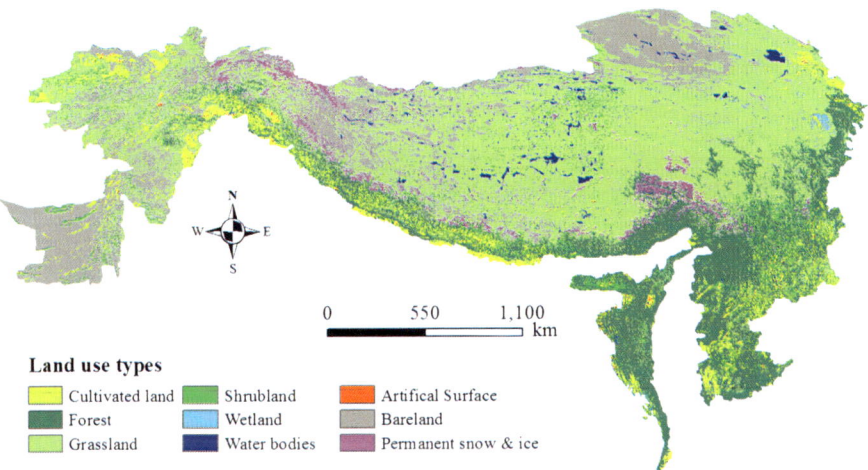

Fig. 3.2 Land use types in the Hindu Kush Himalayan region in 2010

tropical (<500 m) to alpine ice-snow (>6000 m), with the principal vegetation regime comprising tropical and subtropical rain forests, temperate broadleaf deciduous or mixed forests and temperate coniferous forests, and also include high-altitude cold shrub or steppe and cold deserts (Chettri et al. 2008). According to GlobeLand30–2010 data, approximately 43% of the HKH is comprised of grasslands, 20% forest, 3% shrub land and 7% cultivated land. The remaining 27% includes barren land, tundra, built-up areas, snow cover and bodies of water (Fig. 3.2). The HKH has the largest cryosphere in the world and is the source of ten major river basins, including the Brahmaputra, Ganges, Indus, Mekong, Yangtze, and Yellow Rivers. A population of approximately 211 million people lives in the region and more than 1.3 billion people live in downstream basins of the ten large rivers (Table 3.2).

Table 3.2 The ten major river basins of the Himalayan region

River	Basin area (km²)	Countries covered	Population (million)
Amu Darya	534,739	Afghanistan, Tajikistan, Turkmenistan, Uzbekistan	20.86
Brahmaputra	651,335	China, India, Bhutan, Bangladesh	118.64
Ganges	1,016,124	India, Nepal, China, Bangladesh	408.97
Indus	1,081,718	China, India, Pakistan	178.48
Irrawaddy	413,710	Myanmar	32.68
Mekong	805,604	China, Myanmar, Laos, Thailand, Cambodia, Vietnam	57.20
Salween	271,914	China, Myanmar, Thailand	5.98
Tarim	1,152,448	Kyrgyzstan, China	8.07
Yangtze	1,722,193	China	368.55
Yellow	944,970	China	147.42
Total	**8,594,755**	**16**	**1346.85**

The HKH region is undergoing rapid alterations due to climate change, disasters, economic globalization, infrastructure development, migration, and urbanization. As a result, natural resources are being destroyed steadily, making sustainable socio-economic conditions more difficult. Moreover, climate change has undermined water availability, food availability, livelihoods, and health security in the region (Tiwari and Joshi 2015). However, natural regeneration of forests is occurring in some areas due to the increasing trend of rural out-migration and the resulting abandonment of agricultural land.

3.2.2 Data Analysis

3.2.2.1 MODIS GPP

The light use efficiency (LUE) theory was developed by Monteith and the core concept is that vegetation productivity relates to the incident solar energy under the conditions of sufficient water and land fertility (Monteith 1972). Models based on the LUE concept are commonly used to estimate GPP with RS information. An assumption of the LUE model is that a proportional relationship exists between GPP and the amount of APAR. Under favorable environmental conditions, APAR is converted into chemical energy at a fixed rate ε_{max} for a given vegetation type. But the potential LUE should consider environmental stresses in nature and is reduced by a scalar factor (Eq. 3.1).

$$\text{GPP} = \text{PAR} \times \text{FPAR} \times \varepsilon_{max} \times S \tag{3.1}$$

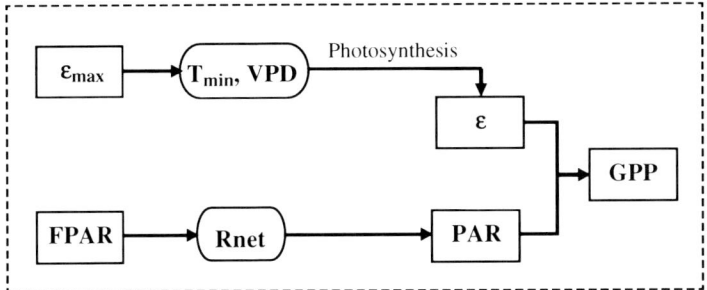

Fig. 3.3 The data flow chart in the daily part of the MODIS GPP algorithm

where PAR is the photosynthetically active radiation reaching the canopy (MJ m^{-2}), FPAR is the fraction of PAR absorbed by vegetation, ε_{max} is the maximum LUE (g C MJ^{-1}) without environmental stresses, and S is a scalar varying from 0 to 1, representing the reduction of potential LUE by environmental stresses. Figure 3.3 shows the flow of the MODIS GPP.

The latest MODIS GPP version 6, MOD17A2H, is a cumulative 8-day composite of values. The GPP parameters are derived empirically from the output of Biome-BGC, a complex ecosystem model, simulations being performed over a gridded global domain using multiple years of gridded global daily meteorological observations (Running et al. 2015). The spatial resolutions of the FPAR (MOD15A2H) and GPP (MOD17A2H) products were improved from 1 km to 500 m and BPLUT (Biome Parameter Look-up Table) and the daily GMAO (Global Modelling and Assimilation Office) meteorological data were also updated for this collection.

It is very difficult to use the annual GPP in the HKH region to evaluate carbon dynamics because of the biodiversity and complex natural factors, as it is a mountainous region with different regional climates. Along with global warming, warming amplifications in high-altitude regions in the HKH are increasing meltwater (You et al. 2017). The vegetation growth on the QTP is particularly sensitive and vulnerable to the rapidly changing climate (Liu et al. 2018; Shi et al. 2018). For example, increased precipitation could stimulate ecosystem carbon fluxes and increase vegetation biomass (Wu et al. 2011; Miao et al. 2015). In the HKH region, most precipitation occurs in July–August and the growing season for plants is from June to September. Heat accumulation before the growing season is the main trigger of vegetation emergence on the QTP; however, it is still unclear which seasonal climate variables dominantly affect this trigger (Shen et al. 2016; Cao et al. 2018). The spatial distribution of GPP depends primarily on climatic conditions; consequently, the MODIS GPP (MOD17A2H) was downloaded on 7.28–9.05 for 2001, 7.27–9.04 for 2004, 7.28–9.05 for 2007, 7.28–9.05 for 2010, 7.28–9.05 for 2013, and 7.27–9.04 for 2016.

3.2.2.2 Digital Elevation Model (DEM) and Land Cover Data

GTOPO30 is a global digital elevation model (DEM) with a horizontal grid spacing of 30 arc seconds (approximately 1 km) that was derived from several raster and vector sources of topographic information. The DEM data are provided by USGS Earth Resources Observation and Science (EROS) Center from the following site: https://e4ftl01.cr.usgs.gov/MEASURES/SRTMGL1.003/2000.02.11/index.html.

The data used to produce GlobeLand30–2010 are multi-spectral images with a main spatial resolution at 30 m, including Landsat TM and ETM+ multi-spectral images and multi-spectral images from HJ-1—Chinese Environmental Disaster Alleviation Satellite. The acquisition time of these images used for GlobeLand30–2010 covered vegetation growing seasons from 2009 to 2010 with cloudless days. The overall accuracy of the land cover data is about 84% and the Kappa indicator is 0.78. The data for the research was provided by the National Geomatics Center of China and the website for the data is http://www.globeland30.org/GLC30Download/index.aspx.

3.3 GPP Variation in HKH

The average daily GPP in the HKH region varied among areas. Higher GPP emerged in areas covered with dense forest (Figs. 3.2 and 3.4), mainly in the southeast and southwest HKH region, especially in Bangladesh, Myanmar, and the Taba mountain range in China. Generally, plant photosynthesis increases with increasing atmospheric CO_2 concentrations, while warming temperature is not conducive to vegetation biomass growth (Cox et al. 2013). In general, the number of extreme cold events decreased, whereas extreme warm events increased in most parts of the HKH during 1961–2015 (Sun et al. 2017). This could explain why the GPP in the southeast HKH region tended to decrease despite the increased air temperature and the ample rainfall. Some areas are deserts or have sparse vegetation and so there was no GPP estimation. Many factors such as land cover change, soil nutrients, CO_2 concentration and air temperature could be responsible for the spatio-temporal variation of GPP in the region. In addition, anthropogenic activities such as farmer out-migration and urban sprawl can also affect the GPP.

3.3.1 The GPP Variation in the Main Part of the QTP in China

The QTP in China is characterized by strong radiation, low air temperature with a large diurnal range, and uneven rainfall distribution. More than 50% of the plateau area is alpine grasslands, which is dominated by low-producing, cold-tolerant perennial plants (Chen et al. 2014). Climate warming and increased precipitation have caused an upward trend of GPP, especially in the southern and northern parts of the

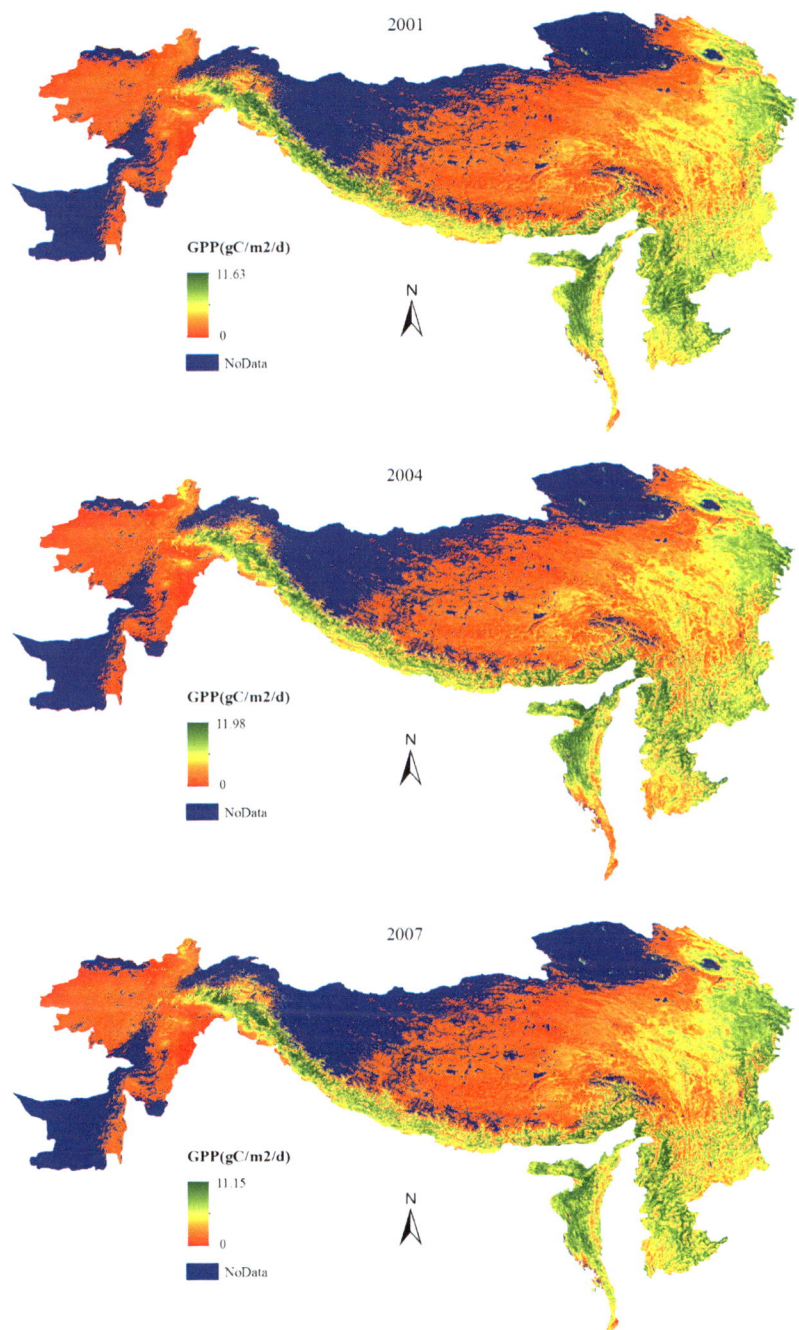

Fig. 3.4 The GPP in the Hindu Kush Himalayan region from 2001 to 2016

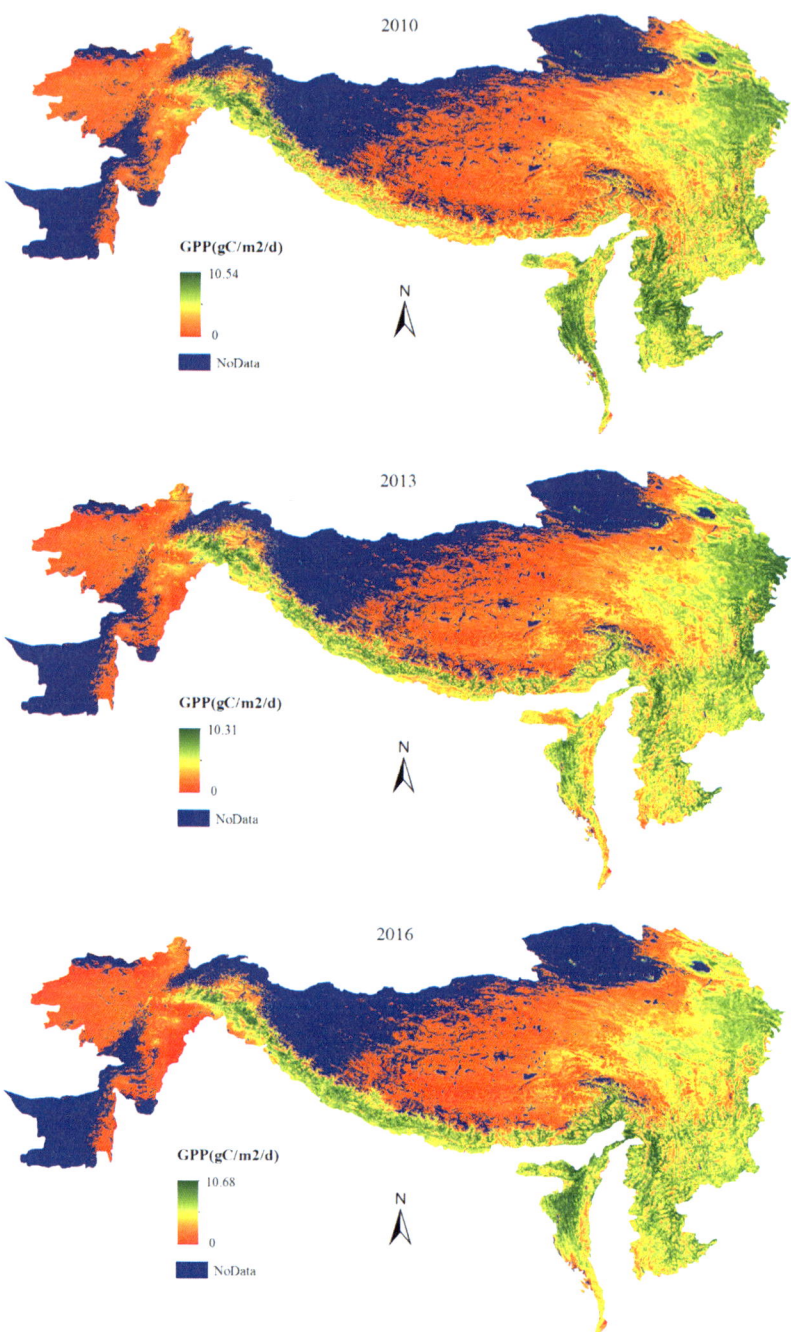

Fig. 4 (continued)

QTP. The areas undergoing vegetation degradation were mainly in the hinterland, Qinghai Lake, Qaidam Basin, Hexi Corridor, Ali region and the southeastern edge of the plateau. Human disturbances such as overgrazing and urbanization, warming temperature and reduced precipitation had negative effects on vegetation, particularly in the fragile alpine grasslands (Peng et al. 2012). Aeolian desertified land is scattered across the QTP, but occurs mainly in the western and northern parts. Ecological protection projects being carried out by the Chinese government and local people and increased rainfall helped reduce the aeolian desertified land (Zhang et al. 2018). The average MODIS GPP of most alpine grasslands was less than 5 g C per m^2 per day, but the total GPP was increasing. In the northwest area, the annual precipitation was less than 50 mm, which lead to a vast expanse of land with little or no vegetation and there was no estimated MODIS GPP. The reason is that these areas are either bare, snow-covered or have sporadic vegetation that is not detected by the GPP algorithm. In the southeast, abundant precipitation and appropriate air temperature increased the GPP between 2001 and 2016, but the distribution of vegetation was scattered. In the middle part of the QTP, the GPP tended to decrease.

3.3.2 The GPP Variation in the North-Western Part: Afghanistan and Pakistan

In Afghanistan, about 56% of the land cover is barren, and forests are located mainly in the east. Most precipitation occurs in winter and the main water source for the cropland is from rivers. Low GPP is common across the country; however, no values are available for the northwestern areas and for the barren land. The average daily GPP decreased between 2001 and 2016 in most areas, with the highest values in the forest areas (Fig. 3.5). The general decrease in GPP was a result of

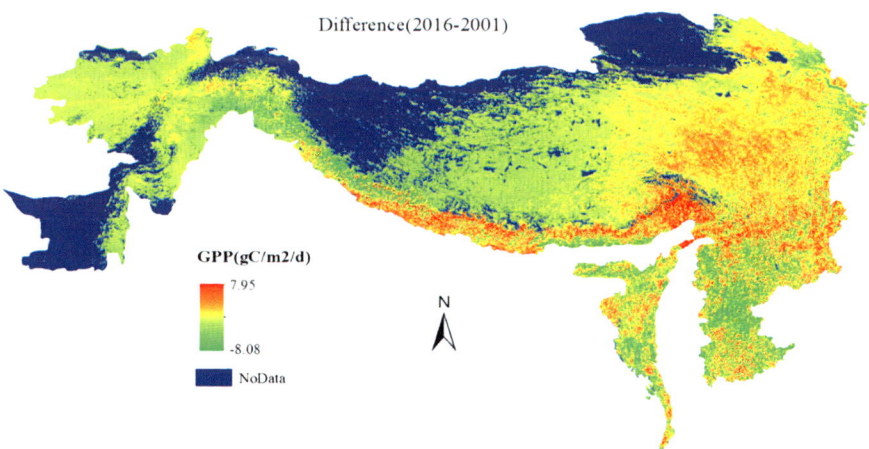

Fig. 3.5 The spatio-temporal variation of GPP difference between 2001 and 2016

Afghanistan being an irrigated, agriculture country, with decades of instability, conflict and destructive war.

Pakistan has a continental climate characterized by extreme variations in air temperature. The land cover types can be summarized as coniferous forests, agriculture lands, rangelands, water bodies, and snow and barren lands. Lands at elevations above 4000 m are sterile and most are often covered with snow and glaciers. Maximum GPP was found in the forest zones, but no estimation of GPP was available for the south-western part of Pakistan. The average GPP between 2001 and 2016 decreased slightly in most areas because of the conversion of forests into agricultural land, the expansion of settlements and infrastructures, and intensive logging.

3.3.3 The GPP Variation in India, Nepal, Bhutan, Bangladesh, and Myanmar

The south-west monsoon season lasts from June to September in India when approximately 80% of the annual rainfall occurs. A large spatial variability in rainfall and the differences in topography results in a large variation in vegetation across India. The average daily GPP in the west and the north, which is dominated by deserts and alpine meadows, was lower than in the south, which is covered by tropical evergreen, and the north-east, which is covered by evergreen forests. Human activities, including urban sprawl and agricultural land transformation, often led to the destruction of vegetation. The total GPP in 2016 was lower than in 2001 and the decreasing trend will continue as India strives for economic development.

Nepal is a predominantly mountainous country and is broadly representative of the land cover diversity. The average daily GPP increased between 2001 and 2016 and reflected land cover types (Fig. 3.5).

Bhutan is a small landlocked country with a rugged, mountainous topography. The settlements are scattered, and more than half of the total population lives in rural areas. The average daily GPP varied considerably and ranged between slightly above 0 to about 10 g C per m^2 during the period from 2001 to 2016. Climate change had a relatively small effect on the GPP, while human activities had a substantial negative effect.

Bangladesh exists on the floodplains of the Ganges, Brahmaputra and Meghna river systems, and small farms and high levels of land fragmentation are common. Monsoon climate occurs from June to September, when precipitation is heavy. About 73% of the land area is covered with forest and nearly 20% of the country is cultivated. The major reason for spatio-temporal change of the GPP could be attributed to deforestation and cropland change.

Myanmar is a tropical country. The highest GPP occurred mainly in the northern, mountainous forested regions during the study period. In the middle part of the country, the GPP was higher than the mean GPP of the plains and rural areas. The GPP was relatively low in the coastal regions because of the sandy coast, however high GPP was scattered in the fertile lands.

3.4 Discussion

The HHK region has sparse ground station data. With the RS-based productivity product, the spatio-temporal variation of MODIS GPP was generally very heterogeneous due to the complex terrain, diverse regional climate, and different human activities. The glaciers in the western part of the Hindu Kush and the western Himalayas are retreating due to climate change. This affects many factors including water availability, biodiversity and ecosystem boundary shifts, which are relevant to the regional ecosystems and even to global feedbacks. In addition, it brings about uncertainties on water supplies and agricultural production for people across Asia (Xu et al. 2009).

The total amount of carbon storage appears to be increasing in the grassland ecosystems in the Three-River Headwaters region (TRHR) due to climate warming and ecological protection programs (Zhang et al. 2015). It has been argued that human activity has little effect on grassland carbon storage (Han et al. 2018), although there is a consensus that urbanization and economic development could destroy the fragile ecosystem (Chen et al. 2014; Zeng et al. 2014; Li et al. 2016; Han et al. 2018). Vegetation restoration and reduction of anthropogenic activities can improve the regional ecological environment and increase regional carbon sequestration.

3.4.1 Use and Accuracy of MODIS GPP

Estimating GPP is a good method to measure atmospheric CO_2 uptake by the terrestrial biosphere. But the current land-use classification driven by RS requires improvement because the different methods often differ at the sub-regional scale (Shim et al. 2014). Moreover, the simulation accuracy of vegetation productivity by RS demonstrated that considerable errors still exist in the current models (Keenan et al. 2012). The biome-specific maximum LUE accounts for the variation in ε between vegetation types, while the scaling with minimum air temperature and vapor pressure deficit accounts for the variation due to climatic conditions (Schubert et al. 2012).

MODIS NPP and GPP products tend to be overestimates at low productivity sites—often because of artificially high values of MODIS FPAR, a critical input to the MODIS GPP algorithm. In contrast, the MODIS products tend to be underestimates in high productivity sites—often a function of relatively low values for vegetation light use efficiency in the MODIS GPP algorithm (Turner et al. 2006). Even after updates, some discrepancies between MODIS GPP and surface-based GPP still exist. The major problems are the uncertainties in meteorological drivers and in FPAR estimates; the accuracy of the LULC product; and the LUE parameterization, mainly in the ε_{max} parameter. When a satellite-based LUE model is recognized as an effective way to assess vegetation productivity, actual LUE is often calculated as a maximum LUE regulated by stresses from environmental factors, such as light intensity, temperature, water, and nutrients (Monteith 1972; Hilker et al. 2008; Chao et al. 2018). LUEmax (maximum light use efficiency), in theory, is a biome-specific variable and is defined as the canopy photosynthesis capacity or maximum conversion rate of absorbed photosynthesis capacity or maximum conversion rate of APAR into biomass under optimal environmental conditions.

Usually, LUEmax is treated as a constant for a specific crop biomass, which can lead to large errors in vegetation productivity estimation. In practice, LUEmax varies with different flora even in the same forest, crop, and spatial scale (Morel et al. 2014; Xin et al. 2015). Another important source of uncertainty in MOD17 GPP is problems with the algorithm at water-limited sites. Other important constraints in the MOD17 GPP algorithm are the difficulties in obtaining accurate measurements of soil nutrients, CO_2 concentrations, and diffuse radiation. The current ε_{max} values used in MOD17 algorithm do not consider the effect of the diffuse PAR or the saturation of photosynthesis under clear skies (Justice et al. 2002; Turner et al. 2006). Furthermore, the use of these variables in many cases may not always be reliable (Propastin and Kappas 2009). The improved MOD17 is a post-reprocessed MODIS GPP/NPP dataset where the contaminated MODIS FPAR/LAI inputs to the MOD17 algorithm have been cleaned. However, the use of the improved MOD17 data are recommended for validation or inter-comparison purposes where study periods cover many years (Running et al. 2015). The spatial resolution of MODIS GPP differs from the resolutions of other data. Scale effects of different datasets should be considered in estimating GPP, especially in heterogeneous areas (Zheng et al. 2018).

With the derived spatially variable LUEmax from satellite RS data, there were significant improvements in biomass estimation accuracy (Dong et al. 2017). The fusion of multi-source data into biophysical simulation models also requires further research in order to better exploit their suitability and transferability. Furthermore, solar-induced chlorophyll fluorescence (SIF), a newly emerging satellite retrieval, can provide a direct, accurate and time-resolved measurement with regard to plant photosynthetic activity. The SIF-based approach accounts for photosynthetic pathways, so more reliable projections of vegetation productivity and climate impact on the ecosystems can be derived. Increased observational capabilities for SIF will improve the mechanistic understanding of vegetation productivity in the HKH, especially in fragmented mountainous areas faced with vulnerability to climate variability and change (Guanter et al. 2014; Guan et al. 2016; Sun et al. 2018).

3.4.2 *Other Constraints to GPP Estimation in HKH Region*

The Asian and Pacific regions have experienced exceptional economic growth over the past two decades, accompanied by a dramatic reduction in extreme poverty. However, moderate forms of poverty remain widespread, especially in rural areas and among households that rely on agriculture. Social developments such as the negative growth of agriculture, instability of economic development and rapid population growth will cause more uncertainty on GPP estimation. The arranged observation station network collects data only from low-lying plains to mountains and plateaus to about 5000 m a.s.l. Most of the HKH region still requires more climate and socio-economic information for robust climate risk assessment, in particular for high-elevation and physically remote locations (Revadekar et al. 2013). The lack of climate and socio-economic data makes the establishment of relevant policies on sustainable development difficult as the fragile ecosystems in the HKH region are vulnerable to climate change and variability. Even when direct photosynthesis measurements can be made, they usually represent small samples in space and time (Anav et al. 2015).

It is difficult to fully understand the effect of climate and human activities on the fragile ecosystem of the HKH. Scientific observation systems, including satellite, airplane, site, and modelling data should be established, which would identify the driving forces and provide more scientific basis for policies and plans (Li et al. 2016; Hua and Wang 2018). Further research should examine vegetation productivity change and the effects of factors such as temperature, precipitation and human activities on the ecosystem.

3.5 Conclusions

There is a large spatio-temporal variation in the GPP due to the complex terrain, diverse regional climate, and different land cover types in the HKH region. Ecosystems are being destroyed mainly in response to population growth and the resultant increased demand for natural resources, economic globalization and rapid urbanization. The MODIS GPP data can be used to track the changing trends of the grassland ecosystem on a large scale and to analyse the carbon storage pattern. The carbon cycle in terrestrial ecosystems is the key to global change. More research is required on the effects of climate and human activities on the fragile ecosystems of the HKH in order to establish policies to improve the livelihood of the local population and the carbon balance in the region.

References

Almeida, C.T.D., R.C. Delgado, L.S. Galvão, et al. 2018. Improvements of the MODIS Gross Primary Productivity model based on a comprehensive uncertainty assessment over the Brazilian Amazonia. *ISPRS Journal of Photogrammetry and Remote Sensing* 145: 268–283.

Anav, A., P. Friedlingstein, C. Beer, et al. 2015. Spatiotemporal patterns of terrestrial gross primary production: A review. *Reviews of Geophysics* 53: 785–818.

Cao, R., M. Shen, J. Zhou, et al. 2018. Modeling vegetation green-up dates across the Tibetan Plateau by including both seasonal and daily temperature and precipitation. *Agricultural and Forest Meteorology* 249: 176–186.

Chao, Z., P. Zhang, and X. Wang. 2018. Impacts of urbanization on the net primary productivity and cultivated land change in Shandong province, China. *Journal of the Indian Society of Remote Sensing* 46: 809–819.

Chen, B., X. Zhang, J. Tao, et al. 2014. The impact of climate change and anthropogenic activities on alpine grassland over the Qinghai-Tibet Plateau. *Agricultural and Forest Meteorology* 189-190: 11–18.

Chettri, N., B. Shakya, R. Thapa, et al. 2008. Status of a protected area system in the Hindu Kush-Himalayas: An analysis of PA coverage. *The International Journal of Biodiversity Science & Management* 4: 164–178.

Cox, P.M., D. Pearson, B.B. Booth, et al. 2013. Sensitivity of tropical carbon to climate change constrained by carbon dioxide variability. *Nature* 494: 341–344.

Dong, T., J. Liu, B. Qian, et al. 2017. Deriving maximum light use efficiency from crop growth model and satellite data to improve crop biomass estimation. *IEEE Journal of Selected Topics in Applied Earth Observations and Remote Sensing* 10: 104–117.

Elalem, S., and I. Pal. 2015. Mapping the vulnerability hotspots over Hindu-Kush Himalaya region to flooding disasters. *Weather and Climate Extremes* 8: 46–58.

Frazier, A.E., C.S. Renschler, and S.B. Miles. 2013. Evaluating post-disaster ecosystem resilience using MODIS GPP data. *International Journal of Applied Earth Observation and Geoinformation* 21: 43–52.

Guan, K., J.A. Berry, Y. Zhang, et al. 2016. Improving the monitoring of crop productivity using spaceborne solar-induced fluorescence. *Global Change Biology* 22: 716–726.

Guanter, L., Y. Zhang, M. Jung, et al. 2014. Global and time-resolved monitoring of crop photosynthesis with chlorophyll fluorescence. *Proceedings of the National Academy of Sciences of the United States of America* 111: E1327–E1333.

Han, Z., W. Song, X. Deng, et al. 2018. Grassland ecosystem responses to climate change and human activities within the Three-River Headwaters region of China. *Scientific Reports* 8: 9079. https://doi.org/10.1038/s41598-018-27150-5.

Hilker, T., N.C. Coops, M.A. Wulder, et al. 2008. The use of remote sensing in light use efficiency based models of gross primary production: A review of current status and future requirements. *Science of the Total Environment* 404: 411–423.

Hua, T., and X. Wang. 2018. Temporal and spatial variations in the climate controls of vegetation dynamics on the Tibetan Plateau during 1982–2011. *Advances in Atmospheric Sciences* 35: 1337–1346.

Justice, C.O., J.R.G. Townshend, E.F. Vermote, E. Masuoka, et al. 2002. An overview of MODIS Land data processing and product status. *Remote Sensing of Environment* 83: 3–15.

Keenan, T.F., I. Baker, A. Barr, et al. 2012. Terrestrial biosphere model performance for interannual variability of land-atmosphere CO2 exchange. *Global Change Biology* 18: 1971–1987.

Li, Q., C. Zhang, Y. Shen, et al. 2016. Quantitative assessment of the relative roles of climate change and human activities in desertification processes on the Qinghai-Tibet Plateau based on net primary productivity. *Catena* 147: 789–796.

Lin, S., J. Li, Q. Liu, et al. 2018. Effects of forest canopy vertical stratification on the estimation of gross primary production by remote sensing. *Remote Sensing* 10: 1329. https://doi.org/10.3390/rs10091329.

Liu, D., Y. Li, T. Wang, et al. 2018. Contrasting responses of grassland water and carbon exchanges to climate change between Tibetan Plateau and Inner Mongolia. *Agricultural and Forest Meteorology* 249: 163–175.

Miao, F., Z. Guo, R. Xue, X. Wang, Y. Shen, C. Cooper, 2015. Effects of Grazing and Precipitation on Herbage Biomass, Herbage Nutritive Value, and Yak Performance in an Alpine Meadow on the Qinghai–Tibetan Plateau. *PLOS ONE* 10(6):e0127275

Monteith, J.L. 1972. Solar radiation and productivity in tropical ecosystems. *Journal of Applied Ecology* 9: 747–766.

Morel, J., A. Bégué, P. Todoroff, et al. 2014. Coupling a sugarcane crop model with the remotely sensed time series of fIPAR to optimise the yield estimation. *European Journal of Agronomy* 61: 60–68.

Peng, J., Z. Liu, Y. Liu, et al. 2012. Trend analysis of vegetation dynamics in Qinghai-Tibet Plateau using Hurst Exponent. *Ecological Indicators* 14: 28–39.

Propastin, P., and M. Kappas. 2009. Modeling net ecosystem exchange for grassland in central Kazakhstan by combining remote sensing and field data. *Remote Sensing* 1: 159–183.

Ren, G., and A.B. Shrestha. 2017. Climate change in the Hindu Kush Himalaya. *Advances in Climate Change Research* 8: 137–140.

Revadekar, J.V., S. Hameed, D. Collins, et al. 2013. Impact of altitude and latitude on changes in temperature extremes over South Asia during 1971-2000. *International Journal of Climatology* 33: 199–209.

Running, S., Q. Mu, M. Zhao, 2015. MOD17A2H MODIS/Terra Gross Primary Productivity 8-Day L4 Global 500m SIN Grid V006 [Data set]. NASA EOSDIS Land Processes DAAC. doi: 10.5067/MODIS/MOD17A2H.006

Sánchez, M.L., N. Pardo, I.A. Pérez, et al. 2015. GPP and maximum light use efficiency estimates using different approaches over a rotating biodiesel crop. *Agricultural and Forest Meteorology* 214-215: 444–455.

Schubert, P., F. Lagergren, M. Aurela, et al. 2012. Modeling GPP in the Nordic forest landscape with MODIS time series data—Comparison with the MODIS GPP product. *Remote Sensing of Environment* 126: 136–147.

Sharma, E., D. Molden, P. Wester, et al. 2016. The Hindu Kush Himalayan monitoring and assessment programme: Action to sustain a global asset. *Mountain Research and Development* 36: 236–239.

Shen, M., S. Piao, X. Chen, et al. 2016. Strong impacts of daily minimum temperature on the green-up date and summer greenness of the Tibetan Plateau. *Global Change Biology* 22: 3057–3066.

Shi, F., X. Wu, X. Li, et al. 2018. Weakening relationship between vegetation growth over the Tibetan Plateau and large-scale climate variability. *Journal of Geophysical Research: Biogeosciences* 123: 1247–1259.

Shim, C., J. Hong, J. Hong, et al. 2014. Evaluation of MODIS GPP over a complex ecosystem in East Asia: A case study at Gwangneung flux tower in Korea. *Advances in Space Research* 54: 2296–2308.

Sun, X., G. Ren, A.B. Shrestha, et al. 2017. Changes in extreme temperature events over the Hindu Kush Himalaya during 1961–2015. *Advances in Climate Change Research* 8: 157–165.

Sun, Y., C. Frankenberg, M. Jung, et al. 2018. Overview of Solar-Induced chlorophyll Fluorescence (SIF) from the Orbiting Carbon Observatory-2: Retrieval, cross-mission comparison, and global monitoring for GPP. *Remote Sensing of Environment* 209: 808–823.

Tiwari, P.C., and B. Joshi. 2015. Local and regional institutions and environmental governance in Hindu Kush Himalaya. *Environmental Science & Policy* 49: 66–74.

Turner, D.P., W.D. Ritts, W.B. Cohen, et al. 2006. Evaluation of MODIS NPP and GPP products across multiple biomes. *Remote Sensing of Environment* 102: 282–292.

United Nations, 2017. World Population Prospects: The 2017 Revision. https://esa.un.org/unpd/wpp/Publications/Files/WPP2017_Wallchart.pdf.

Vitousek, P.M., H.A. Mooney, J. Lubchenco, et al. 1997. Human domination of earth's ecosystems. *Science* 277: 6.

Wu, Z., P. Dijkstra, G.W. Koch, J. Peñuelas, B.A. Hungate, 2011. Responses of terrestrial ecosystems to temperature and precipitation change: a meta-analysis of experimental manipulation. *Global Change Biology* 17(2):927–942.

Xin, Q., M. Broich, A.E. Suyker, et al. 2015. Multi-scale evaluation of light use efficiency in MODIS gross primary productivity for croplands in the Midwestern United States. *Agricultural and Forest Meteorology* 201: 111–119.

Xu, J., R.E. Grumbine, A. Shrestha, et al. 2009. The melting Himalayas: Cascading effects of climate change on water, biodiversity, and livelihoods. *Conservation Biology* 23: 520–530.

You, Q., G. Ren, Y. Zhang, et al. 2017. An overview of studies of observed climate change in the Hindu Kush Himalayan (HKH) region. *Advances in Climate Change Research* 8: 141–147.

Zeng, Y., X. Chen, and W. Jin. 2014. Land use/cover change and its impact on soil carbon in eastern part of Qinghai Plateau in near 10 years. *Transactions of the Chinese Society of Agricultural Engineering* 30: 275–282. (in Chinese).

Zhang, C.L., Q. Li, Y.P. Shen, et al. 2018. Monitoring of aeolian desertification on the Qinghai-Tibet Plateau from the 1970s to 2015 using Landsat images. *Science of the Total Environment* 619-620: 1648–1659.

Zhang, J., C. Liu, H. Hao, et al. 2015. Spatial-temporal change of carbon storage and carbon sink of grassland ecosystem in the Three-River Headwaters region based on MODIS GPP/NPP data. *Ecology and Environmental Sciences* 24: 8–13.

Zhang, Q., J.M. Chen, W. Ju, et al. 2017. Improving the ability of the photochemical reflectance index to track canopy light use efficiency through differentiating sunlit and shaded leaves. *Remote Sensing of Environment* 194: 1–15.

Zhao, M., F.A. Heinsch, R.R. Nemani, et al. 2005. Improvements of the MODIS terrestrial gross and net primary production global data set. *Remote Sensing of Environment* 95: 164–176.

Zhao, M., and S.W. Running. 2010. Drought-induced reduction in global terrestrial net primary production from 2000 through 2009. *Science* 329: 940–943.

Zheng, Y., L. Zhang, J. Xiao, et al. 2018. Sources of uncertainty in gross primary productivity simulated by light use efficiency models: Model structure, parameters, input data, and spatial resolution. *Agricultural and Forest Meteorology* 263: 242–257.

Chapter 4
Livelihood and Carbon Management by Indigenous People in Southern Himalayas

Dil Kumar Limbu, Basanta Kumar Rai, and Krishna Kumar Rai

Abstract The two important anthropogenic drivers for carbon cycle in the Himalayas are land use (rangeland, shifting cultivation and community-based forest management) and fuel use (biomass energy). Shifting cultivation, though illegal in the southern Himalayan regions of India, Bhutan and Nepal, is still widely practiced by indigenous people for their livelihood. This chapter reviews carbon dioxide and methane management in the southern Himalayan region, in particular, Nepal, India and Bhutan. The review will be limited to land and fuel use, as these are the most important anthropometric drivers of the carbon cycle and reduce livelihood security in the Himalayas.

Keywords Southern Himalayas · Indigenous people · Livelihood security · Carbon management · Biomass energy

4.1 Introduction

The Himalayan range extends west-northwest to east-southeast, almost uninterrupted, in an arc for 2500 km, covering an area of 500,000 km^2 (Wadia 1931). Its western and eastern borders are Nanga Parbat and Namcha Barwa, respectively (Fig. 4.1). It is bordered on the north-west by the Karakoram and the Hindu Kush ranges (Valdiya 1998) and on the south by the Indo-Gangetic Plain (Le Fort 1975). Though India, Nepal, and Bhutan have sovereignty over most of the Himalayas, Pakistan and China also occupy parts of the region.

The Himalaya region is among the most vulnerable parts of the world to climate change. The anthropogenic activities that alter the C cycle, and thus negatively affect the climate, can be divided into two broad categories: (1) land use and

---

D. K. Limbu · B. K. Rai (✉)
Central Campus of Technology, Tribhuvan University, Dharan, Nepal

K. K. Rai
Mahendra Multiple Campus, Tribhuvan University, Dharan, Nepal

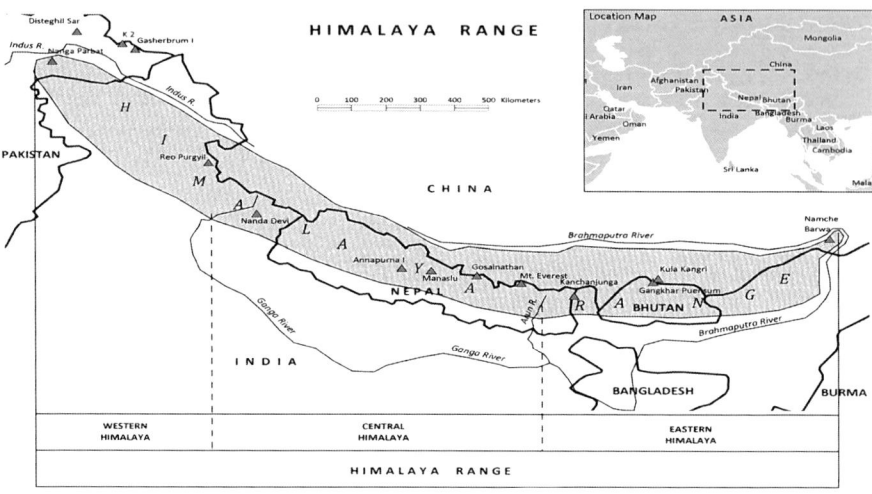

Fig. 4.1 The Himalayan region (Photography by Krishna K. Rai 2018)

management, which is concerned with human choices about how to use/manipulate the land; and (2) fossil fuel usage, which involves human use of C-based fuels (Robinson et al. 2013). The Himalayan region, including the Tibetan Plateau, has shown consistent warming trends during the past 100 years (Yao et al. 2006). However, little is known about the climatic characteristics of the southern Himalayas, partly because of the paucity of observations and partly because of insufficient theoretical attention given to the complex interaction of spatial scales in weather and climate.

Shifting cultivation, though illegal in the Himalayan regions of India, Bhutan and Nepal, is still widely practiced by indigenous people for their livelihood. Pastoralism, another mainstay of livelihood for the mountain people, is also being restricted by governments under the pretext of carbon offset policy. However, because of the general apathy and counter-productive government policies, these guardians of the Himalayas are gradually being deprived of the right they have enjoyed from eons past. Such counter-productive stances on the part of the government may raise issues of livelihood security and resilience among the shifting cultivation-dependent people, who may face reduced provision of ecosystem services because of limited access to land. Rather than decrying the age-old livelihood styles of the mountain indigenous people, the government should develop/enforce policies that have adequate elements of sustainability so that weaning off to sustainable practice is smooth.

4.2 Himalayan People and Their Livelihood

The Himalayas are inhabited by 52.7 million people, spreading across five countries: Nepal, Bhutan, China, India and Pakistan. Of this total, 48.5% reside in western Himalayas, 36.4% in central Himalayas and 15.1% in eastern Himalayas. The indigenous/tribal local people of the Himalayas represent a vast diversity in socio-economic life, cultural heritage and resource use patterns. Despite their habitation in different zones, the tribal people display commonalities in the economic and social life, with variations necessitated to maintain harmonious coordination between resource availability and population structure (Samal et al. 2010).

The Great Himalayan area, which constitutes about 30% of the land, is sparsely populated and contains only 10% of the population. Since the area actually cultivated in the Great Himalaya is hardly 2% of that in the Himalaya, a significant proportion of the population must supplement its income from other sources or else migrate south to the hills and plains (Karan 1987).

Life is very difficult in the infinitely tough Himalayan terrain because the communities are isolated from the world due to climatic and infrastructural adversities. Wrestling with the adversities, the highlanders have learnt to eke out a living from subsistence agriculture, forest utilization and pastoralism/animal husbandry (Naudiyal and Schmerbeck 2017). Women folk are responsible for all the sedentary activities like gathering fodder, fuelwood, farming and cooking, while men are responsible for trade activities and managing the livestock.

4.3 The Indian Context

4.3.1 Physiography and Indigenous People of Indian Himalayas

India, with an area of 3,287,469 km^2, is situated north of the equator between 8°04′ to 37°06′ N latitude and 68°07′ to 97°25′ E longitude (Fig. 4.1). Principal land uses include 161.8 M ha of arable land (11.8% of the world) of which 57.0 M ha is irrigated, 68.5 M ha is forest and woodland, 11.05 M ha is permanent pasture, and 7.95 M ha is for permanent crops.

The indigenous people (or tribal people) of India are referred to by an umbrella term *adivasi* (Gandhi 1968). They include a heterogeneous set of ethnic and tribal groups of the aborigines of India. The Indian Himalayan region (IHR) represents nearly 18.5% of the total tribal population of India. More than 175 of the total 573 known tribes of India inhabit the Himalayan belt stretching through Jammu and Kashmir in the west to Nagaland in the northeast.

4.3.2 Land Use Pattern and Carbon Management in Indian Himalayan Region (IHR)

The adverse impact of climate change on natural resources, including soil, is one of the major issues discussed widely across many parts of the world. Anthropogenic factors and greenhouse gas (GHG) emissions from abrupt land-use change are some of the driving forces contributing to climate change (Meena et al. 2018). Agricultural activities such as land-use change contribute 6–39% of total CO_2 emissions, offsetting of which is possible only through proper land-use specific carbon sequestration approaches (Lungmuana et al. 2018). About two-thirds of the estimated carbon sequestration potential comes from land-use change: shifting cultivated cropland to grass cover for conservation use (e.g., Conservation Reserve Program (CRP) or for use as hay or pasture). On high-tenure farms, three-quarters of sequestration potential is derived from land-use change (Claassen and Morehart 2009). However, there is a paucity of data on linkages between carbon management and land use patterns in the Himalayas. The southern slopes are covered with thick vegetation while the northern slopes are generally barren. Subsistence agriculture, pastoralism, fuel-use pattern and forest utilization are the main components that can be linked to carbon management in the IHR.

In the IHR, forest is the major land use pattern, which covers over 52% of the area, followed by wastelands and arable land (11%). However, dependency on the limited arable land is marginally higher in the IHR than the rest of the Himalayans and the combined cultivators and agricultural laborers comprise about 59% of the total workforce in the region. During the period 1999–2001, whereas India's forest cover recorded a growth of about 6% despite rapid urbanization, the increase in the Himalayan region was only marginal (0.41%), probably due to higher dependence on forest rather than on arable land (Nandy and Samal 2005).

4.3.2.1 Shifting Cultivation in the Indian Himalayas

The people of the IHR, especially those in the northeast, adopted traditional practices of replenishment in the region over the years. One such method was *jhum* cultivation, which is basically the shifting cultivation or the slash-and-burn agriculture (Ramakrishnan 1984).

Simply put, shifting cultivation entails rotation of fields (rather than crops), clearing the cultivation area by means of fire, employment of dibbling stick or hoe and a short period of soil occupancy alternating with long fallow periods. The old cultivated land recovers vegetation cover during the fallow period (Satapathy et al. 2003). The latest review on shifting cultivation can be found in Delang and Li (2013).

This tribal practice enabled regeneration of forests before the same land was cultivated again. The *jhum* cycle that once involved as long as 25 years has been shortened to 4–5 years in recent years (Barthakur 1981; Ramakrishnan 1984). As the *jhum* cycle becomes successively shorter, the rate of soil erosion accelerates. This is a strong indicator of the deteriorating ecological balance of the region, increasing human pressure on land, and growing food needs.

A good number of reviews and research on sifting cultivation in the IHR is available (Ramakrishnan 1984; Nandy and Samal 2005; Choudhary 2012; Sati and Rinawma 2014; Wapongnungsang et al. 2018). Approximately two million tribal people cultivate close to 11 M ha of land under shifting cultivation (Wapongnungsang et al. 2018). In northeast India alone, ~85% of the total cultivation area is covered by shifting cultivation (Rathore et al. 2012) and ~100 tribal farmers (over 0.62 million households) are involved in this practice.

There are both advantages and advantages of shifting cultivation. Shifting cultivation was thought to be beneficial when it emerged (Tomar et al. 2012) for soil fertility rejuvenation, zero tillage, minimal weeds and soil pathogens and release of locked nutrients within the biomass as ash load (Wapongnungsang et al. 2018). In the current context, however, shifting cultivation is considered unsustainable because of the reduced *jhum* cycle (Tomar et al. 2012). The major demerits of shifting cultivation is that it causes deforestation, soil erosion and is not viable on a large scale. Land-use change, particularly soil organic carbon (SOC) loss induced by shifting cultivation is a common land degradation issue in the hilly tracts of the humid tropics (Lungmuana et al. 2018). However, the net per hectare changes in C stock (both biomass and soil) are smaller under shifting cultivation than under permanent cultivation. There is a small net release of other GHG during the cropping cycle but this is still hard to quantify (Tinker et al. 1996). Deforestations for permanent agriculture, plantations and pastures lead to changes in the evapotranspiration, runoff and local climate (Tinker et al. 1996).

Since shifting cultivation is the mainstay of traditional agriculture in the eastern Himalayan region, approaches were implemented to strengthen the existing cultivation practice instead of imposing modern intervention. Against this backdrop, experiments carried out by Kumar et al. (2016) in remote villages of Nagaland deserves mention. The author adopted site-specific agro-based interventions, which proved to be beneficial in augmenting productivity of major crops and livestock, thus ensuring more income, employment and food security. The success achieved in this study could be a model for future policy decisions for sustainable development in wider coverage and assured development in the vast eastern Himalayan region.

According to Bhuyan et al. (2003), land use pattern has a significant effect on soil CO_2 emission in the eastern Himalaya. The authors studied three land use patterns, viz. paddy agro-ecosystem, *jhum* agro-ecosystem and forest. They found significant variation in soil respiration rate, the highest (297 mg CO_2 m^{-2} h^{-1}) being in the forest area and the lowest (136 mg CO_2 m^{-2} h^{-1}) in the *jhum* cultivation. Soil respiration in all sites showed strong seasonal patterns with higher values observed in the wet season. It is not known whether the authors also accounted for CO_2 emission due to slash-and-burn activities. Nevertheless, the rate and amount of *in situ* CO_2 release depends on vegetation types, season, abiotic variables (soil temperature, water content, etc.) and microbial decay of soil organic matter and root (Bhuyan et al. 2003).

4.3.2.2 Rangeland, Grassland, Pasture and Pastoralism in the Indian Himalayas

Grassland, rangeland and pasture are closely related terms and so calls for a distinction among them. While there are many definitions of each of these terms (compiled by Briske 2017), the fundamental differences mentioned by Pandeya (1988) should serve the purpose in the present context. The author referred to 'grassland' as land with more than 80% occupied by grasses, 'rangeland' as vegetation wherein grazing occurs or can occur, and 'pasture' as land in which grasses are grown for feeding. Rangeland is an area where wild and domestic animals graze or browse on uncultivated vegetation.

India's rangelands cover an area of about 121 M ha (~40% of the country's geographical area). Sedentary, semi-migratory, and migratory systems of grazing all occur throughout India. Due to extremities of climate, poor management, and constant grazing, these areas have degraded at an alarming rate. In the Indian Himalayas, alpine grasses and meadows account for 114,250 km^2. These pastures and grasslands range from 300 to 4500 m a.s.l., traversing sub-tropical, temperate, and alpine environments (Singh 1986).

In the north-western part of India, the high altitude sub-alpine and alpine pastures are grazed during the short summer (~4 months) by migratory herds. Rangelands only cover 5.4% and 3.5% of Rajasthan and Gujarat, respectively. In the east, grassland and pastures comprise less than 1% of the total area, even though animal husbandry is an important source of livelihood for local people.

Pastoralism in the IHR rangelands is an important land-use pattern that merits description. The definition of pastoralism varies greatly in terms of purposes and focuses (e.g., intensional, extensional, descriptive, stipulative, etc.) (Dong 2016).

For the pastoralist communities of the IHR, livestock is the sole source of livelihood and is considered as 'the engine and inspiration' of the mountain economy (Maikhuri et al. 2018). However, Indian pastoralism is under-researched and poorly documented. Only a small portion of pastoral groups have been described in some detail—these include some of the larger communities in western India and some of the Himalayan region. This appears natural because worldwide literature on pastoralism is extremely uneven and determined by politics and security issues as much as by the need of empirical data. Be that as it may, there are more than 200 tribes comprising 6% of the country's population engaged in pastoralism (Sharma et al. 2003), contributing 25% of the nation's agricultural GDP and making India one of the world's largest livestock producers (Bhasin 2011).

Pastoralism in the Himalayas is based on transhumant practices and involves cyclical movements of livestock from lowlands to highlands to take advantage of seasonally available pastures at different elevations in the Himalayas (Bhasin 2011; Dong 2016).

In terms of ecological services, accumulated evidence shows that effective animal grazing can contribute to maintaining healthy rangeland vegetation, which generates rich biodiversity, promotes biomass production, captures carbon, reduces erosion, maintains soils, and facilitates water-holding capacity (Frank et al. 1998).

Large pastoral systems represent a great (actual and potential) carbon sink, and pastoralism can effectively promote the potential of rangeland for capturing carbon. It was estimated that grasslands/rangelands store ~34% of the global stock of C. Effective pastoral grazing management can thus be used as tool not only to improve grassland/rangeland biodiversity but also to prevent land degradation and desertification through maintaining rangeland ecosystem integrity (Costanza et al. 1997). However, due to continued growth of human population, agricultural expansion, industrial development, and sedentary livestock farming in recent centuries, pastoralism has been on the decline. The world's few extant grazing ecosystems face large and growing threats (Frank et al. 1998) and pasturelands of IHR are not exceptions. The compounded decline in rangeland quality and quantity (i.e., forage production) associated with climate warming and increased CO_2 concentration may weaken pastoralism but the conversion of cropland or reclamation of mine land into pastureland can mitigate GHG emission by promoting sequestration of carbon in soil (Liebig et al. 2005).

In the Indian Himalaya such as Himachal Pradesh, pastureland-based animal husbandry is quite important, but there is an absence of explicit pastoral policies. The pastoral production systems have been regarded as mal-adaptive and backward and largely overlooked by the policymakers (Sharma et al. 2003). State policies have become increasingly counter-productive by restricting mobile forms of land use adopted by the pastoralists. Decision-makers and politicians are of the view that pastoralism is a threat to the Himalaya because of overgrazing and overstocking (Dong et al. 2016). Thus the pastoralists of IHR are facing uncertain future due to pressure from 'conservation' lobbies that want 'exclusive conservation'. Sustainability has also been threatened by the growing anthropogenic activities and constraints on livestock grazing in many forests (Maikhuri et al. 2018).

4.3.2.3 Community Based Forests (CBFs) and Carbon Management in the Indian Himalayas

CBF regimes can be categorized according to the tenure rights enjoyed by stakeholders. These rights largely determine the extent of empowerment. According to Gilmour (2016), the spectrum of generic types of CBF in order of increasing strength of rights devolved, include: (1) participatory conservation, (2) joint forest management, (3) community forestry with limited devolution, (4) community forestry with full devolution, and (5) private ownership.

There is no denying that forests, when sustainably managed, can play a central role in climate change mitigation and adaptation. Good forest management secures the survival of forest ecosystems and enhances their environmental, socio-cultural and economic functions. It can both maximize forests' contribution to climate change mitigation and help forests and forest-dependent people adapt to new conditions caused by climate change (FAO 2010b). Deforestation and forest degradation account for 12–20% of the annual GHG emissions (Bluffstone et al. 2015).

Understanding the role of CBFs in climate change mitigation is important because 25% of the developing country forests are under effective community management (World Bank 2009). Forests are the key source of carbon sinks and potential GHG emissions, and community forests are about a quarter of the developing country forests, where virtually 100% of net forest biomass loss is taking place.

Important carbon stocks in many forests around the world have been maintained and enhanced due to management practices of local communities, which range from conservation and reforestation to community fire management. Community-based forest management (CBFM), embracing various degrees of community involvement, can significantly contribute to reduce forest emissions and increase forest carbon stocks, while maintaining other forest benefits. Forest-dependent communities are also at the center of climate change adaptation efforts, which must focus on strengthening people's adaptive capacity and resilience. Active participation of communities in all aspects of forest management, taking into account people's needs, aspirations, rights, skills and knowledge, will contribute to the efficiency, sustainability and equity of forest-based measures to tackle climate change (FAO 2010a).

The policies corresponding to the CBFMs in India have not yielded significant results because of the lack of an appropriate tenurial arrangement in favor of the forest managing communities (Nayak 2002). Community-based forest management and natural resource management were under-rated by the policy makers until Joint Forest Management (JFM) policy came into being in the year 2000 (Khawas 2003). Today, about 26–34% of the forests is owned by/reserved for communities and individuals (Molnar et al. 2011; Gilmour 2016), but the literature on community forests in the IHR is rather sketchy. Unlike in Nepal and China, Indian villagers hold only weak tenure rights to forests and enjoy only weak influence over forest management under the country's Joint Forest Management program (Gilmour 2016).

4.3.3 Fuel Use Pattern in the Indian Himalayas

4.3.3.1 Biomass as Fuel

The data given by different authors on global use of biomass energy are contradictory. More than two billion people rely on biomass energy for cooking and heating. Biomass contributes to ~10% of the global energy supply, with two-thirds used in developing countries (IEA 1998; Bensel 2008; Vakkilainen et al. 2013). According to IEA (1998), biomass (then) represented 14% of world's final energy consumption. Around 40% of the world's population depends on fuelwood (Amare 2014).

IEA (2006) predicts that population growth will render 2.7 billion people still relying on plant-based energy forms in the year 2030. In India alone, the population relying on traditional biomass for fuel increased from 740 million in 2004 to 777 million in 2015, with a projected increase to 782 million in 2030. According to a compilation by Bhatt and Sarangi (2010), fuel wood requirements per capita in the eastern-, central-, southern- and Nepalese Himalayas are 3.1–10.4, 1.49, 1.90–2.20 and 1.23 kg/day, respectively.

In the Kanchenjunga Transboundary Conservation Landscape of the eastern Himalaya, over 90% of households remain dependent upon biomass energy (fuel wood) for virtually all domestic uses. Liquefied petroleum gas (LPG) is being adopted only gradually and unevenly. On average, a household in the Conservation Landscape consumes ~4–5 kg/day on an annual basis (Chaudhary et al. 2017). Bhatt and Sachan (2004) estimated an average fuel wood consumption of 3.9–5.81 kg/day for north-east Himalayas.

Fuel wood burning has clear negative consequences for carbon budgets, emissions of GHGs (both locally and globally) and particulate matter, ecosystem health, human health, and livelihoods (Smith 1998; Ekholm et al. 2010). Although there are some variations, consumption of fuel wood, both commercial and domestic, has been described as one of the main drivers of forest degradation throughout the world (Bensel 2008). There are studies that estimate fuel wood extraction resulting in 50–80% deforestation in developing countries (USAID 1989; Bhatt and Sachan 2004). Anthropogenic cause of deforestation is responsible for 17–25% of all GHG emissions.

Wood fuel accounts for over 54% of all global wood harvests per annum, a significant and direct role of firewood extraction in forest degradation (Osei 1993). It has been suggested that firewood collection and shifting cultivation are the major causes of primary deforestation in the region. One means to alleviate fuel wood pressures, where LPG is inaccessible, would be to encourage the widespread adoption of improved cooking stoves. These stoves emit less CO_2 and particulate matter and also have been positively correlated with improved household health status (Bajracharya et al. 2012; Chaudhary et al. 2017).

4.3.3.2 Biogas Option for Carbon Management

Biogas, which refers to a mixture of different gases produced by the breakdown of organic matter (present in agricultural waste, manure, food waste, etc.) in the absence of oxygen, is principally a mixture of methane (CH_4) and CO_2 along with other trace gases. Although CH_4 has a global warming potential (GWP) of 21, burning it (as biogas) is carbon-neutral and does not add to GHG emissions (UF 2018). Replacement of wood fuel with biogas can therefore be an attractive proposition for reduction of GHG in the hills. Cow dung is the most important animal substrate in the rural hills as it is available in almost all households and on average contributes almost 60% of per capita biogas potential (Gross et al. 2017). An abundant supply of animal wastes for biogas production is further ensured by the increase in stall-feeding of animals because of the shrinking grazing area.

In India, the need for alternative sources of energy became apparent with the onset of the oil crisis in the 1970s. Biogas plant soon became the predominant source of renewable energy, particularly in the rural areas where the dung is available in abundance. Today, India has the second largest biogas program in the world after China. By 1999, over 2.9 million family type biogas plants were installed in India, with an estimated potential of 12 million plants. Although biogas has been a success in several states, its presence is merely a token in the north-east and Himalayan region. The most serious limitation of the technology is low gas production during winters (by ~25–30%) because the digestion by microorganisms is slow at lower temperatures (Agrawal 1987).

India has developed a number of models for biogas digesters to suit varying capacities. Bhol et al. (2011) reviewed three successful designs of biogas plants that have been somewhat standardized, and are being widely disseminated in India, viz. (1) KVIC (promoted by Khadi Village Industries Commission of India), (2) Janata, and (3) Deenbandhu (at present the most popular). While the design promoted by KVIC is a floating gas-holder type of plant, both the Janata and Deenbandhu designs are of the fixed-dome type. In colder climate like the lower reaches of the Himalayas, the Deenbandhu fixed model is considered ideal.

The main advantages of using biogas are: (1) methane, one of the gases responsible for global warming, gets burned, (2) households need very little firewood, reducing the pressure on our fragile forests, (3) reduces the drudgery of collecting and carrying the firewood daily for women while allowing time for other activities, (4) it is smokeless and thus has a positive impact on family health, (5) there are no recurring costs, and (6) the dung slurry is composted for manure (Grassroots 2009).

4.4 The Nepalese Context

4.4.1 *Physiography and Indigenous People of Nepal*

The territory of Nepal (Fig. 4.1), with an area of 147,181 km^2, falls in the central portion of the Himalayan arc and extends between 80°04′ and 88°12′ E longitude, and 26°22′ and 30°27′ N latitude. About 80% of Nepal's land is occupied by mountains. Nepal is divided into eight roughly parallel physiographic regions from south to north, viz. (1) Terai, (2) Siwalik Range (or briefly Siwaliks) with dun valleys, (3) Mahabharat Range, (4) Midlands, (5) Fore Himalaya, (6) Great Himalaya, (7) Inner Himalayan valleys, and (8) Tibetan marginal ranges (Dhital 2015).

Indigenous Peoples of Nepal are officially described as Indigenous Nationalities (*Adivasi Janajati*). They number ~8.5 million, or 36% of the country's total population. As many as 59 (previously 61) indigenous communities have been officially and legally recognized by the Nepal government (NFDIN) Act-2002, out of which 17 live in the northern Himalayan region of Nepal. Most of indigenous people live in remote and rural areas and make a living out of subsistence farming, some are nomads (e.g., Raute), some are agro-pastoralists, and some are forest dwellers (e.g., Chepang and Bankariya).

4.4.2 Land Use Pattern and Carbon Management in Nepalese Himalayas

Nepal has very distinct land use patterns, as it is influenced by climatic variation (affected by maritime and continental factors), altitude and land topography. The great variety of micro-climatic conditions in the country results in a diversity of land use and farming practices (Paudel 2015). Land use in the hills differs from that of the plains. The total land cover of Nepal is 14.72 M ha. The land use statistics according to Go-N (2014) is, agricultural land cultivated (21%), agricultural land uncultivated (7%), forest, including shrubs (39.6%), grassland and pasture (12%), water (2.6%) and others (17.8%).

4.4.2.1 The Shifting Cultivation in Nepalese Himalayas

Many indigenous communities (e.g., Chepang, Magar, etc., of poor economic status) in Nepal still use shifting cultivation, locally known as *khoriya* and *bhasme*, in which they clear and cultivate secondary forests in plots of different sizes and leave these plots to regenerate naturally through fallows of medium to long duration (Fujisaka et al. 1996). Shifting cultivation in Nepal in many cases is characterized by 2–4 years of cultivation and 4–9 years of fallow. There is marked tendency to prolong cultivation and shorten the fallow with the increase in population pressure, and a corresponding decrease in the cultivatable land. In some cases, shifting cultivated lands are also being gradually converted to settled farms and regular cropland in some areas (Bajracharya et al. 1993).

The information of the total distribution of shifting cultivation areas in Nepal is not available, but existing studies show that this practice is prevalent in 20 districts of Nepal (Regmi et al. 2005). The Government of Nepal has no specific policy on shifting cultivation, nor is the word/term 'shifting cultivation' acknowledged in any of its land use policies. In essence, the government of Nepal does not support shifting agriculture (Kerkhoff and Sharma 2006). In some cases, since the majority of shifting cultivators do not legally own the land, forest authorities tend to introduce or impose programs that compel them to take up alternative livelihoods.

A case study by Kafle et al. (2009) on shifting cultivation among the resource-poor Chepang people of Gorkha and Tanahu mid-hills describes the relationship among land use transition, climate change and human health. In the past three decades, the fallow period in shifting cultivation lands has been reduced to about 2.5 years because of constraints (including such as population pressure). More than 50% of Chepang farmers now practice annual cropping instead of fallow in the shifting cultivation land (Kafle 2011).

Shifting cultivation is in transition across the world. The characteristics of the shifting cultivation are changing over time. The relationship between the crop yield and fallow period is not clear though it is perceived that shorter fallows result in decreased yield. Altering crop and fallow management practices are widely perceived and practiced to improve the traditional shifting cultivation practices. Studies on shifting cultivation in Nepal are inadequate in concluding current status, distribution, management practices and practical implications in relation to this practice (Kafle 2011).

In the context of changing climate, land use transition on shifting cultivation practice in terms of altering fallow period and agricultural intensification can exacerbate the vulnerability of the resource-poor farmers and mitigation (carbon sequestration) potential of soil in Nepal. There is urgent need for systematic studies on shifting cultivation in Nepal, and based on this, to provide result-based recommendations to the government for its mainstreaming in national plans, policies and priorities (Kafle 2011).

4.4.2.2 Rangeland, Grassland, Pasture and Pastoralism

Rangelands in Nepal comprise an area of 3.33 M ha, accounting for 22.6% of the total land area. More than 80% of the rangeland falls in the Himalayan region. However, livestock utilize only 37% of the available rangeland grass. Depending on the condition of rangeland, annual production of grass is 0.65–360 MT/ha (dry matter) in Nepal (Rangeland Policy of Nepal 2012).

The livelihoods of pastoralists depend greatly on plants, water, animals and other natural resources in the rangelands. Rangelands occupy the single largest proportion of the Nepalese Himalayan region. Grazing lands in the high mountain parks are called *patans* (Figs. 4.2 and 4.3) in western Nepal and *kharka* in eastern Nepal and animal are allowed to graze in specific time on a rotational basis.

The general perceptions of inefficient traditional management of rangelands, confusions over ownership, and conflicts have resulted in a low national priority and neglect of indigenous knowledge of skills and techniques in pastoralism (Sharma et al. 2003). The creation and expansion of protected areas has contributed to the

Fig. 4.2 Rangeland of western Nepal (*patan*) (Photography by Kamal Maden 2017)

Fig. 4.3 Rangeland of eastern Nepal (*kharka*) (Photography by Dil K. Limbu, 2010a)

exclusion of herders from their pasturelands, leading to a decline in pastoral production (Kreutzmann 2012).

Transhumant pastoralism in the mountains of Nepal is an age-old practice. However, due to policy and institutional, governance, and climatic factors, the future of transhumant pastoralism in Nepal is uncertain, and the practice may even disappear (Banjade and Paudel 2008; Aryal et al. 2014). Other constraints to transhumant pastoralism are modifications of livelihood options due to changes in demography, migration, shortage of labor, diversification of agriculture, market influence on rural economy, and privatization and nationalization of rangelands (Namgay et al. 2013). Conflicts (between communities, government, herders and non-herders) in rangeland management and rangeland use have also been identified as a threat to the continuation of transhumant pastoralism in Nepal (Banjade and Paudel 2008).

Increase in population, soil erosion, uncontrolled grazing, wildfire and swidden agriculture have brought about marked loss of biodiversity and productivity and shrinkage of the Himalayan rangelands. Gradual erosion of traditional knowledge on management of rangelands, coupled by lag in adoption of scientific rangeland management practice have posed a challenge to the sustainable management of the rangelands. Climate change due to global increase in air temperature has led to a decrease in productivity. Encroachment of grassland and increase in the number of invasive and/or unpalatable plant species are other challenges facing the development of rangeland. Rangeland Policy 2012 introduced Nepal government aims to address, among other things, measures to minimize environmental degradation, mitigate the effects of climate change, assess the contribution of carbon sequestration, control overharvesting, overstocking, and reduce biodiversity loss.

4.4.2.3 Community Based Forests (CBFs) and Carbon Management

Globally, Nepal appears to be at the forefront of CBFM practice (Ojha et al. 2007). Analysts divide evolution of forestry in Nepal in three distinct phases, viz. (1) privatization (until 1957), (2) nationalization (1957–1970s), and (3) decentralization (late 1970s onward). Forest Act in 1993 was the most significant regulatory development in support of community forests in that it guaranteed the rights of local people in forest management. Nepal became the world's first country to enact such radical forest legislation, allowing local communities to take full control of government forest patches under a community forestry program (Pathak et al. 2017).

Today there are 26,487 CBFs in Nepal, covering 2.3 M ha (38.5% of total forest in the country), and involving more than 3.8 million households. Of the CBFs, community forests alone account for 1.72 M ha, 2.3 million households (Pathak et al. 2017), and 16,000 community forest user groups (CFUGs). Figure 4.4 shows a thriving community forest of eastern Nepal.

The Forestry Act of 1993 and additional forest regulations in 1995 provided authority to CFUGs to manage forests. While the state retained ownership of the forests, it granted community groups the right to manage their forests. Community management of Nepalese forests has resulted in many ecological and economic benefits, including increased crown cover and higher productivity. For instance, a longitudinal 5-year study covering 2700 households from 26 CFUGs in the Koshi Hills showed large-scale improvements of people's livelihoods and food security (Upadhyay 2012). Similarly, Rayamajhi et al. (2012) concluded from their study on economic importance of central Himalayan forests (180 households from lower Mustang district) to rural households that forest dependency level is significant. The average forest income share was found to be 22% of total annual household income.

Fig. 4.4 A thriving community forest of eastern Nepal (Photography by Dil K. Limbu, 2010b)

There is paucity of research data on CBFs of the Himalayas. According to one study, the average carbon in the hills of Nepal is 72.1 tons/ha in the forest managed under community forest program (CFP) to 76.1 tons/ha in non-CFP forests (Bluffstone et al. 2015).

According to Upadhyay (2012), CFUGs effectively managing their forests have been able to provide carbon sequestration and environmental services, including the provision of higher-quality water to downstream communities. Under Reducing Emissions from Deforestation and Degradation (REDD+), an effort is made to quantify such gains so that a proper compensatory mechanism can reward CFUGs according to their contribution in improving the environment. The resources generated from carbon trading and environmental services can be placed into a community forestry fund, which can, in turn, be use to regenerate forests and create livelihood opportunities for users.

In Nepal, six categories of CBFs are recognized, viz. (1) community forest, (2) collaborative forest, (3) pro-poor leasehold forest, (4) religious forest, (5) protected forest, and (6) buffer zone forest. The details of the tenure arrangement and responsibilities regarding these forests are given by Pathak et al. (2017).

4.4.3 Fuel Use Pattern in the Nepalese Himalayas

4.4.3.1 Biomass as Fuel

About 77% of energy consumption of Nepal is met by traditional biomass energy, which includes firewood, dung and agricultural residues. As per the National Census 2011, 64% of the households depend on firewood and another 10.4% on dung for energy, the rest being on LPG and/or electricity.

Firewood consumption rates are influenced by myriad, interrelated factors, such as family size, caste, and season, which can vary greatly. For instance, in a 26-year long survey, Donovan (1981) found firewood consumption to vary by a factor of 67 (when the lowest was compared to the highest). Wood characteristics are, of course, important in its selection for firewood purpose. For example, *Morus laevigata* and *Castanopsis indica* have calorific values (MJ/kg dry weight) of 13.91 and 19.89, respectively (Bhatt and Sarangi 2010). Figure 4.5 shows drying and stacking of firewood in eastern Nepal.

There is paucity of comprehensive data on firewood consumption in Nepal. Fox (1984) carried out survey on firewood consumption in Bhogteni, a village located near Gorkha bazaar (Central Himalaya); a range from 0.95 to 1.07 kg/capita/day was used, based on 107 households with an average of 6.1 individuals per household. The author concludes that firewood consumption patterns provide basic knowledge for designing forest management plans to meet immediate and long-term needs in rural Nepal. Rijal and Yoshida (2002) also carried out a similar study in winter and summer in traditional houses in the Banke, Bhaktapur, Dhading, Kaski and Solukhumbu districts of Nepal and found firewood consumption rates of 0.6–

Fig. 4.5 Drying and stacking of firewood in eastern Nepal (Photography by Dil K. Limbu, 2010c)

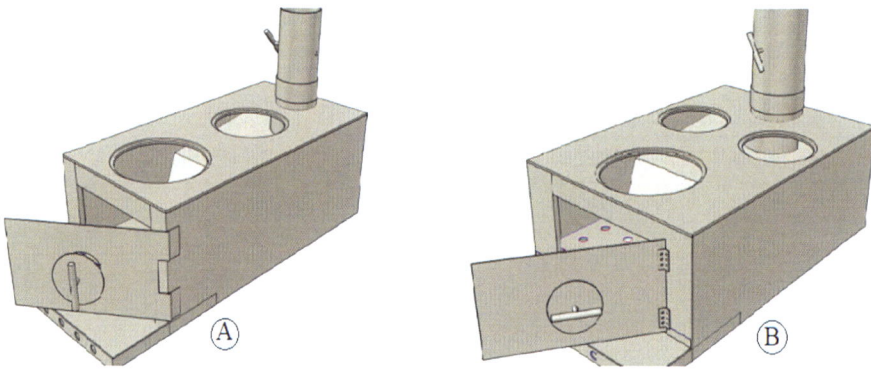

Fig. 4.6 ICS designed by AEPC; (**a**) Two-hole design, (**b**) Three-hole design

4.6 and 0.8–2.6 kg/capita/day in winter and summer, respectively. Cooking required 2.1–5.2 times more firewood than heating. The results also showed that the temperate climate used less firewood than the sub-tropical climate which indicates that proper firewood usage can minimize firewood consumption. It is pertinent at this point to mention the tremendous popularity achieved by locally designed improved cooking stoves (ICS) of Nepal. Figure 4.6 shows some ICS designs offered by AEPC.

As such, ICS was introduced in Nepal in the 1950s and continues to have relevance in the present context. AEPC, together with other government and non-government partners have thus far installed 700,000 ICSs in 63 districts (AEPC 2018).

In a comparative study between ICSs and traditional stoves (involving 12,132 households from three sites), Singh et al. (2012) found significant reduction of $PM_{2.5}$ (particulate matter less than 2.5 μm in aerodynamic diameter) and CO by ICS compared to local stoves. After 1 year of testing, the reduction in PM_{25} and CO was found to be 63.2% and 60.0%, respectively after 1 year.

4.4.3.2 Biogas Option for Carbon Management

Potential of biogas production to reduce firewood consumption in remote high-elevation Himalayan communities in Nepal has been researched by Gross et al. (2017). Biogas Support Program Nepal (BSP-N) has, in partnership with Alternative Energy Promotion Center (AEPC), installed 27,131 biogas plants (anaerobic digesters), with 83,695 beneficiaries. However, BSP-N has worked mostly in the lowlands and hills up to 2100 m a.s.l. It has very little experience at higher elevation areas, where the energy situation is precarious and about 25% of Nepalese population live. Anaerobic digestion at high elevation is challenging due to year-round low air temperature and lower annual biomass production per area compared to lowlands and potentially limited substrate availability. At present, there are approximately 100 biogas companies in Nepal carrying out their construction/promotional activities.

At this juncture it is worthwhile mentioning that India has been very successful with a particular biogas plant named Deenbandhu that works well even in the cold climates of the Himalayan region. Nepal can either adopt this model or modify it to work in the austerity of Nepalese Himalayas.

4.5 The Bhutanese Context

4.5.1 *Physiography and Indigenous People of Bhutan*

The kingdom of Bhutan situated between the Tibetan plateau in the north and the Indian plain in the south. Bhutan's total area is 38,394 km^2 and lies between latitudes 26°N and 29°N, and longitudes 88°E and 93°E. The land consists mostly of steep mountains and deep valleys criss-crossed by a network of swift rivers. This great geographical diversity, combined with equally diverse climate conditions, contributes to Bhutan's outstanding range of biodiversity and ecosystems. However, as is the case in the neighboring Himalayan ranges, the land mass of Bhutan has fragile geology and immature soils.

There are three main ethnic groups in Bhutan. They are not necessarily exclusive: the politically and culturally dominant Ngalop of western and northern Bhutan; the Sharchop of eastern Bhutan; and the Lhotshampa concentrated in southern Bhutan. Other small ethnic groups are Brokpa, Lepcha, and Doya tribes.

4.5.2 Land Use Pattern and Carbon Management in the Bhutanese Himalayas

The Land Use Land Cover (2016) of Bhutan provides useful information on the coverage and distribution of major land cover types to enable management and monitoring of natural resources and changes in forest area and carbon stock.

Bhutan is in fact the only carbon-negative country in the world. Its ability to be a net carbon sink is due to its vast forest cover (~71% of the country) that help remove nearly three times as much CO_2 as it produces (<2.5 million tons of CO_2 each year) (FRMD 2017; Tutton and Scott 2018). Environmental protection is enshrined in the constitution, which states that a minimum of 60% of Bhutan's total land should be maintained under forest cover for all time. The country even banned logging exports in 1999 and by 2025, increased hydroelectricity exports allow the country to offset up to 22.4 million tons of CO_2 per year in the region (Tutton and Scott 2018). More details on Bhutan's commitment to remain carbon neutral for all times is available at TED talks (TED 2016) in which Tshering Tobgay (the then Prime Minister of Bhutan) delivers a very persuasive speech on how Bhutan's forests and hydropower will help achieve the goal. Bhutan's forest resources are also widely touted by many as a significant contributor to rural livelihood. However, most of the forest products are directly consumed by people and thus do not enter the formal economy. Thus, the significance of forest products in general and fuelwood in particular, are often overlooked (Uddin et al. 2007). Although the majority of the total population are subsistence farmers and depend on agriculture and livestock for their livelihoods, agriculture in the country is severely constrained due to the very rugged terrain and extreme climatic conditions (FRMD 2017).

4.5.2.1 The Shifting Cultivation

Farming systems in Bhutan can be classified into three subsystems: (1) pastoral transhumance system, (2) subsistence-level crop and animal husbandry and (3) and early commercial farming, while the cultivation among the subsistence farmers can be divided into three types, viz. (1) valley cultivation, (2) terrace cultivation, and (3) shifting cultivation (*tsheri* in Bhutanese language).

Shifting cultivation is practiced on 32% of the cultivated land to hedge the risk of crop failure and to compensate for food deficit, and almost all farmers practicing it can be regarded as subsistence farmers. Maize represents 68% of crops produced on this land (Upadhyay 1995). The land tenure of shifting cultivation in Bhutan is not the same as in Nepal or India. This is probably because of the Land Act 2007, which is being enforced by the government to convert all shifting cultivation lands to dryland farms, orchards, and wetlands. As a result, the area under shifting cultivation in Bhutan has decreased by 12% (Dorji 2011). However, Bhutan still does not have a clear policy that addresses the issue of shifting cultivation, its tenure, and customary institutions. Despite having achieved success in pilot scales, farmers opine that shifting cultivation is the most suitable land use in sloping areas with little or no irrigation opportunities (ICIMOD 2015). As such, shifting cultivation is illegal in all the three countries described in this chapter, viz. India, Nepal and Bhutan but the manifestation of law varies (Mertz and Bruun 2017).

4.5.2.2 Rangeland, Grassland, Pasture and Pastoralism

High altitude rangeland (*tsa-drog* in Butanese language) and livelihood systems in Bhutan are undergoing changes in resource availability, population and user rights. It is integral to yak (*Poephagus grunniens*) herding and semi-nomadic yak herders' livelihoods (Gyamtsho 2002). High altitude *tsa-drog* includes temperate, sub-alpine and alpine rangelands located between 2500 and 6000 m a.s.l. The yak population of 48,400 heads concerns less than 1400 Bhutanese households, and less than 3% of the market shares of meat, butter and cheese. Yak herders are scattered and isolated in 11 of 20 districts (Derville and Bonnemaire 2010; Tenzing 2018), mostly adopting transhumant grazing (Dong et al. 2016). According to a review by Tenzing (2018), *tsa-drog* degradation is a common problem facing yak herders. The degradation may be attributed to overgrazing, soil erosion, landslides, shrub encroachment, increases in human and livestock populations, family division of assets and climate change with significant rain events.

The Bhutan government introduced the new Land Act of 2007 to promote sustainable governance and incentivize yak herding. The Act includes a nationalization and leasing program, under which herders and livestock farmers are permitted to grow improved pasture and implement maintenance activities hitherto not allowed. However, there is confusion and uncertainty among the herders and farmers as there are no clear mechanisms and guidelines for implementing the leasing program (Gyeltshen et al. 2010).

4.5.2.3 Community Based Forests (CBFs) and Carbon Management

Today, Community Forestry (CF) is a key component of Bhutan's forest policy. It has become an institutionalized part of the system for the sustainable management of Bhutan's rich and diverse forest resources (RGoB 2010). Consequently, CF in Bhutan has gained an unprecedented fast pace in recent years, especially after the revised Forest and Nature Conservation Rules (FNCR) came into being in 2006. By December, 2017, the total number of community forests (CFs) increased to 733 (corresponding to 3% of the total forest area and 30,352 management groups/households) (FRMD 2017). The long term vision for Bhutan's CF is for a future that is sustainable, affordable, makes a significant contribution to rural livelihoods, poverty reduction and improved forest condition which is resilient to climate change (RGoB 2010). Despite some impressive gains in implementing CF across Bhutan in recent years, it is not free from criticisms. In some cases, large scale industries have been forced to close down when calculations of sustainable harvesting limits have proven to be grossly over optimistic.

4.5.3 Fuel Use Pattern in the Bhutanese Himalayas

4.5.3.1 Biomass as Fuel

In Bhutan, fuelwood is the only readily available source of energy for most rural and urban residents. Fuelwood accounts for 78% of total energy consumption and is the primary energy source for most rural and urban residents. The consumption of fuelwood is one of the highest in the world, at 1.2 MT/capita/year (Wangchuk 2011). Fuelwood harvesting in Bhutan is regulated by government laws, which stipulate that a household, irrespective of size, is allowed 16 m^3 of fuelwood/year if it lacks electricity or 8 m^3/year if it has electricity. This amount may not be sufficient for temperate- and alpine residents since Wangchuk et al. (2013) reported that a household required over three times the amount allowed in Wangchuck Centennial Park. When produced and harvested sustainably, fuelwood provides a renewable source of energy with low net carbon emissions (FAO 2010a), but managing this resource in the Bhutanese highlands is a challenge.

4.5.3.2 Biogas as Fuel Option for Carbon Management

For Bhutan, biogas technology is still at an evolving stage. After an initial setback in the late 1980s, its comeback in 2011 with a new formulation has begun to show promise. By the end of 2017, around 3176 family-sized units were installed in 17 districts. The full project (3600 installations) is expected to benefit more than 15,000 people directly. Apart from environmental reasons, this project aims at reducing poverty through reduced costs by replacing fossil fuels, or increasing income via the sale of the digested fertilizer created by the system. The systems have also reduced firewood use by approximately 10,000 MT/year. Since Bhutan has no known fossil fuel reserves, every unit avoided is also one that does not have to be imported, further reducing fossil fuel use, and thereby contributing to environmental protection (Oestereich 2017).

References

AEPC. 2018. *Improved cooking stoves*. Alternative Energy Program Center. Ministry of Energy, Water Resources and Irrigation. Government of Nepal.

Agrawal, S. 1987. Prospects for community biogas plant. In *Rural energy planning for the Indian Himalaya*, ed. T.M.V. Kumar and D.R. Ahuja, 94–117. Rome: ICIMOD and TERI. Wiley Eastern Limited.

Amare, Z.Y. 2014. The role of biogas energy production and use in greenhouse gas emission reduction: The case of Amhara National Regional State, Fogera District, Ethiopia. *Journal of Multidisciplinary Engineering Science and Technology* 1 (5): 404–410.

Aryal, S., T.N. Maraseni, and G. Cockfield. 2014. Sustainability of transhumance grazing systems under socio-economic threats in Langtang, Nepal. *Journal of Mountain Science* 14 (4): 1023–1034.

Bajracharya, K., N.R. Dhakal, and D. Paudel. 2012. *Success stories of ICS promotion*. Nepal: Biomass Support Program (BMSP)/Alternative Energy Promotion Center (AEPC)/Energy Sector Support Program (ESSP.

Bajracharya, K.M., K.B. Malla, and P.B. Thapa. 1993. *Shifting cultivation in Southern Gorkha*. Kathmandu: Forestry and Conservation Technology Services.

Banjade, M.R., and N.S. Paudel. 2008. Mobile pastoralism in crisis: Challenges, conflicts and status of pasture tenure in Nepal Mountains. *Journal of Forest and Livelihood* 7 (1): 36–48.

Barthakur, I.K. 1981. Shifting cultivation and economic change in the north-eastern Himalaya. In *The Himalaya: Aspects of change*, ed. J.S. Lall, 447–460. Delhi: Oxford University Press.

Bensel, T.G. 2008. Fuelwood, deforestation, and land degradation: 10 years of evidence from Cebu Province, the Philippines. *Land Degradation & Development* 19: 587–605.

Bhasin, V. 2011. Pastoralists of Himalayas. *Journal of Human Ecology* 33 (3): 147–177.

Bhatt, B.P., and M.S. Sachan. 2004. Firewood consumption pattern of different tribal communities in Northeast India. *Energy Policy* 24: 1–6.

Bhatt, B.P., and S.K. Sarangi. 2010. Fuelwood characteristics of some firewood trees and shrubs of eastern Himalaya, India. *Energy Sources* 32: 449–474.

Bhol, J., B.B. Sahoo, C.K. Mishra. 2011. *Biogas digesters in India: A review*. National Conference on Renewable and New energy Systems, Odisha, Dec 22–23, 2011.

Bhuyan P., M. Khan, R. Tripathi. 2003. Tree diversity and population structure in undisturbed and human-impacted stands of tropical wet evergreen forest in Arunachal Pradesh, Eastern Himalayas, India. *Biodivers Conserv* 12: 1753–1773.

Bluffstone, R., E. Somanabhan, P. Jha, et al. 2015. Collective action and carbon sequestration in Nepal. *Journal of Forest and Livelihood* 13 (1): 1–7.

Briske, D.D. 2017. Rangeland systems: Foundation for a conceptual framework. In *Rangeland systems: Processes, management and challenges*, ed. D.D. Briske, 5. Switzerland: Springer Nature. https://doi.org/10.1007/798-3-319-46709-2.

Chaudhary, P., R. Seidler, and K. Bawa. 2017. Patterns and determinants of domestic energy use in Kanchenjunga Himalaya. *International Journal of Energy and Environmental Science* 2 (1): 1–11.

Choudhary, V.K. 2012. Improvement of jhum with crop model and carbon sequestration techniques to mitigate climate change in Eastern Himalayan Region, India. *Journal of Agricultural Science* 4 (4): 181–189.

Claassen, R. and M. Morehart. 2009. *Agricultural land tenure and carbon offsets*. Economic Brief Number 14.

Costanza, R., R. d'Arge, R. de Groot, et al. 1997. The value of the world's ecosystem services and natural capital. *Nature* 387 (15): 253–260.

Delang, C.O., and W.M. Li. 2013. *Ecological succession on fallowed shifting cultivation fields: A review of the literature*. New York: Springer.

Derville, M. and J. Bonnemaire, 2010. In *Marginalization of yak herders in Bhutan: Can public policy generate new stabilities that can support the transformation of their skills and organizations? Innovation and Sustainable Development in Food and Agriculture (ISDA)*, France. Retrieved from https://hal.archives-ouvertes.fr/hal-00522045.

Dhital, M.R. 2015. *Geology of the Nepal Himalaya: Regional geology reviews*. Switzerland: Springer.

Dong, S. 2016. Overview: Pastoralism in the world. In *Building resilience of human-natural systems of pastoralism in the developing world—interdisciplinary perspectives*, ed. S. Dong, K-A.S. Kassam, F.F. Tourrand, R.B. Boone, et al, 2, 8–10. Switzerland: Springer Nature.

Dong, S., S.L. Yi, and Z.L. Yan. 2016. Maintaining the human–natural systems of pastoralism in the Himalayas of South Asia and China. In *Building resilience of human-natural sys-*

tems of pastoralism in the developing world—interdisciplinary perspectives, ed. S. Dong, K.-A.S. Kassam, F.F. Tourrand, and R.B. Boone, 93–137. Switzerland: Springer Nature.

Donovan, D.G. 1981. *Fuelwood: How much do we need?* Hanover, NH: Institute of Current World Affairs.

Dorji, M. 2011. *Securing tenureship over tseri land cultivation, Kingdom of Bhutan*. Kathmandu, Nepal: Consultancy Report, Council for Renewable Natural Resources Research of Bhutan, Ministry of Agriculture and Forest, submitted to ICIMOD.

Ekholm, T., V. Krey, S. Pachauri, et al. 2010. Determinants of household energy consumption in India. *Energy Policy* 38: 5696–5707.

FAO. 2010a. *Managing forests for climate change: Supporting countries to manage fragile forest ecosystems*. Food and Agriculture Organization. I1960E/1/11.10.

———. 2010b. *Criteria and indicators for sustainable woodfuels*. FAO forestry paper 160. Rome: Food and Agriculture Organization of the United Nations.

Fox, J. 1984. Firewood consumption in a Nepali village. *Environment Management* 8 (3): 243–250.

Frank, D.A., S.J. McNaughton, and B.F. Tracy. 1998. The ecology of earth's grazing ecosystems. *Bioscience* 48 (7): 629–634.

FRMD. 2017. *Land use and land cover of Bhutan 2016, maps and statistics*. Bhutan: Forest Resources Management Division.

Fujisaka, S., L. Hurtado, and R. Uribe. 1996. A working classification of slash-and-burn agricultural systems. *Agroforestry Systems* 34: 151–169.

Gandhi, M.K. 1968. *The selected works of Mahatma Gandhi: Satyagraha in South Africa*. Navajivan: Navajivan Publishing House. Retrieved Nov 25, 2008.

Gilmour, D. 2016. *Forty years of community-based forestry: A review of its extent and effectiveness*. FAO forestry paper 176. Rome: Food and Agriculture Organization of the United Nations.

Go-N. 2014. *Statistical information on Nepalese agriculture 2014/2015*. Nepal: Government of Nepal, Ministry of Agricultural Development.

Grassroots. 2009. *Bio-gas: Renewable energy for the Himalaya*. Pan Himalayan Grassroots Development Foundation. Retrieved from https://youtube/sMFe%2D%2DbLKdI?list=UUX7rXSJ7oDDpb65drGYSjFg.

Gross, T., A. Zahnd, S. Adhikari, et al. 2017. Potential of biogas production to reduce firewood consumption in remote high-elevation Himalayan communities in Nepal. *Renewable Energy and Environmental Sustainability* 2: 8.

Gyamtsho, P. 2002. Condition and potential for improvement of high altitude rangelands. *Journal of Bhutan Studies, Centre for Bhutan Studies, Thimphu, Bhutan* 7: 82–98.

Gyeltshen, T., N. Shering, K. Tsering, et al. 2010. *Implication of legislative reform under the land act of Bhutan, 2007: A case study on nationalization of Tsamdrog and Sokshing and its associated socioenomic and environmental consequences*. Thimphu: Watershed Management Division, Department of Forest and Park Services.

ICIMOD. 2015. *Shifting cultivation in Bangladesh, Bhutan, and Nepal: Weighing government policies against customary tenure and institutions*. ICIMOD Working Paper 2015/7. Kathmandu: Center for Integrated Mountain Development.

IEA. 1998. *World energy outlook*. International Energy Agency. 1998 Edition. www.iea.org.

———. 2006. *World energy outlook*. Paris: International Energy Agency/Organization for Economic Cooperation and Development. Interscience Publishers.

Kafle, G. 2011. An overview of shifting cultivation with reference to Nepal. *International Journal of Biodiversity and Conservation* 3 (5): 147–154.

Kafle, G., P. Limbu, B. Pradhan, et al. 2009. *Piloting ecohealth approach for addressing land use transition, climate change and human health issues*. NGO group bulletin on climate change. Pokhara: LIBIRD.

Karan, P.P. 1987. Population Characteristics of the Himalayan Region. *Mountain Research and Development* 7 (3): 271–274.

Kerkhoff, E., and E. Sharma. 2006. Debating shifting cultivation in the Eastern Himalayas. In *Farmers' Innovations as Lessons for Policy*. Kathmandu: International Centre for Integrated Mountain Development (ICIMOD).

Khawas, V. 2003. *Joint forest management in India with special reference to Darjeeling Himalaya*. Ahmedabad, India: School of Planning, Center for Environmental Planning and Technology. Retrieved from http://lib.icimod.org/record/11260/files/209.pdf. Retrieved June 30, 2018.

Kreutzmann, H. 2012. Pastoralism: A way forward or back?. In *Pastoral practices in High Asia* (pp. 323–336). Dordrecht: Springer.

Kumar, R., M.K. Patra, A. Thirugnanavel, et al. 2016. Towards the natural resource management for resilient shifting cultivation system in eastern Himalayas. In *Conservation agriculture—An approach to combat climate change in Indian Himalaya*, ed. J.K. Bisht, V. Singh, M. Pankaj, K. Mishra, and A. Pattanayak, 409–436. Singapore: Springer Science + Business Media.

Le Fort, P. 1975. Himalayas: The collided range. Present knowledge of the continental arc. *American Journal Science* 275A: 1–44.

Limbu, D.K. (Photographer). 2010a. Rangeland of eastern Nepal (kharka). [Photograph]. (Author's collection).

———. 2010b. A thriving community forest of eastern Nepal. [Photograph]. (Author's collection).

———. 2010c. Drying and stacking firewood in eastern Nepal. [Photograph]. (Author's collection).

Liebig, M.A., J.A. Morgan, J.D. Reeder, et al. 2005. Greenhouse gas contributions and mitigation potential of agricultural practices in northwestern USA and western Canada. *Soil and Tillage Research* 83: 25–52.

Lungmuana, A.C., B.U. Choudhury, S. Saha, et al. 2018. Impact of post-burn jhum agriculture on soil carbon pools in the north-eastern Himalayan region of India. *Soil Research* 56 (6): 615–622.

Maden, K. (Photographer). 2017. Rangeland of western Nepal. [Photograph]. (Author's collection).

Maikhuri, R.K., L.S. Rawat, P.C. Phondani, et al. 2018. *Livestock—The engine and inspiration of mountain economy*. Bangalore: LEISA-India. Retrieved from https://leisaindia.org/livestock-the-engine-and-inspiration-of-mountain-economy/.

Meena, V.S., T. Mondal, B.M. Pandey, et al. 2018. Land use changes: Strategies to improve soil carbon and nitrogen storage pattern in the mid-Himalaya ecosystem, India. *Geoderma* 321: 69–78.

Mertz, O., and T.B. Bruun. 2017. Shifting cultivation policies in Southeast Asia: A need to work with, rather than against, smallholder farmers. In *Shifting cultivation policies: Balancing environmental and social sustainability*, ed. M. Cairns, 27–31. New York: CABI Publishing.

Molnar, A., M. France, L. Purdy, et al. 2011. *Community-based forest management—The extent and potential scope of community and smallholder forest management and enterprises*. 5th rights + resource anniversary, 7. Washington, DC: The Rights and Resources Initiative.

Namgay, K., J. Millar, R. Black, et al. 2013. Transhumant agro-pastoralism in Bhutan: Exploring contemporary practices and socio-cultural traditions. *Pastoralism* 3 (1): 1–26.

Nandy, S.N., and P.K. Samal. 2005. An outlook of agricultural dependency in the IHR. *ENVIS Newsletter: Himalayan Ecology* 2: 4–5.

Naudiyal, N., and J. Schmerbeck. 2017. The changing Himalayan landscape: Pine-oak forest dynamics and the supply of ecosystem services. *Journal of Forest Research* 28 (3): 431–443.

Nayak, P.K. 2002. *Community-based forest management in India: The issue of tenurial significance*. Paper for the 9th Biennial Conference of the IASCP.

Oestereich, C. 2017. *Green growth initiative in Bhutan: Bhutan biogas project*. Retrieved from https://sdghelpdesk.unescap.org/sites/default/files/2018-03/GG%20-%20Bhutan%20Biogas%20Project.pdf.

Ojha, H.R., N.P. Timsina, C. Kumar, et al. 2007. Community-based forest management programs in Nepal: An overview of issues and lessons. *Journal of Forest and Livelihood* 6 (2): 1–7.

Osei, W.Y. 1993. Wood fuel and deforestation-answers for a sustainable environment. *Journal of Environmental Management* 37: 51–62.

Pandeya, S.C. 1988. *Status of Indian rangelands*, 213. Jhansi, India: Range Management Society if India.

Pathak, B.R., X. Yi, and R. Bohara. 2017. Community based forestry in Nepal: Status, issues and lessons learned. *International Journal of Sciences* 6 (3): 120–129.

Paudel, M.N. 2015. Global effect of climate change and food security with respect to Nepal. *Journal of Agriculture and Environment* 16: 1–20.

Rai, K. K. (Photographer). 2018. The Himalayan Region. [Photograph]. (Author's collection)

Ramakrishnan, P.S. 1984. The science behind rotational bush fallow agriculture system (Jhum). *Proceeding of the Indian Academy of Science (Plant Science)* 93: 379.

Rangeland Policy of Nepal, 2012. Kathmandu, Nepal: Government of Nepal, Ministry of Agriculture and Co-operatives. http://www.nepalpolicynet.com/new/wp-content/uploads/2014/01/2070_-NEPAL-Government_IrrigationPolicy.pdf.

Rathore, S.S., N. Krose, M. Naro, et al. 2012. Weed management through salt application: An indigenous method for shifting cultivation areas, Eastern Himalaya. *Indian Journal of Traditional Knowledge* 11: 354–357.

Rayamajhi, S., C. Smith-Hall, and F. Helles. 2012. Empirical evidence of the economic importance of Central Himalayan forests to rural households. *Forest Policy and Economics* 20: 25–35.

Regmi, B.R., A. Subedi, K.P. Aryal, et al. 2005. *Shifting cultivation systems and innovations in Nepal*. Unpublished report. Pokhara: LIBIRD.

Royal Government of Bhutan (RGoB). 2010. *National strategy for community forestry: The way ahead*. Thimphu, Bhutan: Department of Forests and Park Services, Ministry of Agriculture and Forests.

Rijal, H.B., and H. Yoshida. 2002. *Investigation and evaluation of firewood consumption in traditional houses in Nepal*. In Proceedings: Indoor air (pp. 1000–1006).

Robinson, D.T., D.G. Brown, N.H.F. French, et al. 2013. Linking land use and carbon cycle: Advances in integrated science, management and policy. In *Land use and the carbon cycle*, ed. G.B. Daniel, T.R. Derek, H.F.F. Nancy, and C.R. Bradley. Cambridge, USA: Cambridge University Press.

Samal, P.K., P.P. Dhyani, and M. Dollo. 2010. Indigenous medicinal practices of Bhotia tribal community in Indian Central Himalaya. *Indian Journal of Traditional Knowledge* 9 (1): 140–144.

Satapathy, K.K., B.K. Sharma, S.N. Goswami, et al. 2003. *Developing lands affected by shifting cultivation*. New Delhi: Department of Land Resources, Ministry of Rural Development, Government of India.

Sati, V.P., and P. Rinawma. 2014. Practices of shifting cultivation and its implications in Mizoram, North-east India: A review of existing research. *Nature and Environment* 19 (2): 179–187.

Sharma, V.P., I. Kohler- Rollefson, J. Morton, et al. 2003. *Pastoralism in India: A scoping study. A report from Livestock Production Programme of the United Kingdom*. Department for International Development.

Singh, P. 1986. Status of Himalaya rangeland in India and their sustainable management. In *Proceedings of rangeland and pastoral development in Hindu Kush-Himalayas (5–7 November 1996)*, ed. D.J. Miller and S.R. Craig, 13–22. Kathmandu, Nepal.

Singh, A., K. Tuladhar, K. Bajracharya, et al. 2012. Assessment of effectiveness of improved cook stoves in reducing indoor air pollution and improving health in Nepal. *Energy for Sustainable Development* 16: 406–414.

Smith, K.R. 1998. *Indoor air pollution in India: National health impacts and cost-effectiveness of intervention*. Report Prepared for Capacity 21 Project of India, Mumbai: Indira Gandhi Institute of Development Research.

TED. 2016. Tshering Tobgay. *This country isn't just carbon neutral -it's carbon negative* [Video file]. Retrieved from https://www.ted.com/talks/tshering_tobgay_this_country_isn_t_just_carbon_neutral_it_s_carbon_negative?language=en#t-3411.

Tenzing, K. 2018. Exploring governance structures of high altitude rangeland in Bhutan using Ostrom's Design Principles. *International Journal of the Commons* 12 (1): 428–459.

Tinker, P.B., J.S.I. Ingram, and S. Strue. 1996. Effects of slash-and-burn agriculture and deforestation on climate change. *Agriculture Ecosystem and Environment* 58: 13–22.

Tomar, J.M.S., A. Das, L. Puni, et al. 2012. *Shifting cultivation in Northeastern region of India—Status and strategies for sustainable development*. Dehradun: Central Soil & Water Conservation Research & Training Institute.

Tutton, M., and K. Scott 2018. *CNN: What tiny Bhutan can teach the world about being carbon negative*. Retrieved from http://cnnphilippines.com/world/2018/10/12/bhutan-carbon-negative-greenhouse-gases.html.

Uddin, S.N., R. Taplin, and X. Yu. 2007. Energy, environment and development in Bhutan. *Renewable and Sustainable Energy Reviews* 11: 2083–2103.

UF. 2018. *Biogas—A renewable biofuel*. Florida: University of Florida. Retrieved from http://biogas.ifas.ufl.edu/FAQ.asp.

Upadhyay, K.P. 1995. *Shifting cultivation in Bhutan: A gradual approach to modifying land use patterns: A case study from Pema Gatshel District, Bhutan. Community forestry case study, series 11*. Food and Agriculture Organization of the United States (FAO).

Upadhyay, S. 2012. Community based forest and livelihood management in Nepal. In *The wealth of the commons: A world beyond market and state*, ed. D. Bollier and S. Helfrich, 362–364. Amherst: Levellers Press.

USAID. 1989. *Sustainable natural resources assessment—Philippines, prepared by Dames and Moore International, Louis Berger International and Institute for Development Anthropology*. Philippines: USAID/Manila.

Vakkilainen, E., K. Kuparinen, and H. Jussi. 2013. *Large industrial users of energy biomass. IEA bioenergy, task 40: International bioenergy trade*, 55. Lappeenranta: Lappeenranta University of Technology.

Valdiya, K.S. 1998. Dynamic Himalaya. Jawaharlal Nehru Center for Advanced Scientific Studies, Hyderabad: Universities Press (India) Limited.

Wadia, D.N. 1931. The syntaxis of the northwest Himalaya: Its rocks, tectonics and orogeny. *Record Geological Survey of India* 65 (2): 189–220.

Wangchuk, S. 2011. *Fuelwood consumption and production in alpine Bhutan: A case study of resource use and implications for conservation and management in Wangchuk Centennial Park*. MSc (Resource Conservation, International Conservation and Development) thesis, University of Montana: Montana.

Wangchuk, S., S. Siebert, and J. Belsky. 2013. Fuelwood use and availability in Bhutan: Implications for National Policy and Local Forest Management. *Human Ecology* 42: 127–135.

Wapongnungsang, C. Manpoong, and S.K. Tripathi. 2018. Changes in soil fertility and rice productivity in three consecutive years cropping under different fallow phases following shifting cultivation. *International Journal of Plant & Soil Science* 25 (6): 1–10.

World Bank. 2009. *Forests sourcebook: Practical guide for sustaining forests in international cooperation*. Washington, DC: The World Bank.

Yao, T., X. Guo, L. Thompson, et al. 2006. δ 18 O record and temperature change over the past 100 years in ice cores on the Tibetan Plateau. *Science in China Series D* 49 (1): 1–9.

Part II
Livestock, Grasslands, Wetlands, and Environment on Carbon Dynamics

Chapter 5
Effects of Different Grassland Management Patterns on Soil Properties on the Qinghai-Tibetan Plateau

Jianjun Cao, Xueyun Xu, Shurong Yang, Mengtian Li, and Yifan Gong

Abstract Change in grassland management pattern has an important effect on ecological function of grassland. Two grassland management patterns were developed after grassland contacted to individual households on the Qinghai-Tibetan Plateau, namely multi-household management pattern (MMP) without fences among households, and single-household management (SMP) with fences between adjacent households. Two representative counties (Maqu and Nagchu) of the Qinghai-Tibetan Plateau were selected to compare variation in soil properties between MMP and SMP, and pH, SOC, STN, and STP were selected as indictors of soil properties. The results showed that in Maqu SOC, STN, and STP were all significantly greater under MMP compared to SMP, in Nagchu, their values were also significantly different between them. All of these suggested that MMP without fences among households is of better soil condition than SMP with fences on the Qinghai-Tibetan Plateau.

Keywords The Qinghai-Tibetan Plateau · Grassland contact policy · Grassland management · Soil properties · Carbon losses

5.1 The Evolution of Two Different Grassland Management Patterns

Grasslands on the Qinghai-Tibetan Plateau (QTP), where pastoral practices date back at least 8800 years (Miehe et al. 2009), covered roughly an area of 1.33×10^6 km², accounting for almost 59% of the total area of the QTP and about 30% of the grasslands in China. For a long time, people who lived on the plateau played a crucial role in the formation and maintenance of their grassland

---

J. Cao (✉) · X. Xu · S. Yang · M. Li · Y. Gong
College of Geography and Environmental Science, Northwest Normal University, Lanzhou, China
e-mail: caojj@nwnu.edu.cn; xuxueyun@nwnu.edu.cn; yangsr923@nwnu.edu.cn; gongyf@nwnu.edu.cn

environment (Foggin 2012). However, over the last 50 years, nomads and pastoralists around the world, including those in China, have been accused of exploiting and misusing natural resources in their fragile environments, due to the socioeconomic developmental processes and the prejudice against mobile populations by more numerically dominant sedentary agriculturalists (Török et al. 2016; Cao et al. 2017).

In China, scientists from the Chinese Academy working in remote sensing and policy-making also believed that overgrazing, associated with communal property rights, was the major driver of grassland degradation (Yeh et al. 2017). Accordingly, a grassland contract policy was introduced to the Qinghai-Tibetan Plateau (QTP) in the 1990s (Yeh and Gaerrang 2011). As grassland contract policy implemented, communes, which is a sustainable measures for grassland resources through collective action, through which multiple landowners can come together to manage common resources such as labor, pasture and food, and then decide upon the allocation of those resources in a way that transcends the geographic lines between their properties (Schutz 2010; Wang et al. 2016), were dissolved and families became responsible for livestock and the marketing of their products (Cao et al. 2011).

At the beginning of the grassland contract system on the QTP, contract rules were clearly defined by governments. Namely, winter grassland was to be contracted to single-households, and summer grassland could be contracted to groups (multi-households) (Yu and Farrell 2016) due to the difficulty of fencing in these grasslands with remote areas and various geographies. However, although winter grassland was contracted to single-households, many herders were unwilling to participate in such an isolative practice because of their historic nomadism and dependence on a collective lifestyle (Cao et al. 2011).

As a result, two distinct grazing management patterns emerged: (1) the multi household grazing management pattern (MMP), in which the grasslands are collectively managed by two or more households without fences between them, and (2) the single household grazing management pattern (SMP), in which the grasslands are managed by individual households with fences (Cao et al. 2011, 2013a). With the implementation of these grassland contracts, the scope and area of the available rangeland were reduced. Currently, most of the MMP households have only one summer pasture and a single winter pasture, and some of the SMP households only have one pasture for year-round use (Cao et al. 2013a).

5.2 Effects of MMP and SMP on Social-Ecological System

Though policy-makers and most scientists still assume that overgrazing and climate change are the key drivers of degradation on the QTP, a small but growing number of scientists have tended instead to argue that a series of transformations of traditional forms of that have reduced mobility and fragmented the grasslands deteriorated socioeconomic development (Cao et al. 2013a, b).

5.2.1 Social-Economic System

By social investigation, Cao et al. (2011) found that significantly greater economic, and social benefits accrued under MMP relative to SMP. For example, households under SMP spent about 3100 ¥ y^{-1} more than under MMP for additional fencing and sheepdog breeding costs; with an informal institutional arrangement, households under MMP revealed greater equality and sustainability in the use of grassland resources and were subject to fewer risks during potentially adverse social and natural events, compared to those under SMP.

Similarly, Gongbuzeren and Li (2016) noted that about 24% of the householders from 60 herders' households in 2012 and 59% in 2014 said that milk production had declined under SMP, while about 30% of respondents in 2012 but only 19% in 2014 said it increased under MMP. Between 2012 and 2014, the average livestock mortality under MMP was nearly 10%, while under SMP it was around 14%. Similar results were also found by Cai and Li (2016). Wang et al. (2016) found that herders who pooled their pastures for communal grazing reduced their expenditures on forage, improved their pasture-use efficiency, conserved pasture quality and spread out climate risks over a greater area by having access to a greater range of pasture types.

After indictors of economic system and social system were selected, respectively, the total resilience of social and economic systems between MMP and SMP were compared by Cao et al. (2018a). In their paper, economic system indicators including: (1) Income and expenditure: to measure the livelihood of herders; (2) Infrastructure: to assess the herder's welfare. Social system indicators including: (1) Equity: opportunities available for herders to access natural resources and to participate in decision-making; (2) Health: the outcome of changes to the herder's lifestyle; (3) Assistance: to assess help available to herders when natural hazards or manmade disasters happen; (4) Social relations: to measure the herder's social network and conflicts among herders; (5) Cultural inheritance: the status of the traditional knowledge vital to the protection of grassland, wildlife, livestock and herder's health; (6) Institutional arrangements: costs and benefits of formal and informal institutions.

The resilience of each system was gauged using a decision support tool known as the Mauri Model, which is based on the economic, environmental, social and cultural well-being of the Māori, the indigenous people of New Zealand (Peacock et al. 2012). Based on this, the total resilience of social and economic system for MMP and SMP were presented in Table 5.1. Furthermore, the larger the total household size, the greater the economic benefits (Chen and Zhu 2015). From Table 5.2, it is clear that a lack of equity, cultural transmission and institutional arrangements may be the most serious problems for SMP. Therefore, the resilience of the social system could be enhanced if the SMP herders would recognize these issues and deliberately improve the situation, especially the issues of cultural transmission and institutional arrangements. However, there is very little space for the SMP to achieve equitable forage and water resources utilization due to the limited grazing area (Cao et al. 2018a).

Table 5.1 Resilience of economic and social systems for MMP and SMP, respectively (Date adapted from Cao et al. 2018a)

Indicators	MMP	SMP	Scoring	
			MMP	SMP
Economic system				
• Income (RMB)	High	Low or average	1	−1
• Expenditure (RMB)	Low	High	1	−1
• Infrastructure	Better	Worse	1	−1
Social system				
• Equity	Fair	Unfair	2	−2
• Health	Good	Average	1	−1
• Assistance	Yes	Absent	1	−1
• Social relations	Good	Average	1	−1
• Culture inheritance	Better	Worse	2	−2
• Institutional arrangement	Yes	Absent	2	−2

Table 5.2 Above-ground biomass, species richness, cover and biomass of functional groups under single- and multi-household use pattern, respectively (Date adapted from Cao et al. 2013a)

	2009			2011		
	SMP	MMP		SMP	MMP	
Biomass (g)	24.16 ± 3.30	32.44 ± 5.12	*	34.47 ± 2.26	42.34 ± 2.38	**
Cover (%)	89.2 ± 0.09	92.5 ± 0.13	ns	87 ± 0.01	91 ± 0.01	*
Species richness	21.0 ± 0.70	22.3 ± 0.65	ns	15.0 ± 0.58	18.35 ± 0.53	***
Sedge (g)	8.27 ± 0.81	11.84 ± 1.03	*	15.23 ± 0.98	22.12 ± 1.48	***
Grass (g)	2.71 ± 0.56	3.08 ± 0.53	ns	4.37 ± 0.89	4.09 ± 0.97	ns
Poisonous weed (g)	12.49 ± 0.87	16.42 ± 1.77	ns	13.07 ± 1.60	15.35 ± 1.48	**
Legume (g)	0.67 ± 0.16	1.09 ± 0.37	ns	0.48 ± 0.10	1.49 ± 0.42	*

Note: ns, *, **, *** means not significant, and significant at $p < 0.05$, $p < 0.01$, and $p < 0.001$, respectively

5.2.2 Vegetation System

In 2009 and 2011, Yang (2012) and Cao et al. (2013a) selected 30 MMP winter pastures and 30 SMP winter pastures as sampled sites to compare the vegetation conditions between these two different grassland management patterns. The results were presented in Table 5.2. Based on Table 5.2, we could found that there were no significant differences in coverage, plant species richness, and group functions except sedge group between MMP and SMP in 2009, but by 2011 significant differences among them had emerged except grass function, suggesting that SMP more easier led to vegetable degradation than MMP over time. Also, Abuman et al. (2012) found that comparison with before grassland contracts (MMP), the average aboveground biomass, coverage, vegetation height, and the number of plant species reduced by 47.0%, 15.8%, 44.1%, 33.2%, respectively, after grassland contracts (SMP). In addition, outside of the QTP, such as the Ningxia Hui Autonomous Region (Yu and

Yi 2012; Yu and Farrell 2013, 2016), Inner Mongolia (Hua and Squires 2015; Yu and Farrell 2016), and Gansu province (Hua et al. 2015) support the finding of greater vegetation degradation under SMP compared to MMP.

5.2.3 Soil System

Although a lot of studies showed that MMP can reduce grassland degradation and socio-economic loss, the differences in soil properties, including soil organic carbon (SOC), soil total nitrogen (STN), soil total phosphorus (STP) and soil pH, between MMP and SMP at a relatively larger scale were explored. Therefore, the aim of this paper was to explore whether a grazing management pattern has an influence on soil properties, which are the key factors determining the soil quality and maintaining the plant growth. In this paper, we will synthesize part of our previous studies to illustrate the reasons for the differences in soil properties between MMP and SMP, including the sample method and the main results. This will help readers to understand the comparison method adopted by ourselves and most of others during exploring the differences between MMP and SMP, including social, economic and vegetable differences.

The investigation was performed in Maqu and Nagchu Counties of the QTP (Fig. 5.1). Maqu County (33–34°N, 101–102°E) is located in the eastern QTP and

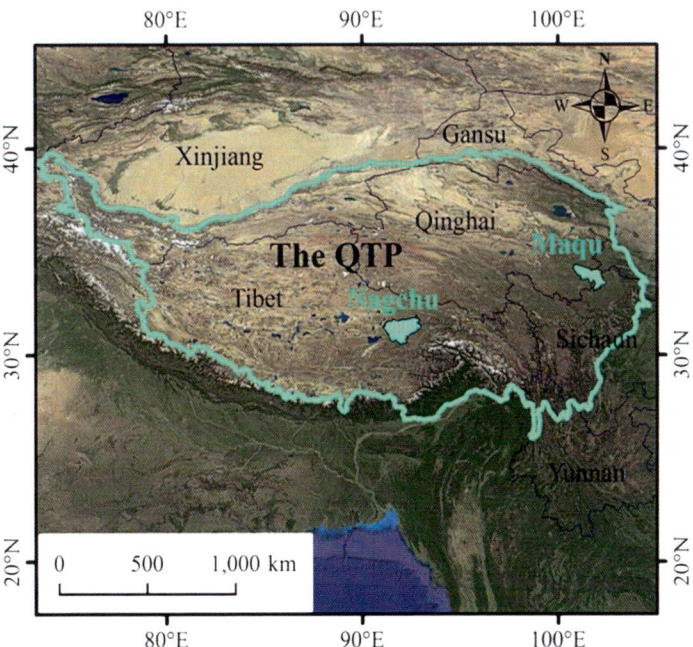

Fig. 5.1 The location of two study cases of Maqu and Nagchu on Tibetan plateau

traverses the boundary of Qinghai and Sichuan provinces in China. The elevation ranges from 2900 to 4000 m with an average annual rainfall of 599.7 mm. The average annual temperature is 1.8 °C with an average monthly low of -10.7 °C in January and a high of 11.7 °C in July. The maximum air temperature during the growing season can reach 29 °C, and there are, on average, 270 frost days annually. The grassland area extends across 8700 km^2: about 60% is considered alpine meadow. The genus *Kobresia* is dominant in this area. This area is commonly known as the 'water tower' of plateau, because the surface runoff from this region accounts for about 58.7% of the total runoff to the Yellow River.

There are seven villages and one town (a sub-administrative unit) in Maqu County with a total population of 57,000 in 2015. About 89% of the people are herdsmen. The annual production value related to animal husbandry is 47 million RMB, constituting about 94% of the agricultural production value in this study area. Currently, about 90% of a herder's income is derived from the trade of livestock and their byproducts; the remaining 10% comes from the sale of medicinal materials and other goods (Cao et al. 2011).

Nagchu County (30–31°N, 91–93°E) is a remote area of northern Tibet with a mean elevation above 4500 m and an average annual rainfall of 407 mm. The Yangtze River, Nujiang River and Lantsang River derives from Nagchu County, where belongs to the semi-arid monsoon climate of the subfrigid zone of the plateau. About 90% of the total land area in this county (16,200 km^2) is covered by *Kobresia*-dominated alpine grasslands and shrubs. The soil is mostly alpine meadow soil, and the special geographical environment and climate conditions make the ecosystem extremely vulnerable in this area.

Before the implementation of the grassland contract policy, which took place in 1996 in Maqu and in 2002 in Nagchu, herders in both counties grazed their livestock on the grassland and frequently shifted among four distinct seasonal grasslands, suggesting that the vegetation and soil conditions were very similar in each seasonal grassland. However, after the implementation of the grassland contract policy, some grasslands have been grazed continuously under MMP, while other grasslands have been grazed continuously under SMP (Cao et al. 2011, 2017). In light of these changes in terms of the kinds of pastoral practices that emerged due to the implementation of the grassland contract policy, the study site presents a unique opportunity to investigate the influence of different grazing management patterns on soil properties on the Qinghai-Tibetan Plateau.

In 2016 and 2017, based on our previous investigation, winter grasslands from 30 MMP and 30 from SMP in Maqu, and 20 winter grassland from MMP and 24 from SMP in Nagchu were taken as sampling sites, respectively. To ensure comparable sample sets in terms of other environmental factors for each management pattern, sampling sites were chosen with the same elevation, aspect, and soil texture and with at least two MMP and two SMP winter grasslands at each of these sites.

In addition, the sampling strategy also had the following stipulations: (1) they had been continuously grazed since the implementation of the grassland contracts in 1996 to ensure that the initial vegetation and soil conditions under the MMP and the SMP were similar, (2) they were predominantly used to graze yak to exclude the effect of type of grazing animal on the soil properties, (3) all grazing livestock on MMP and SMP depended only on foraging from the grassland resources and did not receive any supplementary feed to exclude the effect of nutrient import via excretion on the soil surface, and (4) they had the same stocking rate, which is mandated, monitored, and enforced by the Pastoral Supervisor Stations, as detailed in a previous study (Cao et al. 2013a, 2018b). At each of the sampled winter grasslands, one sample was composited from three plots (10 × 10 m) that were 10 m apart. In each plot, three soil samples (50 × 50 cm) were collected at the ends and midpoint of the diagonal to a depth of 30 cm using soil control sections (0–15; 15–30 cm) (Shang et al. 2014a).

5.2.3.1 Soil Properties Between MMP and SMP

In Maqu, in the 0–30 cm soil depth, pH under the MMP was significantly lower than under the SMP, and at each soil layer (0–15, 15–30 cm) it under the MMP also was significantly lower than its counterpart under the SMP, while in Nagchu, pH between the MMP and the SMP, and at the 0–15 cm soil layer under the MMP and its counterpart under the SMP, were no different; either in Maqu or in Nagchu, the SOC, STN, and STP to the 30 cm soil depth under the MMP were all significantly greater than under the SMP, with approximately 47, 5.0, and 0.77 g kg^{-1}, respectively, for the former, and 43, 4.3, and 0.73 g kg^{-1} for the latter, respectively, in Maqu, and with approximately 84, 6.9, and 0.59 g kg^{-1}, respectively, for the former, and 74, 6.1, and 0.54 g kg^{-1} for the latter, respectively, in Nagchu; both in Maqu and Nagchu, the SOC at each soil layer under the MMP was significantly different to its corresponding layer under the SMP, but the STN and the STP only at the 0–15 cm soil layer under the MMP were significantly different to their counterparts under the SMP in Maqu, while in Nagchu, values of them only at the 15–30 cm soil layer under the MMP were significantly different to their corresponding counterparts under the SMP (Table 5.3).

Soil properties were different between in Maqu and in Nagchu (Table 5.3), suggesting that the differences in ecosystem types are the key factors influencing soil properties (Cao et al. 2017). In Maqu, samples came from the alpine meadow, while in Nagchu, they were from the alpine swamp meadow. However, no matter how much the differences in soil properties between Maqu were and Nagchu, it seems that except pH between the MMP and the SMP being no difference in Nagchu, other soil properties were all better under the MMP than under the SMP. But remarkably, although soil C, N, and P between the MMP and the SMP were different, their stoichiometries between them had no differences (Yang 2018; Li et al. 2018). In this case, stoichiometry, especially for soil C:N may not a good index to compare the soil quality with different management patterns, although Askari and Holden (2014) regarded that it together with SOC and soil bulk density could provide a practical, time and cost effective method for quantitative evaluation of soil quality under temperate maritime grassland management.

Table 5.3 pH, soil organic carbon (SOC), soil total nitrogen (STN) and soil total phosphorus (STP) under the multi-household grazing management pattern (MMP) and the single-household grazing management pattern (SMP)

Property	Maqu		Nagchu	
	MMP	SMP	MMP	SMP
pH				
0–15 cm	6.76a	7.02b	7.21a	7.19a
15–30 cm	6.96a	7.12b	7.33a	7.49b
Average	6.86A	7.06B	7.27A	7.34A
SOC (g kg^{-1})				
0–15 cm	56.17a	49.01b	95.85a	90.94a
15–30 cm	38.38b	34.83a	72.78a	56.21b
Average	47.27A	42.96B	84.31B	73.57A
STN (g kg^{-1})				
0–15 cm	5.57a	5.08b	8.00a	7.74a
15–30 cm	3.62a	3.47a	5.74a	4.39b
Average	4.96A	4.28B	6.87B	6.07A
STP (g kg^{-1})				
0–15 cm	0.82a	0.76b	0.63a	0.60a
15–30 cm	0.71a	0.70a	0.56a	0.47b
Average	0.77A	0.73B	0.59B	0.54A

The letters indicate differences at $p < 0.05$. Capital letters denote differences between MMP and SMP, while lowercase letters denote differences in a particular soil layer between MMP and SMP (Date mainly adapted from Cao et al. 2017, 2018b, c)

5.2.3.2 Soil Carbon Loss Caused by SMP

In Maqu, using Zou et al. (2009) figures of 1.10 g cm^{-3} for SBD under MMP and 1.17 g cm^{-3} for SBD under SMP, the corresponding soil C storage to 30 cm was 156 and 151 Mg ha^{-1}, for MMP and SMP, respectively. In Nagchu, according to Sun et al. (2014), SBD under MMP and SMP can be regarded as 1.55 and 1.74 g cm^{-3}, respectively, and based on this, their corresponding soil C storage to 30 cm was 392 and 384 Mg ha^{-1}, respectively. Given that grasslands have been contracted since 1996 in Maqu, and 2002 in Nagchu, we estimated that about 0.25 Mg C ha^{-1} y^{-1} and about 0.57 Mg C ha^{-1} y^{-1} have been lost in Maqu and Nagchu, respectively.

After considered four policy scenarios for the winter grasslands (Scenario A, Scenario B, Scenario C, and Scenario D) and the summer grasslands (Scenario E, Scenario F, Scenario G, and Scenario H), respectively, across the QTP as below:

Scenario A: Represents an ideal condition in which all winter grasslands are grazed under MMP.

Scenario B: Based on our 2005 investigation in Maqu (Cao et al. 2011), about 80% of winter grasslands were grazed under MMP and remains were grazed under SMP. This could reflect the preferred grassland management pattern adopted by herders.

Scenario C: Based on our investigation in 2015 and 2016 (unpublished) across the QTP, we found that only 30% of the interviewees were willing to adopt MMP in the winter grasslands. This was because unresolved conflicts in MMP had resulted in their breaking up into the smaller SMP units. In addition, a variety of policies, particularly the closing down of village-level schools combined with compulsory education in more distant county towns, and a major government push for education, has led to more and more herders choosing to leave the grasslands, which is facilitated by SMP.

Scenario D: Winter grassland contracted to individual households is still a policy goal so in this scenario all winter grasslands are grazed under SMP.

Scenario E: An ideal condition in which all summer grasslands are grazed under MMP.

Scenario F: Based on our 2005 investigation in Maqu (Cao et al. 2011), about 90% of summer grasslands were grazed under MMP and remains were grazed under SMP. This could reflect the preferred management pattern as in Scenario B.

Scenario G: As described in Scenario C, with the impact of policies of urbanization, marketization, and educational centralization, more and more herders are willing to adopt SMP. About 50% of the interviewees have elected to use SMP in recent years (data unpublished) on the whole QTP, so this scenario can be regarded as the baseline for future summer grassland management unless policymakers recognize the ecological problems associated with SMP and end the contracting policy.

Scenario H: All summer grasslands are grazed under SMP (assuming the various factors described above continue to encourage individual households to take up summer grassland contracts). Based on our findings, this scenario would have the worst ecological impact.

If we assume that half of all alpine grassland is used as summer grassland and half is used as winter grassland, and that the effect of MMP and SMP on SOC in summer grassland is the same as on winter grassland, C losses from the QTP can be calculated for each combination of winter and summer scenarios (Fig. 5.2). Under the combination of Scenarios A and E, soil C would not be lost while the combination of Scenarios D and H would cause the largest soil C loss (6.15×10^7 Mg C y^{-1}) from the QTP.

5.3 The Reasons for Social-Ecologic System Under MMP Being Better Than Under SMP

As described above, social, economic, and ecologic benefits under MMP were better than that under SMP, but there are few studies on exploring the reasons for differences in socioeconomic benefits between them, so we will divide into two parts: one is the institutional reasons for MMP' socioeconomic benefits being better than

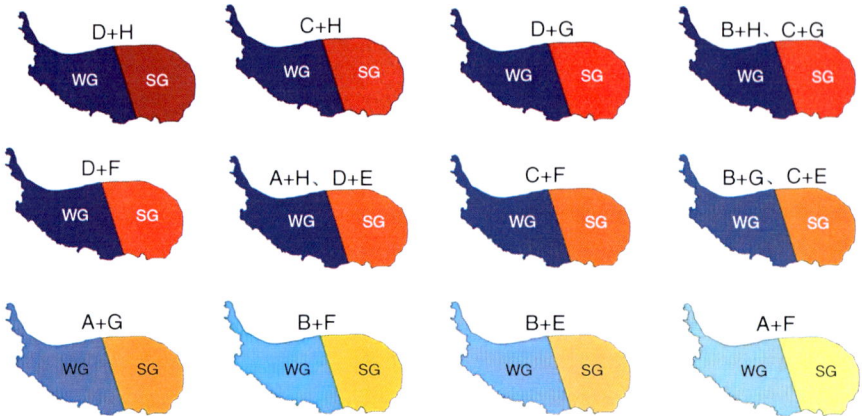

Fig. 5.2 Soil C losses (Mg C y^{-1}) caused by the SMP under different scenario combinations on the QTP. (*WG* winter grassland, *SG* summer grassland. Dark colors indicate greater soil C losses, and light colors indicate less soil C loss. D + H represents the largest soil C loss caused by the SMP (6.15 × 10^7 Mg C y^{-1}), while A + F represents the smallest soil C loss caused by the SMP (0.31 × 10^7 Mg C y^{-1}) from the whole QTP) (Adapted from Cao et al. 2017)

SMP's, and the other is the grazing reasons for MMP' ecologic benefits (vegetation and soil) being better than SMP's.

5.3.1 Institutional Reasons

To explain the institutional reasons, we considered the "Tragedy of the Commons" (Hardin 1968), which discussed the collective dilemma (Elsenbroich and Verhagen 2016) that arose when stakeholders forgot historical strategies for resource utilization, did not adopt proper self-governing institutions, or had little incentive to conserve, resulting in collective benefits being lost and the common good abrogated (Dutta and Sundaram 1993; Heller and Eisenberg 1998; Milinski et al. 2002; Dietz et al. 2003). This type of social dilemma has been studied extensively by political and social scientists, economists and evolutionary theorists (Feeny et al. 1990; Ostrom et al. 1999; Rothstein 2000; Runge 1986; Milinski et al. 2002). While there is no one solution to collective dilemmas, potential resolutions include voluntary small group cooperation through rules and institutions ensuring a shared management of resources (Moritz 2016).

Culture-cognitive elements that involve the creation of shared understandings that are taken for granted (Schermer et al. 2016) are very important for driving environmental change and shaping social behavior and outcomes of natural resource management (Franzén et al. 2015). The common-interest group is a collective governance mechanism, through which multiple landowners can come together to manage common resources such as labor, pasture and food, and then decide upon the allocation of those resources in a way that transcends the geographic lines between their properties (Schutz 2010; Wang et al. 2016). Cooperation is promoted by many mechanisms that human society mainly depends on (Ohtsuki et al. 2006). MMP is a voluntary cooperative action and reflects the influence of cultural norms developed long before privatization, that were designed to balance the supply of ecosystem services for grazing with human survival. In this case, it is particularly important that all households have equal rights of access to use the communal property resources and have the same probability of receiving benefits or sharing losses (Yu and Farrell 2016), i.e., cooperation is symmetrical (He et al. 2015).

It is also important that there are no 'strong' or 'weak' members of the MMP, because such asymmetrical cooperation would result in 'strong' households gaining more than 'weak' households, which might ruin the cooperative effort (Schutz 2010). Furthermore, households under MMP are considerably better at sharing knowledge based on trial and error management practices within the community by cultural transmission, which is a key to many successful resource management decisions (Flanagan and Laituri 2004; Franco and Luiselli 2014).

In the case of the MMP there are many kinds of informal institutions including those are flexible, responsive, multilevel and diverse, all of which can promote the resilience of social-ecological systems (Adger et al. 2005). For example, to balance the relationship between livestock numbers and grassland capacity, MMP households enacted strict regulations to limit stocking rates that each household had to adhere to. Inspections of livestock numbers are made twice each year by MMP representatives (Cao et al. 2011). If one household decides to settle in town, a common phenomenon across the QTP (Du and Zhang 2013; Ptackova 2011; Yeh and Gaerrang 2011), the right to use its grassland must be transferred to households within the MMP first, and the rental prices must be lower than the average level. If Household A has fewer people but has more grassland than Household B, the number of livestock of Household A may not reach the contracted number. Therefore, the grassland from Household A can be rented to Household B. Furthermore, the time of transhumance should be the same and rules regarding the prohibition of long hair, stealing and gambling can greatly influence members within the MMP (Cao et al. 2011). These informal institutions of the MMP perform with high efficiency and at a generally low cost (Cao et al. 2018a).

5.3.2 Grazing Reasons

As there is little deliberate management of soil per se in either MMP and SMP, investment (e.g. money, time, labor) in soil does not explain differences in soil properties. While the animal stocking rates for the MMP and the SMP are the same in both study areas, respectively, compared to MMP, the flexibility and mobility of the livestock under SMP is reduced (Yeh and Gaerrang 2011), thus leading to an intensified trampling and less recovery time between grazing events (Dlamini et al. 2014). Trampling, which, combined with a number of other degrading factors, such as deterioration of soil aggregates, increases in the soil bulk density and decrease in moisture content, water holding capacity and mechanical resistance (Chaudhuri et al. 2015; Herbin et al. 2011; Mei et al. 2013), could reduce the storage capacity and supply of soil nutrients (Christensen et al. 2004), and cause an increase in the C, N and P outputs, mainly through the removal of nutrient-rich clay particles in the topsoil layer (Lu et al. 2014). On the QTP, Luan et al. (2014) found that the effect of trampling on soil C losses can primarily be attributed to the reduced heavy fraction organic carbon and that the effect of trampling on soil N losses is due to an increase in N_2O flux with enhanced soil N transformation rates due to grazing.

Although lack of evidences about vegetation condition between the MMP and the SMP in Nagchu, it reported that with continuous (non-rotational) grazing under the SMP, plants can be smaller, resulting in lower plant cover and biomass, as confirmed in Maqu (Cao et al. 2011, 2013a). In the current study regions and others within the QTP, the lower aboveground biomass and the land cover were presumably the primary limiting factors in improving the soil fertility (Dong et al. 2012). This is because inputs into the soil, which can act to free essential nutrients, could be reduced (Hirsch et al. 2017). In addition to the interactions between the overlying vegetation and soil, which cause lower fertility under the SMP than under the MMP, other factors could also explain this effect. For example, increased grazing intensity within limited areas under the SMP could induce an increase in plant digestibility, nutrient concentrations, and nutrient diversity. This can increase the quality of forage at the expense of losses in both soil carbon stock and nutrient availability (Niu et al. 2016). Meanwhile, as nutrients decline, the population of poisonous plants is likely to increase because such plants can tolerate nutrient-limited soil conditions as found by Cao et al. (2011) in Maqu.

In their study, they found that the species *Ligularia virgaurea* (a poisonous plant) became more common in the grasslands of the SMP compared to grasslands of the MMP. Furthermore, they also found that the biomass of the C4 plant functional group, especially that of Cyperaceae spp. under the SMP, was significantly lower than that under the MMP (Cao et al. 2013a), and thus the soil fertility under the MMP was higher because C4 grasses can potentially improve the above-ground biomass (Wu et al. 2009), and that the biomass of legumes under the SMP was significantly lower than that under the MMP, resulting in the differences in soil fertility between the two management patterns since legumes can fix excess atmospheric N_2 (Peoples et al. 1995).

Overall, with a limited grazing area, the SMP may not only lead to poor soil fertility as found in both study areas, but also result in a less uniform distribution of nutrients compared to the MMP as found in Nagchu (Cao et al. 2018c). This is attributable to the notion that the grazing intervals induced by transhumance under the MMP could improve the distribution and availability of nutrients in grazed pastures (Silveira et al. 2014). However, while this study provides plausible explanations for the degradation of vegetation and soils under the SMP, further studies using well-coordinated multidisciplinary approaches are warranted to identify the effects of the MMP and SMP on soil properties.

5.4 Suggestions on the Sustainable Development on the QTP

Traditionally, warm-season grazing is compatible with conserving species diversity and sequestering nutrients in topsoil, while cold-season grazing is suitable for sequestering nutrients in deeper soil strata; accordingly, periodic cold- and warm-season grazing would constitute a suitable grazing regime to maintain alpine meadow sustainability (Wu et al. 2017). At present, some officials have realized that, under rapid economic development, there are problems associated with such policies, but these problems are unavoidable in a transition process designed for the betterment of future generations (Gongbuzeren et al. 2015).

Along with privatization, to improve the welfare of nomads, the government launched a comprehensive program subsidizing the establishment of stockyards and permanent settlements on winter pastures (Fassnacht et al. 2015). The present livelihood adaption strategies related to sedentary grazing have provided greater convenience to the local population, including providing safe drinking water, improved medical conditions, and economic profitability of the herding livelihood (Wang et al. 2015, 2016). However, the expected benefits from settlements have not been forthcoming, and land degradation and pastoralist poverty have not been eliminated (Yu and Farrell 2016). For example, livestock typically remain in fenced areas near the settlements or overnight campsites, and thus grazing intensity and effects are highest near the settlements, and the effect of trampling on the soil and vegetation is exacerbated (Dorji et al. 2013; Lehnert et al. 2014; Gongbuzeren et al. 2015).

Furthermore, settlements have led to the conversion of marginal croplands and native meadows into pastures (Zhang et al. 2012), while local grazing systems have raised dependence on artificial feeding and inputs coming from outside the grazing system (Wang et al. 2016), and have led to a decline in biodiversity due to the increase in population size and growth rate (Wang et al. 2015).

Today, with the continuing implementation of grassland contracting as well as other policies (e.g. the sedentarization of pastoralists and the Grassland Retirement Program), flexibility in pastoralists' management of pasture and livestock has been considerably decreased (Fu et al. 2012). Wang et al. (2015) proposed that, at a local scale, these unwise spatial and temporal grazing patterns that constrain herds in search of grass and water by moving them when distribution of resources changes (Fu et al. 2012) may be mainly responsible for "overgrazing", and that optimizing seasonal grazing patterns could solve this issue. Therefore, a new management style based on individualized private ownership of properties, such as group and multi-household management, are needed (Wang et al. 2015).

It is acknowledged that there are currently no nomadic management patterns for us to compare with the results of the MMP, based on these two study cases, we could conclude that the SMP has caused significant soil degradation. Therefore, the grassland use policy (which currently advocates the SMP) should be implemented with caution in this important region in the future, as contracting of the grassland to individual households is still a goal of the government. If all grasslands on the QTP are managed using the SMP, it is likely that the losses of soil C, N, and P will become substantial.

Considering the importance of the QTP, especially in terms of regional and global biogeochemical cycles and their roles in stabilizing the underlying environments, the losses of soil nutrients, may lead to regional scarcity in plant productivity. Therefore, appropriate institutional arrangements, which is to retain the nomadic element of pastoral husbandry while increasing the temporal and spatial scale of rotational stocking to reduce pasture degradation, such as put forward by Shang et al. (2014b) must be considered as an important measure to promote sustainable grassland management and to mitigate the negative effects of global warming on rangeland quality on the QTP (Dong et al. 2011).

Except these measures above mentioned, ecological compensation also should be concerned. Ecological compensation can be regarded as the price paid for individuals or regions to sacrifice their development opportunities in order to protect the ecology or the environment. Generally, it can reflect the relationship of relevant groups between environment and economic benefits, and can promote equity and development between regions and social groups, and can also be placed in a border social-economic framework. For example, although there are many causes of degradation (e.g. population growth, sedentarization, market development, construction of infrastructure and mineral exploitation), the increasing number of livestock has probably accelerating a decline in grassland ecological services. However, in pastoral areas of this region, livestock is often regarded as a source of income (the main sources of income and development capital), a subsistence insurance (animals are sold in the event of health emergencies), a fuel source (dung), a lifestyle (via employment) and a measure of social status (more livestock means more respect in the community). Therefore, only a few livestock are traded voluntarily by the herders at market in order to get money, and herders are not willing to sell animals including excess that are over the carrying capacity of grassland.

Since herders are not willing to sell these excessive at market, a non-market method to estimate their value if policy were put in place to encourage herders to give them up in order to balance forage yield and livestock numbers. With non-market method such as contingent valuation, herders are offered a contingent situation, which is the possibility of a compensation for giving up livestock. With this method, Cao et al. (2012) found that the minimum acceptable compensation was estimated as 3717 RMB (~$555) for cattle and 503 RMB (~$75) for sheep at 2005 prices. Based on this, we can provide ecological compensation for locals for giving up livestock in future.

References

Abuman, W.G. Zhang, and M. Chang. 2012. Present situation investigation after grassland contracted by individual households in Gannan pastoral area, Gansu province. *Pratacultural Science* 29 (12): 1945–1950.

Adger, W.N., T.P. Hughes, C. Folke, et al. 2005. Social-ecological resilience to coastal disasters. *Science* 309 (5737): 1036–1039.

Askari, M.S., and N.M. Holden. 2014. Indices for quantitative evaluation of soil quality under grassland management. *Geoderma* 230–231 (7): 131–142.

Cai, H., and W.J. Li. 2016. A study on efficiency and equity of rangeland management on Qinghai-Tibet Plateau in different property right systems. *Journal of Natural Resources* 31: 1302–1309.

Cao, J.J., Y.C. Xiong, J. Sun, et al. 2011. Differential benefits of multi- and single-household grassland management patterns in the Qinghai-Tibetan Plateau of China. *Human Ecology* 39 (2): 217–227.

Cao, J.J., N.M. Holden, Y.Y. Qin, et al. 2012. Potential use of willingness to accept (WTA) to compensate herders in Maqu county, China, for reduced stocking. *Rangeland Ecology & Management* 65 (5): 533–537.

Cao, J.J., E.T. Yeh, N.M. Holden, et al. 2013a. The effects of enclosures and land-use contracts on rangeland degradation on the Qinghai-Tibetan Plateau. *Journal of Arid Environments* 97 (10): 3–8.

———. 2013b. The roles of overgrazing, climate change and policy as drivers of degradation of China's grasslands. *Nomadic Peoples* 17 (2): 82–101.

Cao, J.J., Y.F. Gong, E.T. Yeh, et al. 2017. Impact of grassland contract policy on soil organic carbon losses from alpine grassland on the Qinghai-Tibetan Plateau. *Soil Use and Management* 33 (4): 663–671.

Cao, J.J., M.T. Li, R.C. Deo, et al. 2018a. Comparison of social-ecological resilience between two grassland management patterns driven by grassland land contract policy in the Maqu, Qinghai-Tibetan Plateau. *Land Use Policy* 74 (5): 88–96.

Cao, J.J., X.Y. Xu, R.C. Deo, et al. 2018b. Multi-household grazing management pattern maintains better soil fertility. *Agronomy for Sustainable Development* 38 (1): 6–12.

Cao, J.J., X.Y. Wang, and M.T. Li. 2018c. Effects of different grassland use patterns on soil nutrients and spatial distribution on the Qinghai-Tibetan Plateau, China. *Chinese Journal of Applied Ecology* 29 (6): 1839–1845.

Chaudhuri, S., L.M. Mcdonald, J. Skousen, et al. 2015. Soil organic carbon molecular properties: Effects of time since reclamation in a minesoil chronosequence. *Land Degradation and Development* 26 (3): 237–248.

Chen, H.Y., and T. Zhu. 2015. The dilemma of property rights and indigenous institutional arrangements for common resources governance in China. *Land Use Policy* 42 (1): 800–805.

Christensen, L., M.B. Coughenour, J.E. Ellis, et al. 2004. Vulnerability of the Asian typical steppe to grazing and climate change. *Climatic Change* 63 (3): 351–368.

Dietz, T., E. Ostrom, and P.C. Stern. 2003. The struggle to govern the commons. *Science* 302 (5652): 1907–1912.

Dlamini, P., P. Chivenge, A. Manson, et al. 2014. Land degradation impact on soil organic carbon and nitrogen stocks of sub-tropical humid grasslands in South Africa. *Geoderma* 235–236 (4): 372–381.

Dong, S.K., L. Wen, S.L. Liu, et al. 2011. Vulnerability of worldwide pastoralism to global changes and interdisciplinary strategies for sustainable pastoralism. *Ecology and Society* 16 (2): 85–99.

Dong, S.K., L. Wen, Y.Y. Li, et al. 2012. Soil-quality effects of grassland degradation and restoration on the Qinghai-Tibetan Plateau. *Soil Science Society of America Journal* 76 (6): 2256–2264.

Dorji, T., Ø. Totland, and S.R. Moe. 2013. Are droppings, distance from pastoralist camps, and Pika burrows good proxies for local grazing pressure? *Rangeland Ecology and Management* 66: 26–33.

Du, F.C., and S.H. Zhang. 2013. *Economic transition of pastoralism in western*. China, Beijing: Intellectual Property Publishing House.

Dutta, P.K., and R.K. Sundaram. 1993. The tragedy of the commons? *Economic Theory* 3 (3): 413–426.

Elsenbroich, C., and H. Verhagen. 2016. The simplicity of complex agents: A contextual action framework for computational agents. *Mind and Society* 15 (1): 131–143.

Fassnacht, F.E., L. Li, and A. Fritz. 2015. Mapping degraded grassland on the eastern Tibetan Plateau with multi-temporal Landsat 8 data-where do the severely degraded areas occur? *International Journal of Applied Earth Observation and Geoinformation* 42 (10): 115–127.

Feeny, D., F. Berkes, B.J. McCay, et al. 1990. The tragedy of the commons: Twenty-two years later. *Human Ecology* 18 (1): 1–19.

Flanagan, C., and M. Laituri. 2004. Local cultural knowledge and water resource management: The wind river Indian reservation. *Environmental Management* 33 (2): 262–277.

Foggin, M. 2012. Pastoralists and wildlife conservation in western China: Collaborative management within protected areas on the Tibetan Plateau. *Research Policy and Practice* 2: 17–35.

Franco, D., and L. Luiselli. 2014. Shared ecological knowledge and wetland values: A case study. *Land Use Policy* 41 (4): 526–532.

Franzén, F., M. Hammer, and B. Balfors. 2015. Institutional development for stakeholder participation in local water management-an analysis of two Swedish catchments. *Land Use Policy* 43: 217–227.

Fu, Y., R.E. Grumbine, A. Wilkes, et al. 2012. Climate change adaptation among Tibetan pastoralists: Challenges in enhancing local adaptation through policy support. *Environmental Management* 50: 607–621.

Gongbuzeren, and W.J. Li. 2016. The role of market mechanisms and customary institutions in rangeland management: A case study in Qinghai Tibetan Plateau. *Journal of Natural Resources* 31 (10): 1637–1647.

Gongbuzeren, Y.B. Li, and W.J. Li. 2015. China's rangeland management policy debates: What have we learned? *Rangeland Ecology and Management* 68 (4): 305–314.

Hardin, G. 1968. The tragedy of the commons. *Science* 62: 1243–1248.

He, J.Z., R.W. Wang, C.X. Jensen, et al. 2015. Asymmetric interaction paired with a super-rational strategy might resolve the tragedy of the commons without requiring recognition or negotiation. *Scientific Reports* 5: 7715.

Heller, M.A., and R.S. Eisenberg. 1998. Can patents deter innovation? The anticommons in biomedical research. *Science* 280 (5364): 698–701.

Herbin, T., D. Hennessy, K.G. Richards, et al. 2011. The effects of dairy cow weight on selected soil physical properties indicative of compaction. *Soil Use and Management* 27: 36–44.

Hirsch, P.R., D. Jhurreea, J.K. Williams, et al. 2017. Soil resilience and recovery: Rapid community responses to management changes. *Plant and Soil* 412 (1–2): 283–297.

Hua, L.M., and V.R. Squires. 2015. Managing China's pastoral lands: Current problems and future prospects. *Land Use Policy* 43: 129–137.

Hua, L.M., S.W. Yang, V. Squires, et al. 2015. An alternative rangeland management strategy in an agro-pastoral area in western China. *Rangeland Ecology and Management* 68 (2): 109–118.

Lehnert, L.W., H. Meyer, N. Meyer, et al. 2014. A hyperspectral indicator system for rangeland degradation on the Tibetan Plateau: A case study towards spaceborne monitoring. *Ecological Indicators* 39: 54–64.

Li, M.T., Y.Y. Qin, J.J. Cao, et al. 2018. Effects of different grassland management patterns on soil stoichiometry on the Qinghai-Tibetan Plateau: A case study from Nagqu. *Chinese Journal of Ecology* 37 (8): 2262–2268.

Lu, J., Z. Dong, W. Li, et al. 2014. The effect of desertification on carbon and nitrogen status in the northeastern margin of the Qinghai-Tibetan Plateau. *Environmental Earth Sciences* 71 (2): 807–815.

Luan, J., L. Cui, C. Xiang, et al. 2014. Different grazing removal exclosures effects on soil C stocks among alpine ecosystems in east Qinghai-Tibet Plateau. *Ecological Engineering* 64 (3): 262–268.

Mei, H., G. Zhang, X. Gan, et al. 2013. Carbon and nitrogen storage and loss as affected by grassland degradation in Inner Mongolia, China. *Journal of Food Agriculture and Environment* 11 (3): 2071–2076.

Miehe, G., S. Miehe, K. Kaiser, et al. 2009. How old is pastoralism in Tibet? An ecological approach to the making of a Tibetan landscape. *Palaeogeography Palaeoclimatology Palaeoecology* 276 (1–4): 130–147.

Milinski, M., D. Semmann, and H. Krambeck. 2002. Reputation helps solve the 'tragedy of the commons'. *Nature* 415: 424–426.

Moritz, M. 2016. Open property regimes. *International Journal of the Commons* 10 (2): 688–708.

Niu, K., J. He, and M.J. Lechowicz. 2016. Grazing-induced shifts in community functional composition and soil nutrient availability in Tibetan alpine meadows. *Journal of Applied Ecology* 53 (5): 1554–1564.

Ohtsuki, H., C. Hauert, and E. Lieberman. 2006. A simple rule for the evolution of cooperation on graphs and social networks. *Nature* 441 (7092): 502–505.

Ostrom, E., J. Burger, and C.B. Field. 1999. Revisiting the commons: Local lessons, global challenges. *Science* 284 (5412): 278–282.

Peacock, B.C., D. Hikuroa, and T. Morgan. 2012. Watershed-scale prioritization of habitat restoration sites for non-point source pollution management. *Ecological Engineering* 42 (9): 174–182.

Peoples, M.B., J.K. Ladha, and D.F. Herridge. 1995. Enhancing legume N_2 fixation through plant and soil management. *Plant and Soil* 174 (1): 83–101.

Ptackova, J. 2011. Sedentarisation of Tibetan nomads in China: Implementation of the nomadic settlement project in the Tibetan Amdo area; Qinghai and Sichuan Provinces. *Pastoralism: Research, Policy and Practice* 1: 4–14.

Rothstein, B. 2000. Trust, social dilemmas and collective memories. *Journal of Theoretical Politics* 12 (4): 477–501.

Runge, C.F. 1986. Common property and collective action in economic development. *World Development* 14 (5): 623–635.

Schermer, M., I. Darnhofer, K. Daugstad, et al. 2016. Institutional impacts on the resilience of mountain grasslands: An analysis based on three European case studies. *Land Use Policy* 52: 382–391.

Schutz, A.B. 2010. Grassland governance and common-interest communities. *Sustainability* 2 (7): 2320–2348.

Shang, Z.H., J.J. Cao, R.Y. Guo, et al. 2014a. The response of soil organic carbon and nitrogen 10 years after returning cultivated alpine steppe to grassland by abandonment or reseeding. *Catena* 119: 28–35.

Shang, Z.H., M.J. Gibb, F. Leiber, et al. 2014b. The sustainable development of grassland-livestock systems on the Tibetan Plateau: Problems, strategies and prospects. *Rangeland Journal* 36 (3): 267–296.

Silveira, M.L., F.M. Rouquette, G.R. Smith, et al. 2014. Soil-fertility principles for warm-season perennial forages and sustainable pasture production. *Forage and Grazinglands* 12 (1): 1–9.

Sun, Q.L., Z.B. Sun, and T.P. Liu. 2014. The herdsmen's cognition and behavior about grassland ecological protection—Based on the empirical analysis on 75 villages in Tibet. *Journal of Arid Land Resources and Environment* 28 (8): 26–31.

Török, P., N. Hölzel, R.V. Diggelen, et al. 2016. Grazing in European open landscapes: How to reconcile sustainable land management and biodiversity conservation? *Agriculture Ecosystems and Environment* 234: 1–4.

Wang, P., J.P. Lassoie, S.J. Morreale, et al. 2015. A critical review of socioeconomic and natural factors in ecological degradation on the Qinghai-Tibetan Plateau, China. *The Rangeland Journal* 37: 1–9.

Wang, J., Y. Wang, S.C. Li, et al. 2016. Climate adaptation, institutional change, and sustainable livelihoods of herder communities in northern Tibet. *Ecology and Society* 21 (1): 5–15.

Wu, G.L., G.Z. Du, Z.H. Liu, et al. 2009. Effect of fencing and grazing on a Kobresia-dominated meadow in the Qinghai-Tibetan Plateau. *Plant and Soil* 319 (1): 115–126.

Wu, G.L., D. Wang, Y. Liu, et al. 2017. Warm-season grazing benefits species diversity conservation and topsoil nutrient sequestration in alpine meadow. *Land Degradation and Development* 28 (4): 1311–1319.

Yang, Y.Y. 2012. *The impact studies of Plateau different grazing patterns of grassland degradation.* Master's thesis, Lanzhou University.

Yang, S.R. 2018. *Effects of different grazing patterns on soil nutrients and stoichiometric ratios on maqu grassland.* Master's thesis, Northwest Normal University.

Yeh, E.T., and Gaerrang. 2011. Tibetan pastoralism in neoliberalising China: Continuity and change in Gouli. *Area* 43: 165–172.

Yeh, E.T., L.H. Samberg, Gaerrang, et al. 2017. Pastoralist decision-making on the Tibetan Plateau. *Human Ecology* 1: 1–11.

Yu, L., and K.N. Farrell. 2013. Individualized pastureland use: Responses of herders to institutional arrangements in pastoral China. *Human Ecology* 41 (5): 759–771.

———. 2016. The Chinese perspective on pastoral resource economics: A vision of the future in a context of socio-ecological vulnerability. *Revue Scientifique et Technique* 35 (2): 523–531. (in Chinese).

Yu, L., and J. Yi. 2012. Pastureland governance form a perspective of property right. *Practaculture Science* 29 (12): 1920–1925.

Zhang, Z., J. Duan, S. Wang, et al. 2012. Effects of land use and management on ecosystem respiration in alpine meadow on the Tibetan Plateau. *Soil and Tillage Research* 124 (4): 161–169.

Zou, L.N., Z.Y. Zhou, S.Y. Yan, et al. 2009. Response of soil nutrients to different land utilization types in alpine meadow in Maqu. *Chinese Journal of Grassland* 31: 80–87.

Chapter 6
Carbon Management of the Livestock Industry in the HKH Region

Yu Li, A. Allan Degen, and Zhanhuan Shang

Abstract Human livelihoods in the HKH rely heavily on livestock husbandry, as this sector supplies essential animal products for the local and surrounding areas. This sector also takes on a dominant role in the anthropogenic disturbance in the carbon cycle in this region due, in part, to the extensive grazing and browsing of the livestock. In the HKH, livestock biomass use contributes towards the first step of the human appropriation of net primary production, which has an estimated value (3.69 ton/cap per year) higher than the Asian average. Methane emissions from the main sources, that is, enteric fermentation and manure, amount to 3.77 million tons per year, while the high emission zones are found in the mixed livestock cropping systems. It could be concluded that there is high potential of carbon emission mitigation in the livestock production sector.

Keywords Livestock carbon management · Highland Asia · Hindukush-Karakorum-Himalaya region · HANPP · Carbon emission

---

Y. Li
School of Life Sciences, State Key Laboratory of Grassland Agro-Ecosystems, Lanzhou University, Lanzhou, Gansu, China
e-mail: liyu15@lzu.edu.cn

A. A. Degen
Desert Animal Adaptations and Husbandry, Wyler Department of Dryland Agriculture, Blaustein Institutes for Desert Research, Ben-Gurion University of Negev, Beer Sheva, Israel
e-mail: degen@bgu.ac.il

Z. Shang (✉)
School of Life Sciences, State Key Laboratory of Grassland Agro-Ecosystems, Lanzhou University, Lanzhou, Gansu, China

Key Laboratory of Restoration Ecology of Cold Area in Qinghai Province, Northwest Institute of Plateau Biology, Chinese Academy of Sciences, Lanzhou, Gansu, China
e-mail: shangzhh@lzu.edu.cn

© Springer Nature Switzerland AG 2020
Z. Shang et al. (eds.), *Carbon Management for Promoting Local Livelihood in the Hindu Kush Himalayan (HKH) Region*,
https://doi.org/10.1007/978-3-030-20591-1_6

6.1 Introduction

Livestock industries have emerged to be a significant part of the world carbon cycle. Livestock production directly appropriates about 2 Pg C, or 3% of the global net primary production each year (Imhoff et al. 2004). In addition, it substantially affects the collateral carbon flows via land use (change), animal metabolism and energy use. The diverse livestock production systems occupy over 25% of the terrestrial surface (Asner et al. 2004; Steinfeld and Wassenaar 2007). In terms of carbon losses to the atmosphere, the impact of livestock production has been estimated at 3.2 Pg CO_2-equivalents per year in the world, mainly from livestock grazing and ruminant fermentation products (Asner and Archer 2010).

The Hindukush-Karakoram-Himalayan (HKH) region extends 3500 km from Myanmar in the east to Afghanistan in the west, and is the largest and most diverse mountain region in the world (Joshi et al. 2013). Nearly 60% of the land use/cover system is occupied by rangeland, with pastoralism as the main contribution to the livelihood of the local communities (Kreutzmann 2012). The HKH pastoral communities inhabit mainly high-elevation temperate regions or arid/semi-arid zones to accommodate their grazing livestock, primarily yaks and sheep. Consequently, grazing livestock are the predominent cause of anthropogenic disturbances on the carbon cycle in this region.

The livestock carbon management system could be summarized as material flow, with human interventions, of the biogeochemical processes in the biospheric carbon pool, in which the atmospheric pool is regarded as an external system (Steinfeld and Wassenaar 2007). This system could be described as an input-output model that is composed of inputs (e.g. forage and human labor), anthropogenic carbon management activities (i.e. grassland use, livestock management and energy use) and outputs (e.g. live animals, meat and dairy products). This chapter documents the main issues of livestock carbon management in the HKH, including up-to-date GIS data, and adds more information on carbon management from the livestock industry in highland Asia. Research on carbon management and climate mitigation in the HKH has largely been neglected.

6.2 Livestock Production Systems in the HKH Region

The HKH, covering 4,190,000 km^2, exhibits a high degree of agricultural variability in it's spatial segments. Pastoralism, agropastoralism and agroforestry livestock grazing are the dominant land uses of the vast ranglands of the HKH (Dong et al. 2016). The GIS classification model 'global livestock production systems', engineered by FAO and ILRI, provide sound data support for the determination of spatial land use regimes for livestock (Robinson et al. 2011). In this model, livestock production was mapped in ESRI grids under a decision tree classification method by incorporating data of land cover, population density, length of growing period

(LGP), air temperature and elevation (Robinson et al. 2011). The HKH includes all 14 livestock production systems in the model (Fig. 6.1), and 60% of the land was categorized as livestock-only systems (LG), which is the rangeland has a LGP either less or more than 60 days while the population density is less than 20 persons/km^2. Specifically, the second level system of rangeland-based arid/semi-arid (LGY) zones, distributed centrally in the Tibetan Plateau totalling roughly 1,870,000 km^2, occupies 74% of the LG system (Fig. 6.1). In the south, east and west edge of the HKH, regions having more cultivatable land, the mixed rainfed, irrigated and urban systems account for 15%, 2% and 0.7%, respectively, of the total area.

Pastoral practices are considered to be the major livelihood strategies in the mountains (covers the HKH). The traditional lifestyles, land uses and land management practices are under rapidly increasing pressures from population explosion, modernization and globalization processes (Kreutzmann 2012). More specifically, the 'classical' practices are incorporating novel agricultural and pastoral approaches, and are evolving 'modern' methods that reflect strategies adopted by pastoralists (Kreutzmann 2012; Shang et al. 2014). Changes in the livestock sector and in the utilisation of high pastures were introduced by programs including: 'Combined Mountain Agriculture', 'Detached Mountain Pastoralism', 'Classical Mountain Nomadism', 'Resettlement Project in High Pastures' and 'Agro-Pastoral Resettlement Scheme in Lowland Regions'. In the high elevation areas of the HKH, usually above 3000 m and including Tibet, Qinghai province in China and Bhutan, semi-nomadism is the main practice, while in lower elevation sites, including Afghanistan, India, Nepal and Pakistan, a more sedentary lifestyle is predominant (Wu et al. 2014).

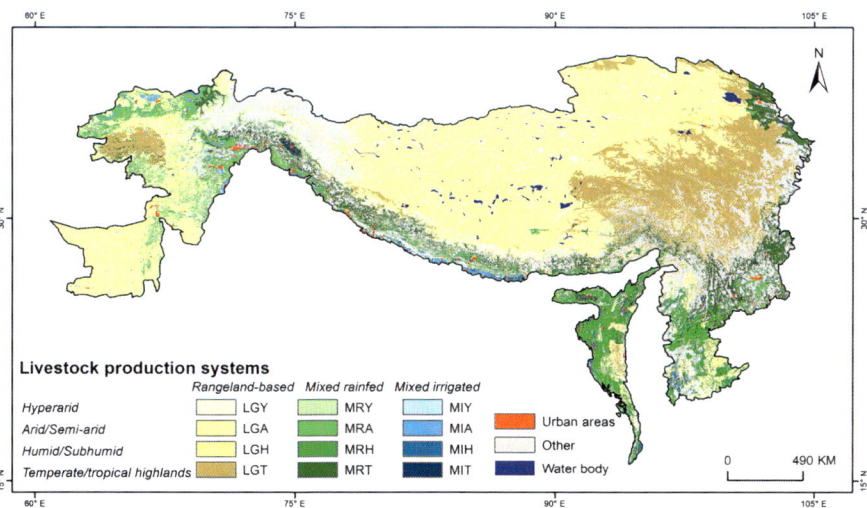

Fig. 6.1 Livestock production systems in the HKH region (Source: Ruminant production systems v.5 by Robinson et al. 2011)

From archaeological and genetic evidence, pastoralism with domesticated animals like yak and sheep has supported human settlements in highland Asia from the late-Holocene period (Rhode et al. 2007; Zhao 2009; Qiu et al. 2015), and has shaped the highland extensive socio-economical development in Eurasia (Frachetti et al. 2017). Typical pastoralist communities in the HKH include: Tibetans in China, Gaddis and Gujjars in India, Tamangs in Nepal, Baloch in northern Pakistan, Kuchi in Afghanistan and Brokpas in Bhutan. Besides, the lower mobility livestock production systems in the HKH include: cattle and pig husbandry in Chittagong of Bangladesh and mixed crop (forestry)—livestock production in west and north-east Myanmar (Goletti 1999; Kreutzmann 2004; Tapper 2008; Halim et al. 2010; Dong et al. 2016). The livestock in the HKH are composed mainly of ruminants: yaks, sheep, goats, buffaloes, and zebu for the use of high pasture resources, while the monogastric animals like pigs, chickens and ducks are generally raised in the lowland settlement area. Yaks and their hybrids have become the symbol of pastoralism in the HKH.

6.3 Livestock Distribution in the HKH Region

Animal distribution, density and population data were taken from FAO (Gilbert et al. 2018). To take advantage of the vast pastoral sources in the HKH, cattle, goats and sheep are raised at high densities, 13–15 head/km^2 (Fig. 6.2), which is higher than the average in Asia, about 11 head/km^2, and the global average, about 2 head/km^2. These high densities in the HKH were not reported (Singh et al. 2011) when discussing the global carbon mitigation issues by the Intergovernment Panel on Climate Change (IPCC). Buffaloes in the HKH are distributed mainly in northern India, south-west China and Nepal, away from the high, cold and arid environment. The average monogastric animal densities in the HKH (6 pigs/km^2, 66 chickens/km^2 and 5 ducks/km^2) are generally lower than the average in Asia (13 pigs/km^2, 250 chickens/km^2 and 26 ducks/km^2), due to the large areas that are unsuitable for these animals; however, the averages in the HKH are several-fold higher than the global averages (Wu et al. 2014).

The numbers of animals in each species are presented in Table 6.1. The HKH region of China, with the largest area (2,395,105 km^2, 26% of country's area), has the highest number of most animal species. This region provides about 15% of the ruminant livestock for the country. India, which has a much smaller area in the HKH than China (404,701 km^2, 13% of country area), has the second largest number of most animals species; however, the numbers are generally lower than 10% of its domestic production, except for horses (36%) and pigs (30%). About half of the territories of Afghanistan, Pakistan and Myanmar are in the HKH (in total nearly 400,000 km^2). The production ratios of livestock in the HKH compared to the total in these countries are all above 70%; in Afghanistan, the HKH is the key livestock production region, more so than in Pakistan and Myanmar. Nepal and Bhutan, which are totally inside the HKH, both have high livestock densities. Bangladesh

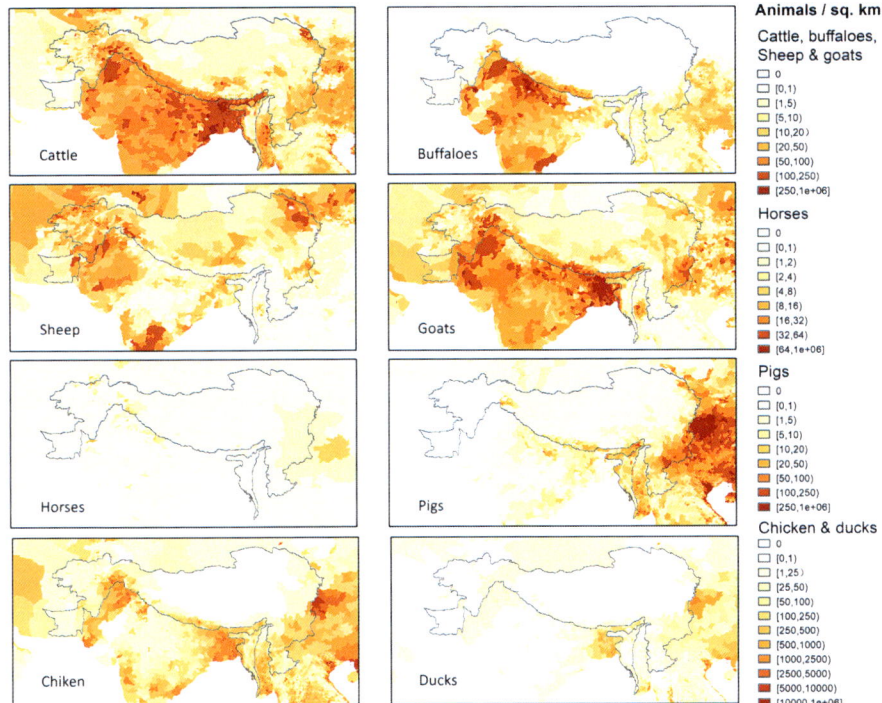

Fig. 6.2 Distribution of the main livestock species in and around the HKH region (Source from FAO Gilbert et al. 2018)

has the smallest area in the HKH (15,543 km², 11% of country area) and relatively low animal numbers. However, the pig density is high as the Bangladesh-HKH is an urban area with a high demand for pork (Halim et al. 2010).

6.4 Livestock Biomass Use in the HKH Region

Feed biomass is what links livestock to land use, and is a key carbon source in the livestock production materials flow (Fig. 6.3). There are four types of feeds: (1) grass for grazing and for silage; (2) grain, usually used as concentrate feed stuff; (3) occasional feeds; and (4) stovers (fibrous crop residue). Grasses constitute an important part of high altitude terrestrial carbon pools as nearly 100% of feed biomass of ruminants is composed of grass (Wang et al. 2010; Herrero et al. 2013) (Fig. 6.1). Consequently, grass biomass is a significant part of the livestock sector affecting the carbon cycle in the HKH.

The FAO sub-national livestock databases enabled us to calculate the biomass use of specific livestock species in the cross-national HKH region. In this calculation, parameters of animal feed intake in south and central Asia were used from peer

Table 6.1 Main livestock species number in the countries' areas within the HKH in 2010

Species	AFG-HKH	BGD-HKH	BTN-HKH	CHN-HKH	IND-HKH	MMR-HKH	NPL-HKH	PAK-HKH
Cattle	4,957,470	550,881	381,611	17,134,195	11,928,360	2,971,110	7,321,860	8,077,874
Buffaloes	3927	58,306	1537	1,378,784	2,070,362	1,359,151	4,908,587	1,621,196
Sheep	10,121,468	26,184	14,482	32,655,768	4,715,141	150,808	799,097	13,843,156
Goats	4,785,893	1,179,778	56,129	24,897,560	6,471,851	482,915	8,947,710	18,434,484
Horses	180,600	4632	17,655	1,457,824	220,034	71,340	45,929	137,538
Pigs	NA	140,356	23,349	18,191,304	3,152,750	2,839,921	1,073,489	NA
Chicken	10,606,645	7,982,598	455,447	91,954,874	28,725,225	33,999,526	25,982,787	78,749,284
Ducks	320,763	556,739	338,697	15,464,382	2,417,245	1,411,188	386,934	1,014,892

AFG Afghanistan, *BGD* Bangladesh, *BTN* Bhutan, *CHN* China, *IND* India, *MMR* Myanmar, *NPL* Nepal, *PAK* Pakistan (Source from FAO Gilbert et al. 2018)

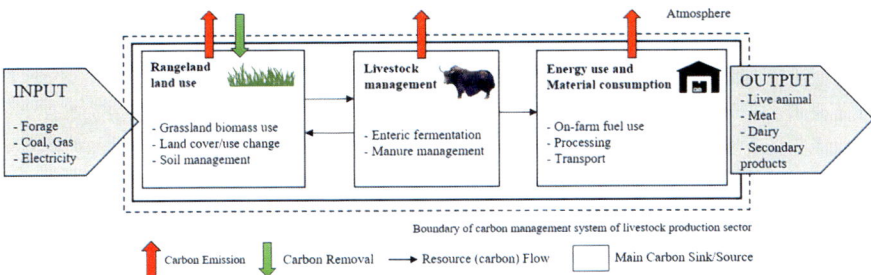

Fig. 6.3 A conceptual framework of carbon management system of the rangeland livestock production sector (Source by author in reference to Steinfeld and Wassenaar 2007)

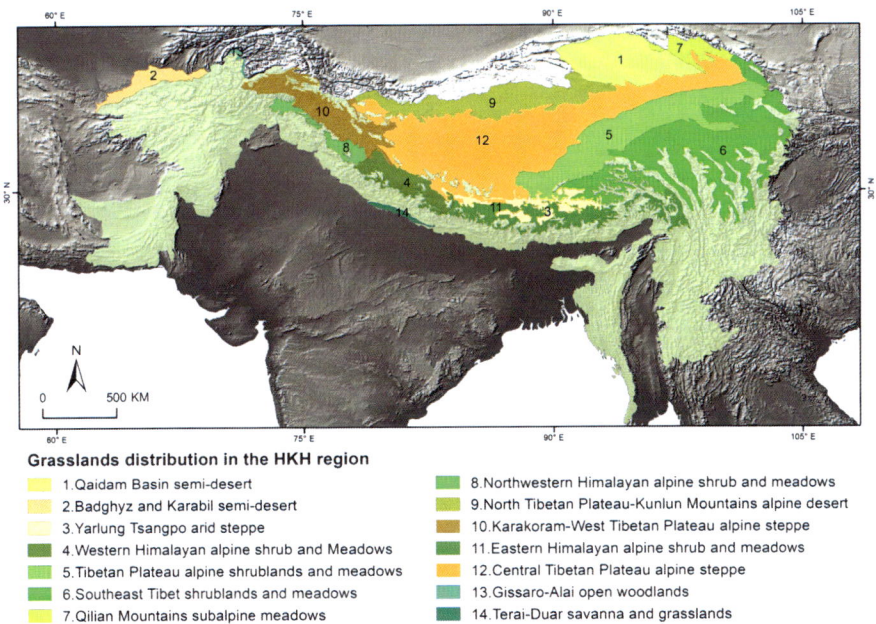

Fig. 6.4 Grasslands distribution in the HKH region (Source from World Wildlife Fund, Ripple et al. 2014)

reviewed papers (Xue et al. 2004, 2014; Haberl et al. 2007). In the global and national levels of bovine statistics, the yak was generally classified with cattle. These statistics could not be used to calculate biomass intake of ruminants for the HKH, as the dominant species, the yak, differs substantially in its production indices (live weight, diet quality, feed intake and feed-use efficiency) from lowland cattle. To account for these differences, we differentiated the original cattle data into yak and cattle by using the grassland divisions suggested by the International Vegetation Classification (IVC) (Fig. 6.4) (Dixon et al. 2014). The HKH region has 14 grassland divisions under the IVC, including the Central Asian Alpine Scrub,

Forb Meadow and Grasslands (3–12 in Fig. 6.4), the Eastern Eurasian Grassland and Shrubland (13–14 in Fig. 6.4) and the Eastern Eurasian Cool Semi-Desert Scrub and Grasslands (1–2 in Fig. 6.4). These grasslands are generally above 3000 m where mainly yaks are raised. From an extensive literature review of suitable habitats for yak and cattle (Wiener et al. 2018; Han et al. 2003; Rhode et al. 2007; Kreutzmann 2012; Wiener 2013) and, concomitantly confirming the yak distribution in the HKH IVC grasslands map (Kreutzmann 2012; Dixon et al. 2014), we were able to determine the boundary splitting the yak and cattle habitats.

Livestock production is a low efficient process in terms of carbon fixing from the atmosphere to the biosphere, as the net primary production consumed by rangeland livestock converted to secondary production is low (in the order of 2%) (Asner and Archer 2010). Besides, livestock land use (specifically grazing) contributes nearly 30% of human appropriation of net primary production (HANPP), regardless of its linkage to cropping activities (49.8% of HANPP) (Haberl et al. 2007). Thus, there may be large amounts of carbon consumed by the livestock in the HKH. In this chapter, carbon appropriation by livestock feed biomass intake was calculated by assuming an average carbon content of dry matter (DM) biomass of 50% (Krausmann et al. 2013). For the whole HKH region in 2010, cattle and buffaloes consumed the largest amount of biomass and carbon content (11.9 million ton C per year, 5.94 million ton C per year), followed by sheep and goats (4.65 million ton per year, 2.32 million ton C per year), yaks (2.48 million ton per year, 1.24 million ton C per year), pigs (0.84 million ton/year, 0.42 million ton C per year), horses (0.78 million ton C per year, 0.39 million ton C per year) and poultry (0.55 million ton C per year, 0.27million ton C per year) (Table 6.2). The HANPP in the livestock sector is generally composed of biomass harvest for feed and the NPP change due to soil degradation induced by grazing activities (Krausmann et al. 2013). These measurements are important because of the dominant role of pastoralism and the uncertainty of the effect of grazing on land cover in the HKH (Wang et al. 2010; Lu et al. 2015; Zhang et al. 2015; Huang et al. 2016). By assuming that the livestock feed intake was a

Table 6.2 Estimation of livestock feed biomass consumption and carbon appropriation in the HKH in 2010

Livestock	Quantity (head)	Feed intake (kg DM/head/day)	Biomass intake (million ton per year)	Carbon appropriation (million ton per year)
Yak	13,568,103	5[a]	2.48	1.24
Cattle and buffaloes	50,873,176	6.4[b]	11.88	5.94
Sheep and goats	127,327,013	1[b]	4.65	2.32
Horses	2,136,284	10[b]	0.78	0.39
Pigs	25,652,325	0.9[b]	0.84	0.42
Poultry	299,244,225	0.05[b]	0.55	0.27

[a]Data from Xue et al. (2004)
[b]Data from Haberl et al. (2007)

direct measure of HANPP in the HKH, and using the population density data in 2010, we calculated that the average HANPP amounted to 3.69 ton C/cap per year, which was more than double the average in Asia in 2005 (1.3 ton C/cap per year). This high value compared to the Asian average could be attributed to the high ruminant density in the HKH. Moreover, from the data of Zhao and Running (2010), we calculated that the rate of livestock appropriation of carbon to the land amounted to 2.5% of the NPP in 2010.

6.5 Carbon Emission of Livestock Production Sector in the HKH Region

Sources of carbon emissions in the livestock production system include soil, enteric fermentation, manure and energy use (Fig. 6.3). In the extensive rangeland system, where the local feed resource-based production is dominant, CH_4 emission is the main component of total GHG emission in the livestock sector. Yaks generally produce less methane than cattle, which may be attributed to the co-evolution of rumen microbiomes toward energy-saving strategies in adapting to the high altitude environment (Zhang et al. 2016).

We calculated livestock CH_4 emission from enteric fermentation and manure from the main species in the HKH in 2010 by using the Tier 1 method in the IPCC guideline (Table 6.3). It emerged that livestock produced 3.77 million tons of CH_4 per year, of which 96% was from enteric fermentation and 4% from manure. The bovine was the main source of carbon, and contributed 64% of the total CH_4 by livestock species (Table 6.3). By calculating livestock carbon appropriation and emission in the HKH (Fig. 6.5), the most intensive activity of the carbon cycle by livestock was in the south (northern India and northeast Pakistan), followed by northeast Qinghai, Gansu, Sichuan and Shigatse-Tibetan in China. The high

Table 6.3 CH_4 emission factors of livestock in HKH in 2010

Livestock	Enteric fermentation CH_4 factors (kg per year)	Manure CH_4 factors (kg per year)	Enteric fermentation (CH_4 million ton per year)	Manure (CH_4 million ton per year)	Total (CH_4 million ton per year)
Yak	40	1	0.54	0.01	0.56
Cattle and buffaloes	47	1	2.39	0.05	2.44
Sheep and goats	5	0.1	0.64	0.01	0.65
Horses	18	1	0.04	0.00	0.04
Pigs	1	2	0.03	0.05	0.08
Poultry	0	0.01	0.00	0.00	0.00
Total			3.63	0.13	3.77

Emission factors from IPCC (2006) and Xue et al. (2014)

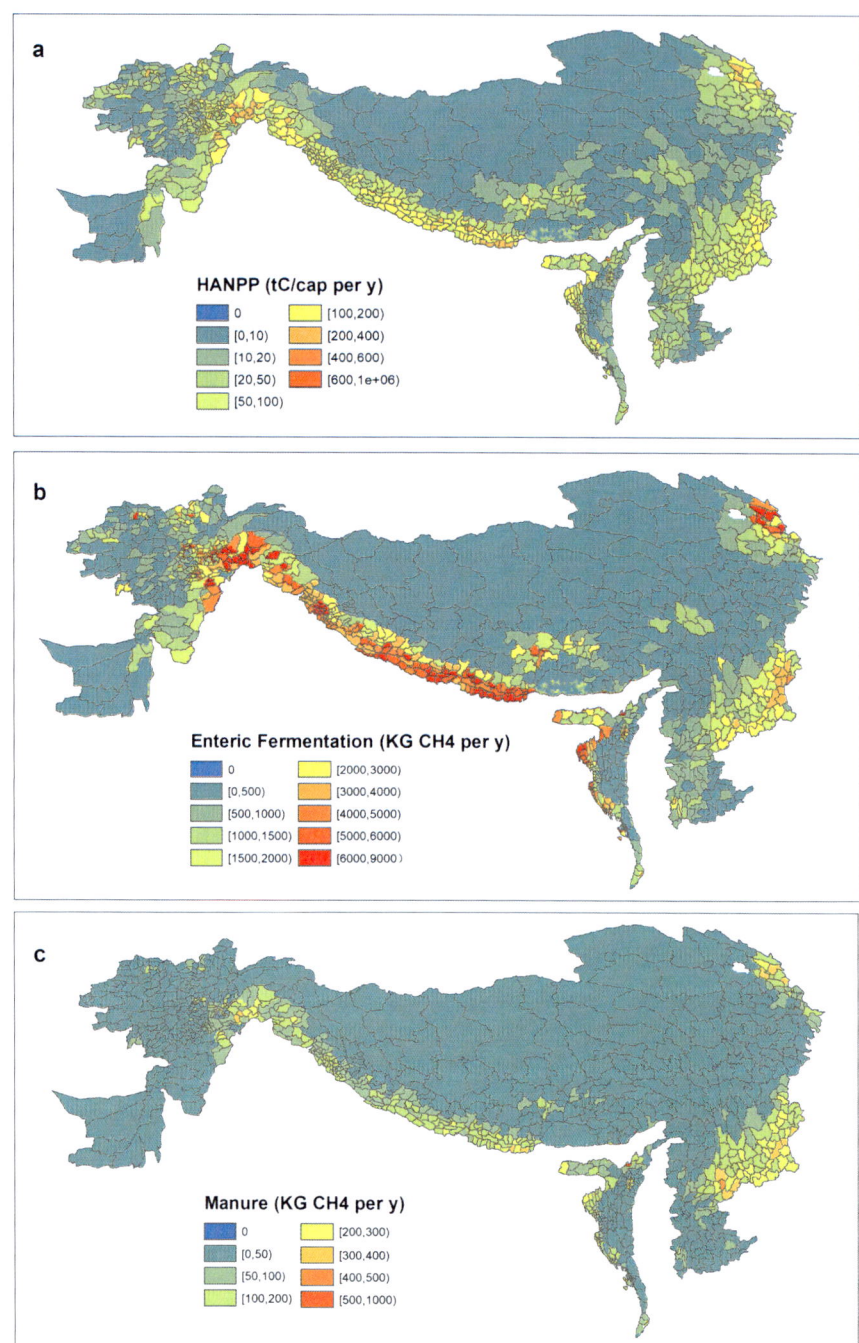

Fig. 6.5 (**a**) HNAPP in the HKH in 2010 by assuming the livestock as the only carbon appropriate sector; (**b**) CH_4 emission from livestock enteric fermentation in the HKH in 2010; (**c**) CH_4 emission from livestock manure in the HKH in 2010

emission areas were found to be mostly in the mixed rainfed systems (MR) (Fig. 6.1). In comparing animal species in CH_4 emission and the HANPP, poultry had the highest intake efficiency (the C input/C output was 0.3%), and yaks and cattle/buffaloes the lowest (the C input/C output was 12% and 11%, respectively).

On a time scale, the pastoral system in the HKH is under the process of shifting from traditional ruminant grazing and browsing to intensive industrial production (Kreutzmann 2012; Bai et al. 2018). The livestock production mobility change on the Tibetan Plateau is being driven mainly by government policies, including the projects *Ecological Resettlement*, *Turning Pastureland into Grassland*, and *Nomadic Settlement*. Numerous local pastoral communities have increased their livestock production systems in feedlots (Figs. 6.6 and 6.7), while the previous grazing lands have been fenced for ecological restoration. These changes can be summarized under the term 'industrialization', which is limiting livestock grazing while increasing infrastructures dealing with storage, processing, transport and energy. These transformations have changed the carbon emissions, as on the one hand, feedlots ensure the quantity and quality of livestock feed in winter and as a consequence, the emission to production efficiency increases (Ding et al. 2010), but, on the other hand, industrialization converts the energy use pattern in the livestock food chain from solar energy to high carbon equivalent fossil fuel (Steinfeld and Wassenaar 2007). For example, in the pastoralist household, the on-farm fuel is mainly biofuel of yak dung and wood with low heating value (Rhode et al. 2007), while for the industrial production system, energy sources are mainly electricity, gas and coal, which have higher CO_2 emission coefficients. These changes could increase the total livestock production emission as has been evaluated by Zhuang et al. (2017).

Fig. 6.6 Grazing yak on the Qilian mountains of the Tibetan Plateau

Fig. 6.7 Industrial feedlot yak production on the Tibetan Plateau

From another perspective, industrialization entails land use intensification, which in the HKH has important ecological significances, as the dominant ecotypes, alpine and desert grasslands, are highly fragile, and are sensitive to human disturbances like grazing and land cultivation (Singh et al. 2011). Regarding the carbon cycle, carbon emission from soil degradation could be extremely high, as the soil carbon pool generally accounts for more than 90% of the total carbon in the ecosystem (Wen et al. 2013), and once degraded, becomes a carbon source emitted to the atmosphere. The rangeland use intensification could reduce the pressure of grazing in the low valued grasslands while it maximizes the use of high productive land. Such a practice could reduce soil degradation from inappropriate grazing, and thus maintain the land carbon sequestration capacity. However, intensification implies higher inputs per unit land. In agriculture and livestock mixed systems, this includes fertilizers, seeds and land management activities, which can increase emissions. For example, there would be an increase in CH_4 release from the breakdown of fertilizer and in emissions related to land use of forage seeds production, to soil organic matter oxidation/erosion due to cultivation and to respiration from organisms due to irrigation. As has been estimated, there is great potential of carbon sequestration in the dryland pasture soil (maximum achievable of 1 billion ton C per year) due to its large storage capacity, which is far from being saturated (Lal 2004). This would entirely offset livestock-related emissions from agriculture, and therefore, industrialization to a certain extent, could result in a carbon-negative practice by 'sequestration through intensification' (Steinfeld et al. 2006). Soil quality improvement as a result of increased soil carbon could bring important social and economic benefits to the drylands of the HKH, where the world's poorest communities are living.

Some practical examples could be observed on the Tibetan Plateau, where overgrazing and climate change has exacerbated soil degradation extensively (Du et al. 2004; Wang et al. 2012), while in recent decades, restoration practices (e.g. artificial grass and animal exclosures) with subsidy policies (e.g. payment for ecosystem services) have been carried out with good environmental and social outcomes (Bryan et al. 2018).

6.6 Conclusion

The HKH region is the home to the world highest altitude pastoralist communities, whose livelihood is tied closely to animal production. The anthropogenic carbon disturbance of livestock production in the HKH involves livestock HANPP and carbon emission, which have been described in detail in this chapter. The quantitative results of the HANPP and emissions indicate that there is a large carbon mitigation potential in the livestock sector in the HKH. The current rapid transformation of livestock production to industry should be considered, as these changes will require developing novel strategies for carbon management. Policies should be directed towards combining livestock production systems with infrastructure improvement and technological advances while, concomitantly, taking into account the political, economical and cultural aspects of the area.

References

Asner, G.P., and S.R. Archer. 2010. Livestock and the global carbon cycle. In *Livestock in a changing landscape: Drivers, consequences and responses*, ed. H. Steinfeld, H. Mooney, F. Schneider, and L.E. Neville, 69–82. Washington, DC: Island Press.

Asner, G.P., A.J. Elmore, L.P. Olander, et al. 2004. Grazing systems, ecosystem responses, and global change. *Annual Review of Environment and Resources* 29: 261–299.

Bai, Z., W. Ma, L. Ma, G.L. Velthof, et al. 2018. China's livestock transition: Driving forces, impacts, and consequences. *Science Advances* 4: eaar8534.

Bryan, B.A., L. Gao, Y. Ye, et al. 2018. China's response to a national land-system sustainability emergency. *Nature* 559: 193–204.

Ding, X.Z., R.J. Long, M. Kreuzer, et al. 2010. Methane emissions from yak (*Bos grunniens*) steers grazing or kept indoors and fed diets with varying forage: Concentrate ratio during the cold season on the Qinghai-Tibetan Plateau. *Animal Feed Science and Technology* 162: 91–98.

Dixon, A.P., D. Faber-Langendoen, C. Josse, et al. 2014. Distribution mapping of world grassland types. *Journal of Biogeography* 41: 2003–2019.

Dong, S., S. Liu, and L. Wen. 2016. Vulnerability and resilience of human-natural systems of pastoralism worldwide. In *Building resilience of human-natural systems of pastoralism in the developing world*, ed. S. Dong, K.A. Kassam, J. Tourrand, and R. Boone, 39–92. Cham: Springer.

Du, M., S. Kawashima, S. Yonemura, et al. 2004. Mutual influence between human activities and climate change in the Tibetan Plateau during recent years. *Global and Planetary Change* 41: 241–249.

Frachetti, M.D., C.E. Smith, C.M. Traub, et al. 2017. Nomadic ecology shaped the highland geography of Asia's Silk Roads. *Nature* 543: 193–198.

Gilbert, M., G. Nicolas, G. Cinardi, et al. 2018. Global distribution data for cattle, buffaloes, horses, sheep, goats, pigs, chickens and ducks in 2010. *Scientific Data* 5: 180–227.

Goletti, F. 1999. *Agricultural diversification and rural industrialization as a strategy for rural income growth and poverty reduction in Indochina and Myanmar.* MSS Discussion Paper (No. 596-2016-40031).

Haberl, H., K.H. Erb, F. Krausmann, et al. 2007. Quantifying and mapping the human appropriation of net primary production in earth's terrestrial ecosystems. *Proceedings of the National Academy of Sciences of the United States of America* 104: 12942–12947.

Halim, M., M. Kashem, J. Ahmed, et al. 2010. Economic analysis of Red Chittagong Cattle farming system in some selected areas of Chittagong district. *Journal of the Bangladesh Agricultural University* 8: 271–276.

Han, X.T., A.Y. Xie, X.C. Bi, et al. 2003. Effects of altitude, ambient temperature and solar radiation on fasting heat production in yellow cattle (Bos taurus). *British Journal of Nutrition* 89: 399–407.

Herrero, M., P. Havlik, H. Valin, et al. 2013. Biomass use, production, feed efficiencies, and greenhouse gas emissions from global livestock systems. *Proceedings of the National Academy of Sciences of the United States of America* 110: 20888–20893.

Huang, K., Y. Zhang, J. Zhu, et al. 2016. The influences of climate change and human activities on vegetation dynamics in the Qinghai-Tibet plateau. *Remote Sensing* 8: 876.

Imhoff, M.L., L. Bounoua, T. Ricketts, et al. 2004. Global patterns in human consumption of net primary production. *Nature* 429: 870–873.

Intergovernmental panel on climate change (IPCC). 2006. 2006 IPCC guidelines for National Greenhouse Gas Inventories. Agriculture, forestry and other land use (AFOLU), Vol. 4, Eggleston, S., L. Buendia, K. Miwa, T. Ngara, and K. Tanabe (eds.). Prepared by the National Greenhouse Gas Inventories Programme, Institute for Global Environmental Strategies, Hayama, Japan. Available online at www.ipcc-nggip.iges.or.jp/public/2006gl/vol4. html; last accessed Jan. 10, 2010.

Joshi, L., R.M. Shrestha, A.W. Jasra, et al. 2013. Rangeland ecosystem services in the Hindu Kush Himalayan region. *High-altitude Rangelands Their Interfaces Hindu Kush Himalayas* 2013: 157–174.

Krausmann, F., K.H. Erb, S. Gingrich, et al. 2013. Global human appropriation of net primary production doubled in the 20th century. *Proceedings of the National Academy of Sciences of the United States of America* 110: 10324–10329.

Kreutzmann, H. 2012. Pastoral practices in transition: Animal husbandry in high Asian contexts. In *Pastoral practices in high Asia. Advances in Asian Human-Environmental Research*, ed. H. Kreutzmann, 1–30. Dordrecht: Springer.

Kreutzmann, H. 2004. Pastoral practices and their transformation in the North-Western Karakoram. *Nomad People* 8: 54–88.

Lal, R. 2004. Carbon sequestration in dryland ecosystems. *Environmental Management* 33: 528–544.

Lu, X., Y. Yan, J. Sun, et al. 2015. Short-term grazing exclusion has no impact on soil properties and nutrients of degraded alpine grassland in Tibet, China. *Solid Earth* 6: 1195–1205.

Qiu, Q., L. Wang, K. Wang, et al. 2015. Yak whole-genome resequencing reveals domestication signatures and prehistoric population expansions. *Nature Communications* 6: 10283.

Rhode, D., D.B. Madsen, P.J. Brantingham, et al. 2007. Yaks, yak dung, and prehistoric human habitation of the Tibetan Plateau. *Developments in Quaternary Sciences* 9: 205–224.

Ripple, W.J., P. Smith, H. Haberl, et al. 2014. Ruminants, climate change and climate policy. *Nature Climate Change* 4: 2–5.

Robinson, T. P., Thornton, P. K., Franceschini, G., et al. 2011. *Global livestock production systems*, 2011. Rome, Italy: FAO and ILRI.

Shang, Z.H., M.J. Gibb, F. Leiber, et al. 2014. The sustainable development of grassland-livestock systems on the Tibetan plateau: Problems, strategies and prospects. *Rangeland Journal* 36: 267–296.

Singh, S.P., I. Bassignana-Khadka, B.S. Karky, et al. 2011. *Climate change in the Hindu Kush-Himalayas: The state of current knowledge*. Katmandu: International Centre for Integrated Mountain Development (ICIMOD).

Steinfeld, H., and T. Wassenaar. 2007. The role of livestock production in carbon and nitrogen cycles. *Annual Review of Environment Resources* 32: 271–294.

Steinfeld, H., P. Gerber, T.D. Wassenaar, et al. 2006. *Livestock's long shadow: Environmental issues and options*. Rome, Italy: FAO.

Tapper, R. 2008. Who are the Kuchi? Nomad self-identities in Afghanistan. *Journal of Royal Anthropological Institute* 14: 97–116.

Wang, L., K.C. Niu, Y.H. Yang, et al. 2010. Patterns of above- and belowground biomass allocation in China's grasslands: Evidence from individual-level observations. *Science China-Life Sciences* 53: 851–857.

Wang, S., J. Duan, G. Xu, et al. 2012. Effects of warming and grazing on soil N availability, species composition, and ANPP in an alpine meadow. *Ecology* 93: 2365–2376.

Wen, L., S. Dong, Y. Li, et al. 2013. The impact of land degradation on the C pools in alpine grasslands of the Qinghai-Tibet Plateau. *Plant and Soil* 368: 329–340.

Wiener, G. 2013. The yak, an essential element of the high altitude regions of Central Asia. *Études mongoles et sibériennes, centrasiatiques et tibétaines* 43–44:2-10.

Wiener, G., J.L. Han, R.J. Long, et al. 2018. Yak. In *Encyclopedia of Animal Science (two-volume set)*, 1121–1124. New York: CRC Press.

Wu. N., Ismail, M., Joshi, S., et al. 2014. Livelihood diversification as an adaptation approach to change in the pastoral Hindu-Kush Himalayan region. *Journal of Mountain Science* 11(5): 1342–1355.

Xue, B., X.Q. Zhao, and Y.S. Zhang. 2004. Feed intake dynamic of grazing livestock in nature grassland in Qinghai-Tibetan Plateau. *Ecology of Domestic Animal* 25 (4): 21–25.

Xue, B., L.Z. Wang, and T. Yan. 2014. Methane emission inventories for enteric fermentation and manure management of yak, buffalo and dairy and beef cattle in China from 1988 to 2009. *Agriculture Ecosystem & Environment* 195: 202–210.

Zhang, T., Y. Zhang, M. Xu, et al. 2015. Light-intensity grazing improves alpine meadow productivity and adaption to climate change on the Tibetan Plateau. *Scientific Reports* 5: 15945.

Zhang, Z., D. Xu, L. Wang, et al. 2016. Convergent evolution of rumen microbiomes in high-altitude mammals. *Current Biology* 26: 1873–1879.

Zhao, M., and S.W. Running. 2010. Drought-induced reduction in global terrestrial net primary production from 2000 through 2009. *Science* 329: 940–943.

Zhao, Z. 2009. Eastward spread of wheat into China: New data and new issues. *Chinese Archaeology* 9: 1–9.

Zhuang, M., Gongbuzeren, and W. Li. 2017. Greenhouse gas emission of pastoralism is lower than combined extensive/intensive livestock husbandry: A case study on the Qinghai-Tibet Plateau of China. *Journal of Cleaner Production* 147: 514–522.

Chapter 7
Wetlands as a Carbon Sink: Insight into the Himalayan Region

Awais Iqbal and Zhanhuan Shang

Abstract The Hindu-Kush Himalaya region's land cover is comprised of 54% rangeland, 25% agricultural land, 14% forest, 5% permanent snow and 1% water bodies. The Himalayans contain some of the largest water reservoirs, which are critical for HKH countries. Amidst these, wetlands have remained important to ecosystem services and the overall water cycle of the basins. Beside their cultural and provisioning amenities, wetlands are important carbon reservoirs, accounting for 20–30% of the global carbon pool. They act as a sink for atmospheric carbon, thus can influence GHG emissions, especially CH_4, and, thus, should be managed properly. However, substantial data gaps remain in quantifying carbon sequestration and the potential of CH_4 emission. Furthermore, studies on CH_4 fluxes in high-altitude wetlands, particularly in remote areas, remain inconclusive. Hence, more research is required to understand the role of wetlands in term of GHG emissions and carbon sequestration.

Keywords Wetland · Methane · Hindu-Kush Himalaya · Carbon sink

---

A. Iqbal
School of Life Sciences, State Key Laboratory of Grassland Agro-Ecosystems, Lanzhou University, Lanzhou, Gansu, China

Z. Shang (✉)
School of Life Sciences, State Key Laboratory of Grassland Agro-Ecosystems, Lanzhou University, Lanzhou, Gansu, China

Key Laboratory of Restoration Ecology of Cold Area in Qinghai Province, Northwest Institute of Plateau Biology, Chinese Academy of Sciences, Lanzhou, Gansu, China
e-mail: shangzhh@lzu.edu.cn

7.1 Introduction

Wetlands are considered to be one of the oldest territories of the planet. Initially they posed a hindrance to human exploration, however once people learnt the art of boat building, what was once a hindrance became a convenient means for travel and exploration. Subsequently, wetlands became prolific habitats, and a vital food resource. In general, wetlands may seem as abrasive and threatening, with shallow standing water, strong odors, infinite swarms of mosquitoes and deep mud. The word wetland might seem simply as an area that is wet; nevertheless, describing a wetland is more challenging than one would anticipate. Wetlands comprise all regions where shallow water, both salty or fresh, moves or stands. Overall, oceans, seas and deep lakes are usually omitted from the definition of a wetland, yet the shallow boundaries of seas and lakes are considered as wetlands (Moore and Garratt 2006).

Wetlands are occasionally termed as "the kidneys of the landscape" since they play an important role as the downstream receivers of waste and water from human and natural resources. Given this background, wetlands consume CO_2, filter pollutants, protect against flooding, hold and gradually release storm water runoff and generate much oxygen. In addition, wetlands have been called "ecological supermarkets" because of their rich biodiversity and vast food chain. Relating to its importance, wetlands are now valued worldwide and have led to wetland regulations, conservation, management plans and protection laws. The Ramsar Convention on 'Wetlands of International Importance' (Ramsar Convention Secretariat 2013) endorsed a definition of wetlands that embrace all types of aquatic habitats: "Wetlands are areas of marsh, fen, peatland or water, whether natural or artificial, permanent or temporary, with water that is static or flowing, fresh, brackish or salty, including areas of marine water the depth of which at low tide does not exceed six meters." The convention in its second article further stated that wetlands "may incorporate riparian and coastal zones adjacent to the wetlands, and islands or bodies of marine water deeper than six meters at low tide lying within the wetlands."

Wetlands have championed the cause *célebre* for administrations across the world, because of the disappearance of water resources in natural habitats and loss of economy. To overcome these problems, scientists, engineers, lawyers and government officials are studying wetland ecology and wetland management to comprehend, preserve, and recreate these vulnerable ecosystems (Mitsch et al. 2009). Development of wetlands entails the interaction between living (animals and plants) and non-living components of the habitat (water, chemical, particles and rock). Collectively, the living and non-living elements co-exist as an integrated ecosystem (Fig. 7.1).

7 Wetlands as a Carbon Sink: Insight into the Himalayan Region

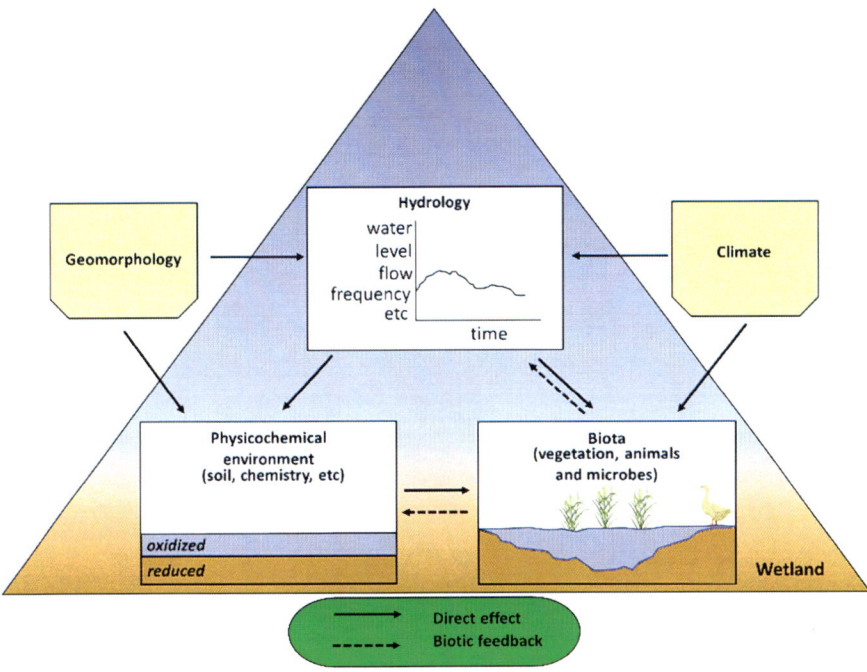

Fig. 7.1 Living and non-living components of an ecosystem working together to form a wetland (Mitsch et al. 2009)

7.2 Hindu Kush-Himalaya

The Hindu Kush-Himalayan (HKH) region covers an area greater than four million km^2 (Gurung et al. in preparation), approximately 18% of the global mountain areas and almost 2.9% of the global land area. These gigantic mountain ranges contain the world's largest peaks, including K2, Mount Everest, Lhotse, Kangchenjunga, Makalu, Cho Oyu, Annapurna and Dhaulagiri, and the headwaters of the ten foremost river systems, including Brahmaputra, Amu Darya, Ganges, Indus, Irrawaddy, Mekong, Yangtze, Yellow, Salween, and Tarim. In addition, these mountain regions support the livelihood of 210 million inhabitants while providing services and goods for 1.3 billion people residing downstream. Collectively, around three billion people benefit from food and energy yielded by the river basin. The HKH region includes China, Pakistan, India, Bangladesh, Myanmar, Afghanistan and all of Nepal and Bhutan (Fig. 7.2).

The land cover throughout the HKH region comprises 54% rangeland and scrubland, 25% agricultural land (including areas with a mixture of natural vegetation), 14% forest, 5% permanent snow and 1% water bodies. The water bodies of the Himalaya consist of rivers, lakes and wetlands of various shapes and sizes, making

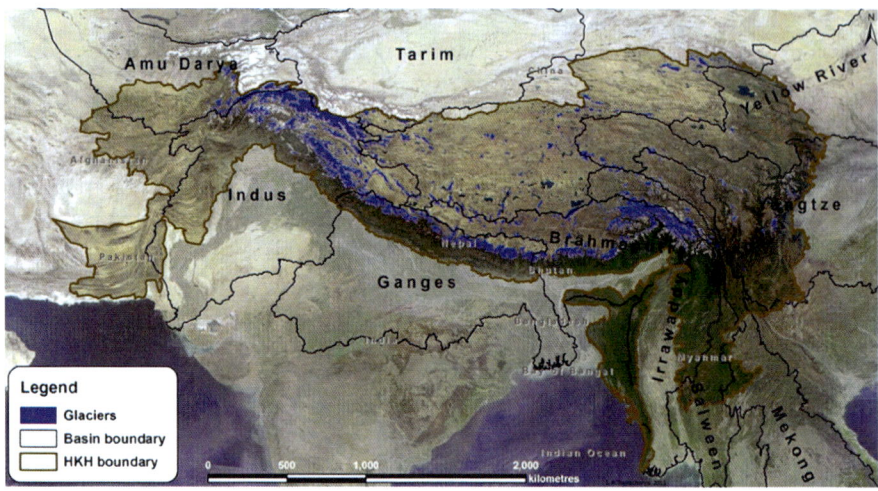

Fig. 7.2 Regions included in the HKH and Major river basins

it an ecosystem that fulfils a crucial function in the overall water cycle of the basins. Wetlands and lakes have received little attention in terms of water management and conservation; however, they are receiving more attention owing to their importance due to the probable consequences of global climate change. Table 7.1 shows some of the international important wetlands/regions, including those in the HKH.

7.3 Wetland Types

Wetlands differ among themselves in chemical, biological, hydrological and sedimentation characteristics, as well as in in size and shape. Table 7.2 represents all types of wetlands that exist among the HKH countries. Numerous classification schemes have been employed in distinct regions (Finlayson and Van der Valk 1995). US Fish and Wildlife Service divide wetlands into two basic groups: freshwater and saltwater. Freshwater wetlands include marshes (prairie potholes, wet meadows, playa lakes, vernal pools), swamps (mangrove, shrub, forested), and bogs (common, fen, pocosin), while saltwater wetlands include tidal marshes and mangrove swamps. Estuarine/coastal wetlands account for 10% of the global wetlands, whereas freshwater wetlands make up the remaining 90%. The Ramsar convention primarily implemented a simple classification that identified 22 wetland types (Table 7.3) but was revised later to recognize more groups.

Table 7.1 Some internationally important Wetlands/regions of the Hindu Kush-Himalaya (Boere et al. 2006; Gujja 2007; Jha 2009; Byomkesh et al. 2009; Gopal et al. 2010; Sherab et al. 2011; Tan et al. 2011; Li et al. 2014; Khan and Arshad 2014; Uddin et al. 2015; Zhu et al. 2015; Meng et al. 2017; Paudel et al. 2017; Qamer et al. 2008)

Afghanistan	
Himalayan regions	1. Pamir-i-Buzurg (wakhan) 2. Imam Sahib (kunduz) 3. Khulm, Ajar (Baghlan) 4. Nursitan, band-i-amir (Bamiyan) 5. Ab-i-Estada (Ghazni) 6. Ab-i-nawar (Dasht-i-nawar)
Ramsar sites	N.A.
Altitude	2000–4800 m
Total area/total number of wetlands	N.A.
Bangladesh	
Himalayan regions	1. Sundarbans mangrove wetlands 2. Chandabill-baghiar bill 3. Bildakatia, Atadanga Boar 4. Marijat Baor (Ganges Delta)
Ramsar sites	2
Wetlands type	Mangrove wetlands, Oxbow lakes
Total wetlands/total area	Number: NA Area: 7–8 million (ha)
Bhutan	
Himalayan regions	1. Torsa (Amo Chu) 2. Raidak (Wong Chu) 3. Sankosh (Mo Chu) 4. Manas (Gangri) 5. Phobjikha 6. Khalong Chu 7. Jigme Darji wildlife sanctuary
Ramsar sites	3
Total area/total number of wetlands	Number: 3027 Area: 10,200 ha Glacial lakes: 2674 (2000–4500 m) and 60 (3500 m)
China	
Himalayan regions	1. Tibet autonomous region {Pumqu (Arun) 2. Poiqu (Bhote-Sun Koshi) 3. Rangxer (Tama Koshi) 4. jilongcanbu (Humla karnali)} 5. Yunnan region (Bitahai and Napahai wetland) 6. Zoige 7. Sichuan 8. Yangzte 9. Mekong 10. Salween 11. Namucuo lake 12. Yanzongyongcuo lake 13. Lhasa
Ramsar sites	57
Average altitude	4500
Total area/total number of wetlands	Number: NA Area: 53.42×10^6 ha Glacial lakes: 824

(continued)

Table 7.1 (continued)

India		
Himalayan regions	1. Tripura 2. Meghalaya 3. Assam 4. Arunachal 5. Nagaland 6. Manipur 7. Mizoram 8. Darjeeling 9. Jalpaiguri 10. Coochbehar 11. Sikkim 12. Ladakh 13. Hokar Sar wetland	
Ramsar sites	26	
Total wetlands/total area	Natural: 67,429 (lakes, marsh, ponds, reservoirs and oxbow lakes, tanks) Area: 4.1 million (ha)	
Myanmar		
Himalayan regions	1. Irrawaddy-Chindwin 2. Sittaurg 3. Salween	
Ramsar sites	5	
Total wetlands/total area	Numbers: 99 (lakes, marsh, ponds, reservoirs and paddy fields) Area: N.A.	
Nepal		
Himalayan regions	1. Terai region (Koshi Tappu Wildlife Reserve) 2. Shey Phoksundo 3. Lake Rara 4. Lake Tilicho 5. Khumbu Himal region 6. Gokyo lake 7. Phuksundo 8. Gosaikunda lakes 9. Dhaap lake 10. Beeshazar lakes 11. Pokhara (8 lakes)	
Ramsar sites	10	
Total wetlands/total area	Number: 242 Area: 7,43,563 ha Glacial lakes: 2323 lakes (above 3000 m) and 90 lakes (4, 500 and 5645 m)	

(continued)

Table 7.1 (continued)

Pakistan	
Himalayan regions	1. Sheosar lake (4250 m) (deosai) 2. Saucher lake 3. Karumbar lake (deosai, 4150–4250 m) 4. Rupal (shaigiri) 5. Mankial lake 6. Swat, jabba (falak sher) 7. Lulusar wetland complex 8. Uttar lake 9. Handrap
Ramsar sites	19
Altitude	4200
Wetland types	Inland, delta marshes, mangroves, lakes, reservoir, paddies
Total area/total wetlands	Number: 225 Area: 189,089.4 (ha)

7.4 Value of Wetlands

The wetlands are characterized by huge biodiversity and beautiful landscapes. Regardless of any importance, everyone wants to know "what use it provides?" However, there is no difficulty in demonstrating the worth of wetlands. People ascribe 'value' to many ecosystem functions, where functions are the product of the relations among the biological, chemical and physical elements of the wetlands. Instant, primary or secondary functions are important since they provide products such as fiber, food, fuel, timber and fodder. The Ramsar Convention stated that these functions and values are "products, functions and attributes" while the Millennium Ecosystem Assessment (MEA) mentioned "the assistance people gain from ecosystems" or simply "ecosystem services" (MEA 2005) and divided them into four categories: (a) providing services such as food and water; (b) controlling services such as regulations of land degradation, disease, flood and drought; (c) assisting services such as nutrient cycling and soil formation; and (d) social services such as spiritual, recreational, religious and additional non-material benefits (Table 7.4). In general, the value of Himalayan wetlands relies on their characteristics, location and the economic and socio-cultural status of the peoples nearby. Many glacial lakes at these high altitudes are considered to be a rich source of water along with headwaters of various rivers. Moreover, wetlands, the rivers to be precise, in the Himalayan region are important for their hydropower potential (Agrawal et al. 2010). Most of the glacial lakes and accompanying marshes and additional high-altitude wetlands are principal habitats of many wildlife species such as migratory waterfowl and cold-water fish. Some of these lakes, e.g. Lake Tsomgo and Lake Guru Dongmar in Sikkim, are also sacred according to local communities (Gopal et al. 2008). However, at middle and lower altitudes, wetlands are usually valued for their provisioning services—forage for wildlife and domestic animals and fish for human

Table 7.2 Types of wetlands in the Hindu Kush-Himalaya

Country	Types of wetlands	Reference
Afghanistan	HAW, Lakes, Rivers and Marshes	Bridge et al. (2006)
Bangladesh	HAW, Lakes, Mangroves, Back-swamp (Haor), Oxbow (Baors), Paddy fields and rivers etc.	Islam and Gnauck (2008), Byomkesh et al. (2009)
Bhutan	HAW, Alpine wetlands, Paddy terraces, water bodies, Alpine marshes, Bogs, Riverbeds, Perpetual Snow cover, Rivers (chu) etc.	Sherab et al. (2011)
China	HAW, Alpine wetlands, Mires, Lakes, Paddy fields, Inland Marshes/Swamps, Snow Covers, Glacial lakes, Peat lands etc.	Zhao and He (n.d.), Meng et al. (2017)
India	HAW, Glacial lakes, Mangroves, Swamps, Peat lands, Coastal lagoons etc.	SAC (2013), Bassi et al. (2014)
Nepal	HAW, Lakes, Ponds, Marshes, Swampy lands, Paddy fields etc.	Jha (2009), Bhandari (2009)
Pakistan	HAW, Lakes, Paddy fields, Mangroves, Delta marshes etc.	Qamer et al. (2008), Asad and Sana (2014)
Myanmar	HAW, Lakes, River Basin, Marshes, Mangroves, Paddy fields etc.	"Ye Htut Deputy Director Nature and wildlife Conservation Division I. Water is Life: Too Much or Too Little, Every Drop Counts," (2013)

HAW high altitude wetlands

consumption. It is believed that most of the local communities depend on these wetlands (Barik and Katiha 2003). The efficient output of swamps, marshes, and streams are evident that these wetlands harbor huge amount of nutrients and act as significant systems for household waste discharged into the rivers prior to treatment.

7.5 Wetland Losses and Consequences

Globally, 1052 sites in Europe; 359 sites in Africa; 289 sites in Asia; 211 sites in North America; 175 sites in South America; and 79 sites in the Oceana region have been recognized as Ramsar sites or wetlands of International prominence (Ramsar Convention Secretariat 2013). Wetlands are vital to the livelihoods of almost 250 million people residing on the valley floors and plateau areas of the Himalayas (Trisal and Manihar 2004). The environmental meanings and amenities of the Himalayan wetlands in controlling river drifts and maintaining grasslands are indispensable for more than 10 countries in Asia. Furthermore, these countries have about 140 million people residing in high-altitude areas and 1.3 billion people downstream of the river basins (Xu et al. 2009).

Table 7.3 Ramsar classification system for natural and artificial wetlands (Ramsar 2006)

Types	Examples
Inland wetlands	
Permanent inland deltas	N.A.
Permanent rivers/streams/creeks	waterfalls
Seasonal/intermittent/irregular rivers/streams/creeks	N.A.
Permanent freshwater lakes (over 8 ha)	Large oxbow lakes
Seasonal/intermittent freshwater lakes	Floodplain lakes
Permanent saline/brackish/alkaline lakes	N.A.
Seasonal/intermittent saline/brackish/alkaline lakes and flats	N.A.
Permanent saline/brackish/alkaline marshes/pools	N.A.
Seasonal/intermittent saline/brackish/alkaline marshes/pools	N.A.
Permanent freshwater marshes/pools; ponds (below 8 ha), marshes and swamps on inorganic soils; with emergent, vegetation water-logged for at least most of the growing season	N.A.
Seasonal/intermittent freshwater marshes/pools on inorganic soils	Sloughs, potholes, seasonally flooded, meadows, sedge marshes
Non-forested peatlands	Shrub or open bogs, swamps, fens
Alpine wetlands	Alpine meadows, temporary waters from snowmelt
Tundra wetlands	Tundra pools, temporary waters from snowmelt
Shrub-dominated wetlands	Shrub swamps, shrub-dominated freshwater marshes, shrub, alder thicket on inorganic soils
Freshwater, tree-dominated wetlands	Freshwater swamp forests, seasonally flooded forests, wooded swamps on inorganic soils
Forested peatlands	Peat swamp forests
Freshwater springs	Oases
Geothermal wetlands	
Karst and other subterranean hydrological systems	Inland
Human-made wetlands	
Aquaculture	Fish/shrimp ponds
Ponds	Farm ponds, stock ponds, small tanks (generally less than 8 ha)
Irrigated land	Irrigation channels and rice fields
Seasonally flooded agricultural land	Intensively managed or grazed wet meadow or pasture
Salt exploitation sites	Salt pans, saline
Water storage areas	Reservoirs/barrages/dams/impoundments (generally greater than 8 ha)
Excavations	Gravel/brick/clay pits; borrow pits, mining pools
Wastewater treatment areas	Sewage farms, settling ponds, oxidation basins
Canals and drainage channels, ditches	N.A.
Karst and other subterranean hydrological systems	Human-made

Table 7.4 Value and services of wetlands (Millennium Ecosystem Assessment 2005)

Services	Examples
Provisioning	
Food	Fish, wild game, fruit, and grain
Freshwater	Storage and retention of water for domestic, industrial, and agricultural use
Fibre and fuel	Production of logs, fuelwood, peat, and fodder
Biochemical	Extraction of medicines and other materials from biota
Genetic materials	Genes for resistance to plant pathogens, ornamental species, and so on
Regulating	
Climate regulation	Source of and sink for greenhouse gases; influence on local and regional temperature, precipitation, and other climatic processes
Water regulation (hydrological flows)	Groundwater recharge and discharge
Water purification and waste treatment	Retention, recovery, and removal of excess nutrients and other pollutants
Erosion regulation	Retention of soils and sediments
Natural hazard regulation	Flood control and storm protection
Pollination	Habitat for pollinators
Cultural	
Spiritual and inspirational aspects of wetlands	Source of inspiration; many religions attach spiritual and religious value to wetlands
Recreational	Opportunities for recreational activities
Aesthetic	Many people find beauty or aesthetic value in wetland ecosystems
Educational	Opportunities for formal and informal education and training
Supporting	
Soil formation	Sediment retention and accumulation of organic matter
Nutrient cycling	Storage, recycling, processing, and acquisition of nutrients

Human activities and climate change have resulted in degradation of many wetland environments in the Himalayas (Erwin 2009). As a consequence, loss of vegetation cover and degradation of wetlands expose the soil surface which ultimately increases CH_4 emission into the atmosphere (Melton et al. 2013), hence influencing global climate. Studies over the past 50 years attributed people for changing ecosystems more rapidly than in any parallel period. Summary of a recent conclusion of Working Group II of the Intergovernmental Panel on Climate Change forecasted that "if existing warming rates are continued, Himalayan glaciers possibly decay at very prompt rates, decreasing from the current 500,000 square kilometers to 100,000 square kilometers by 2030s." Shrinking of glaciers, reducing water flow in rivers, gradual rain failure during monsoons, hefty and unpredictable rain in shorelines and climate fluctuations are some examples of the climate change predicted under the Indian Prospective (Thomas et al. 2007). Amidst the innumerable factors cited as influencing the changes in wetland, land use has been categorized as the sole factor accounting for their degradation in the western Himalayas. In addition, serious loss has also been related to unplanned urban sprawl, unregulated agricultural development and inflow of fertilizers, silt, solid waste and pesticides into the wetland.

Hokar Sar is in the Doodhganga watershed of the western Himalayas in the extreme northern part of Indian occupied Kashmir, where it plays a significant role in the overall water cycle of the basin. This wetland is also a wildlife reserve and was named as a Ramsar site in 2005. Current studies have suggested that this wetland is under serious threats regarding land cover use and climate change. Based on the China National Land Cover Database, from 1990 to 2010, China lost 2883 km^2 of wetlands to urbanization, of which approximately 2394 km^2 were in the eastern regions (North China, Northeast China, South China, and Southeast China). Overall, four urbanization-induced wetland losses in China were recognized, including the Qinghai Tibetan plateau, Yangtze river delta, one of the important Himalayan regions of China (Mao et al. 2018). Similarly, threats to Nepal's wetlands can be broadly classified as habitat degradation and destruction, loss of environmental integrity and reduction of species abundance and diversity.

With a population of nearly 30 million people, Nepal has a growth rate of 2.5% per year. About 81% of the population depends on agriculture, and consequently the conversion of wetlands to agricultural land, specifically to paddy fields, will continue to rise, exerting more pressure on wetlands. Various other factors such as disposal of untreated industrial effluents, use of agro-chemicals for high value crops, construction of dams thus inundating important habitats; disturbance of nutrient dynamics, reduction of downstream water flow and alteration of local microclimates are some the issues which need to be addressed by Nepal.

7.6 Insight into Wetlands

A group of individuals of the same species collaborating is called a population while organisms interacting with other living organisms belonging to different species, is called a community. A community interacting with the non-living world is called an ecosystem. Generally, all ecosystems have a pattern of energy flowing through them. Sunlight, which is utilized mostly by plants for photosynthesis, is the basic source of energy for most ecosystems. However, there are certain microorganisms which photosynthesize, while others acquire energy from non-solar sources by reacting with inorganic materials, for example the oxidation of iron.

Based on this energy flow, we can identify groups of organisms which have distinct roles. Plants, being the primary producers, fix solar energy into organic matter; and are autotrophic, which means they can make their own food. Herbivores are primary consumers; depending on plants for energy. Predators are categorized in the heterotrophic group as they rely on plants indirectly, feeding on herbivores or animals that consume herbivores. They are referred to as secondary and tertiary consumers. They are listed in different spots in a hierarchy of feeding, occasionally regarded as a food web in the ecosystem. The waste products of living organisms and dead parts/left-overs which remain unconsumed by predators are utilized by decomposers. Decomposers of the ecosystem are mostly bacteria and fungi, which consume all the energy-rich materials that remain. Energy remains constant—nothing is wasted, nothing is lost. Energy entering the ecosystem is consumed and is released as heat during respiration. Chemical elements are recycled as the energy flows through the ecosystem.

At a glance, the carbon atoms (gaseous form CO_2) that are taken up by plants are fixed either into carbohydrates or fats or converted into other compounds by incorporating elemental nitrogen (soil), or possibly modified into other forms by the addition of phosphorus to synthesize phospholipids. Animals in turn consume the biomass stored in plants and eliminate a proportion as waste. Respiration releases carbon back into the atmosphere as CO_2 gas, whereas nitrogen and phosphorus along with other elements are returned to the soil through the process of decomposition. Hence, they are recycled and can be utilized again. This rotating wheel of element motion is known as a nutrient cycle.

Many different types of wetlands exist, each of which can be regarded as an ecosystem, with its own pattern of energy flow and nutrient cycle. Wetlands have many features in common which make them different from other ecosystems. The most evident feature is their water, which influences the pattern of energy flow and storage. The foremost barrier posed by the water is the low availability of oxygen. Every organism, except for some fungi and bacteria, requires oxygen to respire. The energy trapped in the organic compound upon which they feed can only be retrieved if oxygen is available, as they convert sugars into CO_2.

The supply of oxygen is abundant in air (~21%) while it decreases under water, as it dissolves. Therefore, fast flowing streams might be rich in dissolved oxygen, whereas in still water, oxygen relies on the course of diffusion where the molecules flow from high density to regions of low density. Diffusion of oxygen molecules follows from the surface layers, where water interacts with air, descending deeper, where oxygen is utilized by the decomposers at the bottom layer of wetlands. However, oxygen diffuses about 10,000 times more slowly in water as in air, thus oxygen consumed by the respiration of organisms in still water is replaced very slowly. The slow diffusion of dissolved oxygen results in the incomplete decomposition of dead organic matter. Generally, all residual matter of the ecosystem that enters the soil is ultimately decomposed and is lost, but in the case of wetlands, the slow decomposition may create an imbalance in the flow of energy, hence developing a layer of energy-rich organic matter deposited in the sediments of the ecosystem.

Due to this imbalance, wetlands appear as a "sink" for atmospheric carbon and as an aid in the assimilation of carbon released by the combustion of fossils fuels. Considering that wetlands form through the process of succession, the energy pool within the sediments increases, and this is the foundation of the growing mass of peat in bogs. It is important to mention that most of the energy does not recycle, but often follows a complicated route through an ecosystem, which ecologists have termed food webs instead of food chains. It must also be noted that energy is reduced as it travels through the ecosystem, as some of the energy is absorbed, hence only 10% of the energy (in aquatic environment) present at one feeding level (known as trophic level) is passed to the next level. The remainder is handled by detritivores and decomposers.

Decomposition occurs mostly in the surface layers of peat and sediments, where oxygen diffuses easily. The upper, loosely compacted layer of the peat deposit, where movement of water and decomposition occur rapidly, is termed as acrotelm, while the deeper layer, where decomposition and movement of water is comparatively slow, is termed as catotelm. Poor oxygen diffusion under water and waterlogging are, apparently, the key factors in regulating the rate of decomposition; however, some materials are less palatable to fungi and bacteria then other materials. In summary, the atmosphere is common to all terrestrial ecosystems, thus offering a medium through which elements can shift from one ecosystem to another.

Looking at elemental shifts, carbon is one element that is obtained from the atmosphere by plants for building biomass. Importantly, it is said to be the critical element for all living organisms, as almost all energy transfer and storage encompass compounds built utilizing carbon atoms. Wetlands and carbon are discussed in synchrony, as wetlands act as "sinks" for atmospheric carbon. Moreover, it also enters the ecosystem in a dissolved form in rainfall and rivers or streams. This dissolved CO_2 is made available for photosynthesis by phytoplankton and submerged aquatic plants. Other means of carbon entry into the wetlands is via hydrogen carbonate ions, HCO_3^-. This may have originated from carbonic acid, by the chemical reaction of CO_2 with water, or from lime (calcium carbonate) dissolving in water. Aquatic plants of the wetlands can once more utilize this hydrogen carbonate ion for photosynthesis. Most of the carbon fixed by photosynthesis is lost in respiration, while some is deposited in the sediments that accumulate, particularly in the peat deposits of the temperate bogs.

Current dilemma concerns the entry of carbon into the atmosphere in that about 5.5 billion tons of carbon enters the atmosphere every year due to human combustion, of which over half remains in the atmosphere, increasing every year and causing climate change. Eventually, some is recruited by forests and some is dissolved in oceans. There is a "missing sink" for carbon that has yet to be identified. The absorption of carbon by peat land is probably a portion of this sink for atmospheric carbon. Peat deposits grow slowly, so the total carbon taken up is limited, yet every carbon sink is the key in precluding the buildup of CO_2 in the atmosphere.

7.7 Carbon Sink and Gaseous Emissions

Predominant biogeochemical processes aid wetlands to be one of the most important carbon reservoirs, accumulating approximately 20–30% of the global carbon pool. Carbon deposits in the wetland sediments are very sensitive to environmental changes such as in temperature, flood, microbial activity and nutrient regime. The wetlands provide a major source of carbon. Especially greenhouse gasses (GHG) such as CH_4, CO_2 and nitrous oxide (N_2O) (Mitra et al. 2005).

From 1750 to 2011, the amount of CO_2 increased from 278 to 390.5 ppm, N_2O from 271 to 324.2 ppb and CH_4 from 722 to 1803 ppb (IPCC 2013). "Wetlands contain 12% of the global carbon reservoir with a carbon density of 723 t ha^{-1} (Mitra et al. 2005), storing nearly 2500 Pg 1 Pg = 10^{15} g) of the earth's carbon pool" (Mitsch et al. 2013). The main carbon pool of wetlands includes: (a) particulate organic carbon; (b) plant biomass carbon; (c) dissolved organic carbon; (d) gaseous end products such as CO_2 and CH_4; and (e) microbial biomass carbon. Plant remnants are the major organic matter in the wetlands, comprising 45–50% of carbon (Kayranli et al. 2010), where it goes through a series of complex aerobic and anaerobic processes. The main aerobic processes entail respiration while the main anaerobic processes undergo fermentation, methanogenesis and sulfate, iron and nitrate reduction.

The wetlands function as a carbon sink, but they also contribute substantial GHG emissions (Mander et al. 2014). Among all GHGs, CH_4 is considered the most prominent and potent gas in terms of global warming. Methane is the product of methanogenesis in anoxic environments and low redox potential of $\leq$150 mv by transmethylation or decarboxylation of acetic acid and by the reduction of CO_2 (Segers 1998; Singh et al. 2000), whereas, N_2O is the net result of: (a) nitrate or nitrite-reducing processes of denitrification; (b) nitrate ammonification; (c) ammonia oxidation; and (d) nitrifier denitrification (Baggs 2011). The escalation of the GHGs is possibly the main cause of the rise in ocean and land surface temperatures.

Being one of the most significant GHGs exchanged amidst terrestrial ecosystems and the atmosphere, CH_4 as compared to CO_2 has 28 times more global warming potential on a 100-year time horizon (IPCC 2013). Therefore, any change in atmospheric CH_4 concentration can contribute to significant climate impacts because of strong greenhouse effects (Bridgham et al. 2013). Currently, worldwide atmospheric CH_4 concentration is about 1810 parts per billion (ppb), a rise of over 1000 ppb since pre-industrial level (Myhre et al. 2013). Hence, increases in atmospheric concentration of CH_4 have prompted significant interest in calculating CH_4 emission from various sources.

Studies have revealed that CH_4 fluctuations are associated to biological (plant life cycle related with microbes) and physical (wind, temperature and water table) variables at various temporal and spatial scales (Song et al. 2015). However, there remain large uncertainties while estimating future and current CH_4 emission from wetlands for many reasons. Until now, many studies on CH_4 fluxes have focused largely on high-latitude wetlands, since these wetlands store vast amounts of carbon in waterlogged sediments.

The HKH contain most of the high-altitude wetlands, and there is a paucity of research on these wetlands, especially in remote areas. Most of the studies on Himalayan wetlands were carried out by Chinese and Nepalese researchers. Hirota et al. (2004) measured CH_4 emission from the Luanhaizi alpine wetlands in four vegetation zones on the Qinghai-Tibetan Plateau: three emergent plant zones, including *Hippuris*-dominated, *Scirpus*-dominated and *Carex*-dominated zones and one submerged zone, *Potamogeton*-dominated. The study revealed that the lowest

CH_4 flux (seasonal mean was 33.1 mg CH_4 m^{-2} day^{-1}) was observed in the *Potamogeton*-dominated zone while the highest CH_4 flux (seasonal mean was 214 mg CH_4 m^{-2} day^{-1}) was in the *Hippuris*-dominated zone (Hirota et al. 2004). Similarly, CH_4 fluxes were also calculated in regions such as Maduo (Qinghai), Hongyuan (Sichuan) and Zoige (Qinghai) (Table 7.5).

Changes in CH_4 emission were mostly due to variation in vegetation, seasons and type of wetlands. A recent study compared Beeshazar and Dhaap, two wetlands located in the southern slopes of Himalayan Nepal, for day-time based CH_4 emission (Zhu et al. 2015). The authors correlated plant community height, standing water depth and soil temperature to CH_4 emission from these wetlands. They suggested that their findings could be valuable in filling the gap and in generating the emission pattern of CH_4 at regional and global scales. Sundarbans mangroves are the largest biosphere reserves in the world, which stretches over 1000 km^2 covering West Bengal, India and Bangladesh. During the summer season, the mean concentration of CH_4 was 1682 ppb while the daily CH_4 flux was 150 mg m^{-2} day^{-1}. Many studies have suggested that the global warming status of Himalayan wetlands and the specific quantification of CH_4 from these regions are not fully understood and need more research data.

7.8 Conclusion and Recommendations

The HKH region is home to various types of wetlands and the services they provide are essential for people as well as the environment. They also act as reservoirs for carbon in their biomass, peat, sediment and litter. They pose a serious threat to the environment by emitting GHGs, especially CH_4. Wetland loss due to commercial development, extraction of minerals and peat, intensive agriculture, drainage schemes, pollution, construction of dams, over-fishing and tourism are causing serious damage to the wetland ecosystem. These threats have affected the overall biogeochemical cycle and biodiversity and have caused rising floods and droughts, nutrient runoff, water pollution and mounting global warming.

To maintain the wetlands, several organizations are focusing to conserve and manage these water bodies and relieve the environmental pressure. In this regard, the WWF (World Wildlife Fund) has initiated projects such as the conservation of high-altitude wetlands. The project has produced socio-economic, technical, scientific and biodiversity related information which is being utilized to protect and conserve these unique places in the interest of local communities and biodiversity. The organization is also working with Ramsar for collecting scientific information essential for wetland management. ICIMOD (International Center for Integrated Mountain Development) aims to assist sustainable and resilient mountain development for better livelihoods via knowledge and regional cooperation. However, most of the structural policies have led to the implementation of countrywide plans that were not particularly designed for mountain people.

Table 7.5 Studies on methane emission from wetlands located in the HKH

Location	Wetland type	Measurement period	Season	Vegetation	Methane emission mg CH_4 m^{-2} day^{-1}	Reference
China						
Luanhaizi, Qinghai	Peatland	4 Jul.–15 Sep. 2002	N.A.	Carex allivescer	196	Hirota et al. (2004)
				Scirpus distigmaticus	99.5	
				Hippuris vulgaris	214	
				Potamogeton pectinatus	33.1	
Maduo, Qinghai	Peatland	27 Jul.–8 Aug. 1997	N.A.	Kobresia humilis	1.01–147	Jin et al. (1998)
				Hippuris vulgaris	−4.38 to 2.81	
				Kobresia tibelica	12.9–88.8	
				Kobresia humilis	3.82–10	
				Batrachium trchophyllum	3.58–18.8	
				Kobresia humilis	−6.49 to 75.6	
				Batrachium trchophyllum	−19.4 to 63.3	
				Hippuris vulgaris	−5.76 to 188	
				Carex muliensis	−4.39 to 347	
				Batrachium trchophyllum	0.28–23.9	
Hongyuan, Sichun	Peatland	May–Oct. 2001	N.A.	Carex muliensis	12.2–197	Ding et al. (2004)
				Carex meyeriana	8.64–241	
		May–Oct. 2002		Carex muliensis	3.84–138	
				Carex meyeriana	20.6–214	
Zoige, Qinghai	Alpine	2011	N.A.	Kobresia littledalei	144.3 µg m^{-2} h^{-1}	Wei et al. (2015)
				Carex moorcroftii	67.6 µg m^{-2} h^{-1}	

7 Wetlands as a Carbon Sink: Insight into the Himalayan Region

Zoige, Qinghai	Natural Peatland	2013	N.A.	Potamogton pusittus	19.13 mg m^{-2} h^{-1}	Zhou et al. (2017)
	Drained peatland			Utricularia vulgaris	0.14 mg m^{-2} h^{-1}	
Nepal						
Beeshazar Lake, Chitwan National Park	Freshwater lake	June 10th, 2013	Monsoon season	Kans grass	4.46 ± 4.80	Zhu et al. (2015)
				Water pepper	11.45 ± 6.29	
		December 31st, 2013	Dry season	Kans grass	2.43 ± 5.18	
				Water pepper	1.13 ± 3.68	
Dhaap Lake, Shivapuri-Nagarjun National Park	Freshwater lake	July 23rd, 2013	Monsoon season	Water moss	0.44 ± 0.71	
				Sweet flag	1.60 ± 1.39	
		January 21st, 2014	Dry season	Water moss	0.40 ± 0.90	
				Sweet flag	0.12 ± 0.31	
India (West Bengal)						
Sundarbans	Mangrove	April-May, 2012	NA	NA	150.22 ± 248.87	Jha et al. (2014)

Consequently, the HKH mountain region lags behind in conservation-linked management. Conservation is just not placing fences around areas and keeping people and livestock out. The local people must get involved with organizations and management to conserve biodiversity. Such conservation and management must be based on a combination of literature surveys, workshops, field research and incentive mechanisms to know how these wetlands function and to meet the challenges of climate change, mainly in relation to adaptation, biodiversity conservation and restoration of wetlands as carbon reservoirs in the HKH.

References

Agrawal, D.K., M.S. Lodhi, and S. Panwar. 2010. Are EIA studies sufficient for projected hydropower development in the Indian Himalayan region. *Current Science* 98: 154–161.
Asad, A.K., and A. Sana. 2014. Wetlands of Pakistan: distribution, degradation and management. *Pakistan. Geographical Review* 69 (1): 28–45.
Baggs, E.M. 2011. Soil microbial sources of nitrous oxide: Recent advances in knowledge, emerging challenges and future direction. *Current Opinion in Environmental Sustainability* 3 (5): 321–327.
Barik, N.K., P.K. Katiha, 2003. Management of fisheries of floodplain wetlands: Institutional issues and options for Assam. In *A Profile of People, Technologies and Policies in Fisheries Sector in India*, 141–158.
Bassi, N., M.D. Kumar, A. Sharma, et al. 2014. Status of wetlands in India: A review of extent, ecosystem benefits, threats and management strategies. *Journal of Hydrology: Regional Studies* 2: 1–19.
Bhandari, B. 2009. Wise use of wetlands in Nepal. *Banko Janakari* 19 (3): 10–17.
Boere, G.C., C.A. Galbraith, D.A. Stroud, et al. 2006. *Waterbirds around the world*, 960pp. Edinburgh, UK: The Stationery Office.
Bridge, L. K., Stroud, D., Galbraith, C. A., and Boere, G. 2006. Waterbirds around the world. The Stationery Office.
Bridgham, S.D., H. Cadillo-Quiroz, J.K. Keller, et al. 2013. Methane emissions from wetlands: Biogeochemical, microbial, and modeling perspectives from local to global scales. *Global Change Biology* 19: 1325–1346.
Byomkesh, T., N. Nakagoshi, and R.M. Shahedur. 2009. State and management of wetlands in Bangladesh. *Landscape and Ecological Engineering* 5 (1): 81–90.
Ding, W.X., Z.C. Cai, and D.X. Wang. 2004. Preliminary budget of methane emissions from natural wetlands in China. *Atmospheric Environment* 38: 751–759.
Erwin, K.L. 2009. Wetlands and global climate change: The role of wetland restoration in a changing world. *Wetlands Ecology and Management* 17 (1): 71–84.
Finlayson, C.M., and A.G. Van der Valk. 1995. Wetland classification and inventory: A summary. *Vegetatio* 118 (1–2): 185–192.
Gopal, B., A. Chatterjee, and P. Gautam. 2008. *Sacred waters of the India Himalaya*. New Delhi: WWF-India.
Gopal, B., R. Shilpakar, E. Sharma, et al. 2010. *Functions and services of wetlands in the Eastern Himalayas: Impacts of climate change*. Climate change impact and vulnerability in the Eastern Himalayas-Technical Report 3. Kathmandu: ICIMOD.
Gujja, B. 2007. Conservation of high-altitude wetlands: Experiences of the WWF network. *Mountain Research and Development* 27 (4): 368–371.
Hirota, M., Y.H. Tang, Q.W. Hu, et al. 2004. Methane emissions from different vegetation zones in a Qinghai-Tibetan Plateau wetland. *Soil Biology & Biochemistry* 36: 737–748.

IPCC, W.G.I. 2013. *Contribution to the IPCC fifth assessment report. Climate change*, 36.
Islam, S.N., and A. Gnauck. 2008. Mangrove wetland ecosystems in Ganges-Brahmaputra delta in Bangladesh. *Frontiers of Earth Science in China* 2 (4): 439–448.
Jha, S. 2009. Status and conservation of lowland terai wetlands in Nepal. *Our Nature* 6 (1): 67–77.
Jha, C.S., S.R. Rodda, K.C. Thumaty, et al. 2014. Eddy covariance-based methane flux in Sundarbans mangroves, India. *Journal of Earth System Science* 123 (5): 1089–1096.
Jin, H.J., G.D. Cheng, B.Q. Xu, et al. 1998. Study on CH_4 fluxes from alpine wetlands at the Huashixia permafrost, Qinghai-Tibetan plateau. *Journal of Glaciology and Geocryology (in Chinese)* 20: 172–174.
Kayranli, B., M. Scholz, A. Mustafa, et al. 2010. Carbon storage and fluxes within freshwater wetlands: A critical review. *Wetlands* 30 (1): 111–124.
Khan, A.A., and S. Arshad. 2014. Wetlands of Pakistan: Distribution, degradation and management. *Pakistan Geographical Review* 69 (1): 28–45.
Li, Z., J. Xu, R.L. Shilpakar, et al. 2014. Mapping wetland cover in the greater Himalayan region: A hybrid method combining multispectral and ecological characteristics. *Environmental Earth Sciences* 71 (3): 1083–1094.
Mander, Ü., G. Dotro, Y. Ebie, et al. 2014. Greenhouse gas emission in constructed wetlands for wastewater treatment: A review. *Ecological Engineering* 66: 19–35.
Mao, D., Z. Wang, J. Wu, et al. 2018. China's wetlands loss to urban expansion. *Land Degradation and Development* 29 (8): 2644–2657.
MEA. 2005. *Ecosystems and human well-being: wetlands and water*. World Resources Institute.
Melton, J.R., R. Wania, E.L. Hodson, et al. 2013. Present state of global wetland extent and wetland methane modelling: Conclusions from a model intercomparing project (WETCHIMP). *Biogeosciences* 10: 753–788.
Meng, W., M. He, B. Hu, et al. 2017. Status of wetlands in China: A review of extent, degradation, issues and recommendations for improvement. *Ocean and Coastal Management* 146: 50–59.
Mitra, S., R. Wassmann, and P.L. Vlek. 2005. An appraisal of global wetland area and its organic carbon stock. *Current Science* 88 (1): 25–35.
Mitsch, W.J., J.G. Gosselink, L. Zhang, and C.J. Anderson. 2009. *Wetland ecosystems*. Wiley.
Mitsch, W.J., B. Bernal, A.M. Nahlik, et al. 2013. Wetlands, carbon, and climate change. *Landscape Ecology* 28 (4): 583–597.
Moore, P., and R. Garratt. 2006. *Biomes of the earth. Wetlands*. Warsaw: Livro.
Myhre, G., D. Shindell, F.M. Bréon, et al. 2013. Anthropogenic and natural radiative forcing. In *Climate change 2013: The physical science basis. Contribution of working group I to the fifth assessment*.
Paudel, N., S. Adhikaril, and G. Paudel. 2017. Ramsar lakes in the foothills of Himalaya, Pokhara-Lekhnath, Nepal: An overview. *Janapriya Journal of Interdisciplinary Studies* 6: 134–147.
Qamer, F.M., F.M. Qamer, M.S. Ashraf, et al. 2008. Pakistan wetlands GIS-a multi-scale national wetlands inventory. *Wetlands* 10: 3–6.
Ramsar, C. M. 2006. The Ramsar convention manual: a guide to the convention on wetlands. In T, editor. Ramsar convention secretariat (pp. 6–8).
Ramsar Convention Secretariat. 2013. The Ramsar convention manual. In *The Ramsar convention manual: A guide to the convention on wetlands (Ramsar, Iran, 1971)*, Vol. 109, 6th ed. http://www.ramsar.org.
Segers, R. 1998. Methane production and methane consumption: A review of processes underlying wetland methane fluxes. *Biogeochemistry* 41 (1): 23–51.
Sherab, N. Wangdi, N. Norbu 2011. *Inventory of high altitude wetlands in Bhutan. The Wetlands Sky High: Mapping Wetlands in Bhutan*.
Singh, S.N., K. Kulshreshtha, and S. Agnihotri. 2000. Seasonal dynamics of methane emission from wetlands. *Chemosphere-Global Change Science* 2 (1): 39–46.
Song, W., H. Wang, G. Wang, L. Chen, Z. Jin, Q. Zhuang, and J.S. He. 2015. Methane emissions from an alpine wetland on the Tibetan Plateau: Neglected but vital contribution of the non-growing season. *Journal of Geophysical Research: Biogeosciences* 120 (8): 1475–1490.

Space Applications Centre (SAC). 2013. *National wetland inventory and assessment high altitude Himalayan lakes.* http://www.sac.gov.in.

Tan, Y., X. Wang, Z. Yang, et al. 2011. Research progress in cold region wetlands, China. *Sciences in Cold and Arid Regions* 3 (5): 441–447.

Thomas, D.S., C. Twyman, H. Osbahr, et al. 2007. Adaptation to climate change and variability: Farmer responses to intra-seasonal precipitation trends in South Africa. *Climatic Change* 83 (3): 301–322.

Trisal, C.L., and T.H. Manihar. 2004. *The atlas of Loktak lake.* New Delhi: Wetlands International-South Asia, and Loktak Development Authority.

Uddin, K., S.M. Wahid, M.S.R. Murthy, et al. 2015. *Mapping of koshi basin wetlands using remote sensing.* In *5th International Conference on Water & Flood Management (ICWFM-2015)*

Wei, D., T. Tarchen, D. Dai, et al. 2015. Revisiting the role of CH_4 emissions from alpine wetlands on the Tibetan Plateau: Evidence from two in situ measurements at 4758 and 4320 m above sea level. *Journal of Geophysical Research: Biogeosciences* 120 (9): 1741–1750.

Xu, J., R.E. Grumbine, A. Shrestha, et al. 2009. The melting Himalayas: Cascading effects of climate change on water, biodiversity, and livelihoods. *Conservation Biology* 23 (3): 520–530.

Ye Htut Deputy Director Nature and wildlife Conservation Division I. 2013. *Water is life: Too much or too little, every drop counts.*

Zhou, W., L. Cui, Y. Wang, et al. 2017. Methane emissions from natural and drained peatlands in the Zoigê, eastern Qinghai-Tibet Plateau. *Journal of Forestry Research* 28 (3): 539–547.

Zhu, D., N. Wu, N. Bhattarai, et al. 2015. A comparative study of daytime-based methane emission from two wetlands of Nepal Himalaya. *Atmospheric Environment* 106: 196–203.

Chapter 8
Milk and Dung Production by Yaks (*Poephagus grunniens*): Important Products for the Livelihood of the Herders and for Carbon Recycling on the Qinghai-Tibetan Plateau

A. Allan Degen, Shaher El-Meccawi, and Michael Kam

Abstract The Qinghai-Plateau, characterized by an extremely harsh environment, namely, severe cold, low air oxygen content and strong ultraviolet radiation, accounts for one-third of the total grasslands in China. Yaks (*Poephagus grunniens*) have adapted well, both physiologically and anatomically, to the high altitude and extreme cold. They are vital for the livelihoods of the herders, providing milk and meat for consumption, dung for fuel, fertilizer and building material and fiber for clothing, tent material and rope. It is estimated that, on an annual basis, each lactating female yak produces an average of 288 kg milk, 19 kg butter, 17 kg hard dried curds (*qula*) and 84 kg yoghurt and each yak produces an average of 786 kg of dung organic dry matter. This paper discusses the importance of milk and dung production by yaks for the livelihood of the herders and the role of these products in carbon recycling.

Keywords Yaks · Milk production · Dung production · Carbon recycling · Qinghai-Tibetan Plateau · Extreme climate · Sedentarization

---

A. A. Degen (✉) · M. Kam
Desert Animal Adaptations and Husbandry, Wyler Department of Dryland Agriculture, Blaustein Institutes for Desert Research, Ben-Gurion University of the Negev, Beer Sheva, Israel
e-mail: degen@bgu.ac.il; mkam@bgu.ac.il

S. El-Meccawi
Achva Academic College, Arugot, Israel

8.1 Introduction

The Qinghai-Tibetan Plateau, which extends over 1.29×10^8 ha, is the highest and largest plateau in the world, accounting for one-third of the total grasslands in China (Wiener et al. 2003). The plateau occupies more than a third of the Hindu-Kush Himalayan (HKH) region and one eighth of the entire area of China and has an average altitude greater than 4000 m above sea level (Tashi and Partap 2000). The area is characterized by an extremely harsh environment, namely, severe cold, low air oxygen content and strong ultraviolet radiation (Wiener et al. 2003). These factors, in particular long-term exposure to ultraviolet light, can be sources of oxidative stress for Tibetans inhabiting the plateau (Zheng and Huang 2008).

8.2 Carbon Recycling

Rangelands produce a large proportion of total land carbon and sequester large quantities of carbon above and below ground. The immense rangelands on the Tibetan plateau account for 2.4% of the world's soil carbon storage (Wang et al. 2014) and, consequently, "could have widespread effects on regional climate and global carbon cycles". In fact, "alpine meadow and alpine steppe range, found primarily on the Tibetan plateau, comprise 40 percent of all carbon stored on China's rangelands, indicating that these ecosystems have a significant and long-lived effect on global carbon cycles" (Miller 2003). This can be explained, at least in part, by the low air temperatures on the plateau, which slow down organic decomposition rates and increase the mean retention time of the carbon. The biomass carbon stock of the Qinghai-Tibetan Plateau's grasslands has been increasing over the past 20 years, suggesting that alpine grasslands are storing carbon (Gao et al. 2012).

From the pre-industrial period to 2010, the atmospheric concentration of CO_2 has increased by 39%, from 280 to 390 ppm. This enrichment in atmospheric CO_2 concentration, along with other greenhouse gases, including methane (CH_4) and nitrous oxide (N_2O), "may accentuate radiative forcing and alter the Earth's mean temperature and precipitation" (Lal undated). To minimize the risks of global warming, it was recommended at the 2009 Copenhagen Accord that atmospheric CO_2 concentrations should be contained below 441 ppm by 2100 (Ramanathan and Xu 2010) and that this should be accomplished "by reducing CO_2, CH_4 and N_2O emissions and by offsetting emissions through sequestration of carbon in soils and other terrestrial and inland aquatic ecosystems" (Lal undated).

The 390 ppm of atmospheric CO_2 has a total mass of 3030 or 825 Pg of carbon. Anthropogenic emissions through activities such as animal husbandry and burning of biomass total 9.9 Pg carbon per year, of which 4.2 Pg is absorbed by the atmosphere. Consequently, human and livestock activities are important components of CO_2 emission in the carbon cycle (IPCC 2007; WMO 2010).

Livestock are an important source of income in the HKH, more so with increasing elevation. Most farmers practice a mixed crop-livestock farming system (Tulachan 2000); but even in the "lower valleys where cropping and forestry are possible, livestock is still the main activity and major means of sustaining livelihoods and food security. This means that nearly half of the population of Tibet is deriving some livelihood from farm animals and that nearly 30% depend almost entirely on livestock. Animal products from rangeland areas meet more than 50% of the people's food and agricultural needs" (Tashi and Partap 2000). The main livestock raised on the Qinghai-Tibetan Plateau are yaks (*Poephagus grunniens*) and Tibetan sheep (*Ovis aries*). This chapter will focus on the yak and its production of milk and dung.

8.3 Sedentarization and Land Enclosure

In raising their livestock, Tibetans practiced a nomadic lifestyle in which grazing lands were communally controlled. Mobility was a key feature in their livestock production system, in which winter and summer grazing lands were designated. However, since the 1980s, there have been substantial and rapid changes in pastoralism, in lifestyle and in livelihood of the Tibetans on the Qinghai-Tibetan Plateau (Levine 2015). In fact, two recent journals (Himalaya, volume 30, issue 1, 2011, and Nomadic Peoples, volume 16, issue 1, 2015) devoted special issues to the dramatic changes that have been taking place.

In essence, the policy of the Chinese government has been to settle the nomadic Tibetans. Bauer (2015) commented on the changes as: "among other rationale, resettlement is of symbolic value to the state, as authorities see it as a measure of Tibetans' integration into Chinese society and of 'modern' development. Underlying these policies is the assumption that nomads are 'backward' and practice 'inefficient' methods of land and livestock management that are associated with poverty and environmental degradation (Goldstein et al. 1990; Williams 1997). Nyima (2011) added that "the new programs were to 'modernize' farmers, and 'stabilize' backward villages in order to raise family incomes. The underlying assumption is that local villagers are of 'low quality' (*suzhidi*) and have a 'backward mentality' (*sixiang luohou*), which have been singled out as the principal causes for the aforementioned problems. In accordance with this discursive representation, the government thus launched the 'three rurals' (*sannong*) program to end the 'chronic problems of poverty' facing rural villagers."

The primary goal of the government was to use the grazing land more intensively, by the herders: (1) possessing a durable house for living at winter sites; (2) growing fodder crops for hay in fenced-in plots; and (3) providing animal shelters due to the severe winters (Levine 2015). Communal grazing lands were privatized by the Chinese government and these lands were distributed among the members at the household level. The households were provided with long-term contracts and the acquired land was enclosed by the household, often with barbed wire, which prevented mobility as was practiced in the past (Bauer and Nyima 2011). In addition, under the Eco-Migration Policy initiated in 2003, large tracts of grasslands were lost to grazing.

The government was concerned with environmental damage and argued that due to widespread rangeland degradation as a result of high livestock stocking rates, grazing must cease or be reduced substantially in large tracts of land to allow the rangelands to recover. Consequently, national natural preservation zones were created in the Three River Headwaters Region, designated as the National Sanjianyuan Nature Reserve Project. Pastoralists in these areas were relocated to ecological migrant villages as either: (1) entire village migration, in areas where there would be zero grazing; and (2) partial migration, where a portion of the pastoralists were relocated to reduce grazing pressure. These relocated households were obligated to sell their livestock (Bessho 2015); however, they managed to retain some animals. Ptackova (2015) reported that "inhabitants of the settlement have retained more livestock than has been reported. Although many people reduced the size of their herds significantly after moving into the settlement, many animals were also redistributed to other family members. Nevertheless, local herds are comparatively small and therefore, in the majority of cases, animal husbandry cannot serve as the main or only source of income anymore." Levine (2015) added that "over the last decade, under government sponsorship, large numbers of households have abandoned animal herding and moved to modern houses in newly-built towns and settlement areas."

Interestingly, although the total number of livestock has been reduced substantially on the plateau through these changes, the population of yaks remained relatively constant. Tibetans were reluctant to sell or slaughter yaks, due in part to cultural reasons, and managed to conceal them from the authorities. The number of Tibetan sheep, however, has been reduced drastically to the point where Sulek (2011) commented on the "disappearing sheep" phenomenon. This author pointed out that the reduction in sheep numbers was concomitant with the emergence of new, non-pastoral sources of income, in particular, income from gathering the caterpillar fungus (*Ophiocordyceps sinensis*). The caterpillar fungus trade has become very profitable in recent years, with a massive demand from China's eastern provinces, and income could be realized not only from the harvesting of the fungus but also from the sale and the leasing of land to others for that purpose (Cencetti 2011). Many Tibetans abandoned agriculture and employment in agriculture dropped from 76% of the population in 1999 to 56% in 2008 (Fischer 2011). In general, the non-educated Tibetans continued practicing pastoralism while the educated Tibetans accessed different sources of income (Iselin 2011).

8.4 The Yak

The yak (*Poephagus grunniens*) belongs to the family *Bovidae*, together with bison (*Bison*), buffalo (*Bubalus*) and cattle (*Bos*), and is the only species of its genus. Initially, the yak was named *Bos grunniens* due to its relationship to cattle, but was then placed into its own genus, *Poephagus*, because of its distinctness from other bovines (Zhao 2000a). However, there is still some confusion concerning its nomenclature and many authors still refer to the yak as *Bos grunniens* (Zi 2003), while some simply use both genus names interchangeably (Han et al. 2002).

Yaks originated in high mountainous regions bordering the Himalayas. They were first domesticated, most likely, in the Stone Age by the ancient Qiang people on the Qinghai-Tibetan Plateau about 10,000 years ago; archaeological proof of domestication dates back approximately 5000 years (Cai 1996; Luo et al. 1997; Wiener et al. 2003). Today, yaks are found in the mountainous highlands of central Asia at 2500–6000 m above sea level, between 70° and 115° of east longitude and between 27° and 55° of north latitude. Wild yaks still exist in remote areas of the Tibetan plateau and adjacent highlands and a few exist in the Chang Chenmo Valley of Ladakh in Eastern Kashmir, India; however, in 2002, there were likely fewer than 10,000 in total (IUCN 2002). It is estimated that there are about 14.7 million domesticated yaks spread across areas of China, Nepal, Bhutan, Mongolia, Southern Russia, Tajikistan, Kyrgyzstan, Myanmar, Pakistan and Afghanistan, with about 95% (14 million) of these in China alone (Lu et al. 2004).

Yaks are vital for the livelihoods of the herders, providing them with milk and meat for consumption, dung for fuel, fertilizer and building material and fiber for clothing, tent material and rope (Wiener et al. 2003). This ruminant is also used for transportation, as a pack animal and for draught power. In addition, at least in Nepal, yak blood is drunk by some people as it is believed to possess medicinal properties (Vinding 1999; Degen et al. 2007).

The importance of the yak has been emphasized by a number of researchers. Miller (2000) stated that "although Tibetan Nomads also raise other animals, they place so much value on the yak that the Tibetan word for yaks, *nor*, is also translated as wealth. The yak makes life possible for man in one of the world's harshest environments. There is little doubt that the presence of wild yaks, and their later domestication, was the single most important factor in the adaptation of civilization on the Tibetan Plateau". Dong et al. (2003) added that "without yaks, it is doubtful whether man could survive on the harsh, high-altitude grazing lands of the Plateau (Qinghai-Tibetan Plateau of China)". In fact, many Tibetans relied solely on livestock, mainly yaks, for their means of subsistence (Gyal 2015) and held yaks in very high esteem (Figs. 8.1 and 8.2).

Indigenous to the plateau, the yak has adapted excellently, both physiologically and anatomically, to the high altitude and extreme cold. High altitude does not affect energy requirements while low air temperatures actually cause a decrease in energy requirements of yaks (Han et al. 2002). The ability of the yak to survive under extremely low air temperatures is aided by its thick hide, subcutaneous fat and thick long hair. Furthermore, because of its relatively large heart and lungs and high erythrocyte count, the yak can tolerate the low oxygen levels at high altitudes.

Nonetheless, in spite of its adaptation, losses can be very high due to inclement weather. For example, the winter of 1997–1998 was extremely severe in the Tibetan Autonomous Region when an estimated three million head of livestock perished. Unusually early and heavy snowfall was followed by extremely low air tempera-

Fig. 8.1 Statue of golden yaks in Llasa (Photograph by A. Allan Degen)

Fig. 8.2 Skull of a yak in entrance of Tibetan house (Photograph by A. Allan Degen)

tures which prevented the snow from melting. Livestock were unable to reach the forage under the snow and, consequently, many perished. Sheep and goats suffered most but mature yaks were also severely affected. In some townships, 70% of the livestock was lost. Nearly 25% of the million nomads was affected and hundreds of nomad families lost all their livestock (Miller 2000).

Average body mass of male yaks in China is 357 kg and of females is 223 kg. Body length, heart girth, and wither height of males are 144, 197 and 124 cm, respectively, while for females these values are 125, 161 and 108 cm, respectively (Wiener et al. 2003). Yaks raised in Nepal are considerably smaller; the body mass of the male and female yaks barely falls within the range of yaks in China, but body length, heart girth and wither height are all below the range (Joshi 2003) (Table 8.1). There are 12 recognized yak breeds in China, while in Nepal yaks are not separated into breeds. Much more selection has been done for certain traits in yaks in China, which has also resulted in the development of some larger breeds.

A low reproductive rate is the primary limiting factor in yak production, as only 40–60% of mature females reproduce annually (Zi 2003). Yaks are seasonal breeders with mating and conceptions taking place in the relatively warmer months of the year from mid-July to early November, which coincide with the period of peak forage growth. Females are usually mated for the first time at 3–4 years of age, produce four to five calves in a lifetime and calve once every alternate year. Gestation is approximately 258 days (Yu et al. 1993; Zhao 2000b; Zhang 2000).

The number of chromosomes of both yaks and cattle is 60 and cross-breeding between the two is common. Yak bulls crossed with cows (*Bos indicus* or *Bos taurus*) or bulls crossed with female yaks produce sterile males and fertile females. The sterile males are relatively gentle and are used as pack animals, while female crosses produce considerably more milk than female yaks (Sherchand and Karki 1997; Devkota 2002; Kharel et al. 2005).

Table 8.1 Body mass, length and height of adult male and female yaks in China and Nepal

Measurement	China		Nepal	
	Males	Females	Males	Females
Body mass (kg)	357 (235–594)	223 (188–314)	245	215
Body length (cm)	144 (123–173)	125 (114–140)	110	105
Heart girth (cm)	197 (176–236)	161(152–182)	140	130
Wither height (cm)	124 (114–138)	108 (103–117)	105	102

Note: For China—Wiener et al. (2003): values are mean of ten locations and nine breeds and ranges are in brackets; for Nepal—Joshi (2003)

8.5 Yaks, Their Management and Carbon Recycling on the Qinghai-Tibetan Plateau

As a consequence of the cold conditions on the Qinghai-Tibetan plateau, the growing season for vegetation is very short, only 100–150 days during summer and autumn. Emergence of grasses starts in May, with peak biomass in August–September. During winter and spring, that is from October to May, vegetation withers and the available forage is often insufficient to sustain the livestock. Nonetheless, yaks subsist almost entirely on grazing rangelands throughout the year (Fig. 8.3). As a result, yaks generally increase their body mass in summer and autumn and decrease their body mass during winter and spring. The decrease in body mass can be about 25% and mortality can be 30% in extremely cold years.

Customarily, it is the man who makes family decisions and is in charge of yak husbandry, while women are responsible for most of the labor needed for yak production, including milking and processing of the milk (Fig. 8.4), collecting, drying and storing dung (Fig. 8.5) and drying leather (Li and Yang 2005; Shang et al. 2016). Yaks are grazed rotationally on natural pasture often without receiving any supplemental feed except at harsh times in winter, when they are supplemented with harvested, cultivated herbage because of severe feed deficiency. As a strategy to deal with a deficiency in energy intake in winter, pastoralists also harvest forage, mainly *Kobresia tibetica*, in September and October in the wetlands, which are protected from being grazed. Other herbages, comprising mainly *Avena nuda*, *Avena sativa*, and *Pisum sativum*, are also cultivated by pastoralists. In general, only old and weak yaks are provided with supplementary feed.

In alpine grassland ecosystems, carbon is fixed from the atmosphere through photosynthesis and accumulates in the grassland vegetation and soils. While part of the carbon eventually returns to the atmosphere via geochemical cycles, the cold,

Fig. 8.3 Yaks grazing on the Qinghai-Tibetan Plateau (Photograph by A. Allan Degen)

8 Milk and Dung Production by Yaks (*Poephagus grunniens*): Important Products... 153

Fig. 8.4 Woman milking a yak on the Qinghai-Tibetan Plateau (Photograph by A. Allan Degen)

Fig. 8.5 Woman stacking dry yak dung on the Qinghai-Tibetan Plateau (Photograph by A. Allan Degen)

low oxygen environment reduces its ability to decompose. With the consumption of plant carbon by livestock, carbon accumulates at a higher trophic level. Animal carbon can re-enter the carbon cycle via two pathways. The first is through the decomposition of the body directly into the carbon cycle. The second is through its products such as milk, dung, meat and leather. By consumption of livestock products and sale of products outside the grassland ecosystem, and by the use of dung, livestock products become part of the carbon output in the carbon cycle. Dung is an important source of energy in the cold alpine areas that not only reduces fossil fuel imports into the grassland ecosystem and the use of firewood, but also accelerates the decomposition of livestock carbon into the atmosphere or ash into the soil.

From the 1970s, yields of animal products increased throughout the alpine region. The increase in milk products, and hence carbon, was a result of increased livestock. The increased livestock numbers consumed more grass and, therefore, more carbon, which reduced the grassland's regenerative capacity and, inevitably, resulted in grassland degradation. In essence, heavy grazing reduced the net ecosystem exchange (NEE), that is, the net effect of carbon fixation by plants, heterotrophic and autotrophic respiration and soil carbon sequestration, resulting in reduced plant biomass (Zhu et al. 2015). Consequently, nutrition levels for livestock have declined. Zhu et al. (2015) concluded that "reducing stocking rates on heavily grazed grasslands of Northern China to moderate grazing levels would enhance NEE, and benefit biomass and animal production."

8.6 Milk and Milk Products

The milk and milk products from yaks, namely, butter, yoghurt and *qula* (hard, dried curds), are the main food items consumed by the Tibetans and, consequently, provide them with most of their vitamins and nutritional needs. In the past, they did not eat fruits and vegetables for 8 months of the year and, at times, for the whole year, but, despite this, there was no apparent sign of vitamin and mineral deficiencies (Goldstein and Beall 1987; Beall and Goldstein 1993). This is so, at least in part, as yak milk and its derived products "seem to be particularly rich in functional and bioactive components, which may play a role in maintaining the health status of Tibetan nomads. This includes particular profiles of amino acids and fatty acids, and high levels of antioxidant vitamins, specific enzymes, and bacteria with probiotic activity (yoghurt is the main food). Based on that, it is proposed that the Tibetan nomads have developed a nutritional mechanism adapted to cope with the specific challenges posed by the environment of the world's highest plateau" (Guo et al. 2014).

Three milking regimes exist in different localities: once-daily in the mountain areas of the northern plateau, twice-daily on the grasslands of the central plateau and thrice-daily along the valleys of the eastern plateau (Dong et al. 2003). For milking, the female yak is secured tightly with her back legs hobbled and her calf is secured nearby with access to the yak. The calf sucks to start milk letdown, then is removed from the udder and the women milks the animal (Fig. 8.4).

Fresh milk is processed immediately into a variety of indigenous products capable of being stored for long periods. Most milk products are consumed locally, although some are sold outside the ecosystem. Traditionally, fresh yak milk is not consumed but is used for milk tea. Milk tea is a popular drink with Tibetans and, besides fresh milk, also usually contains salt and yak butter. The rest of the milk is used to produce butter, yoghurt and *qula* (dried cheese). To produce butter, some yoghurt is added to the milk, which is then covered with a cloth and allowed to ferment for a day. The milk is then boiled and after cooling to about 40 °C is poured into a wooden churn. By rotating a stick with a paddle in the center of the churn, the fat, after it solidifies, rises to the top. This butter is scooped out by hand and placed in cold water. After the butter is removed from the churned milk, skimmed milk or "milk residue", as it is known locally, remains. This can be drunk or made into *qula*. To make *qula*, the skimmed milk is heated to 50 °C and then sour milk is added to make the milk curdle. This milk is poured into a gauze bag to allow the whey to drip out. The remaining curds are dried and can be stored for long periods. The whey is usually discarded, but can be fed to pigs. Yoghurt is made by boiling skimmed milk and when it has cooled to about 50 °C, sour milk is added and mixed till the temperature declines to about 40 °C and then placed in a bucket. The bucket is kept warm with a blanket for 5 or 6 h in summer and somewhat longer in winter after which time the yoghurt is ready.

Modifications to these processes are possible and are dependent on the amount of yak milk available. For example, fresh milk can be replaced by skimmed milk in milk tea or can be omitted when yak butter is included. Also, yoghurt can be made from fresh milk instead of skimmed milk. Another use of milk and milk products are in the preparation of *tsampa*, a stable high-energy component of the Tibetan diet. It is usually prepared by hand-kneading flour of roasted highland barley or oats or both with butter tea, dried yak cheese, yak butter and sometimes sugar into a dough-like texture. Butter is also used for many purposes other than food, often for religious reasons. Monks fashion sculptures, some as high as two to three stories, out of yak butter mixed with different coloring for religious ceremonies and New Year celebrations. It is used by lamas in sacred lamps and as a fuel in domestic lamps. Women use butter on their skin and hair and, in addition, butter can be used for tanning and for polishing fur coats and as a component of some Tibetan medicines (Wiener et al. 2003). Yak butter is sold in large chunks, often outside monasteries (Fig. 8.6).

We used data from a number of yak breeds in 14 studies to calculate their average milk yield and estimate their milk products production (Table 8.2). Daily milk yield during lactation in these studies ranged between 800 and 3000 g and averaged 1723 ± 810 g, while annual milk production ranged between 148 and 464 kg and averaged 288 ± 115 kg per yak over an average lactation period of 149 ± 49 days. Fat content of the milk ranged between 5.4 and 7.5% and averaged 6.3 ± 0.56%. Yak herders typically use about 10% of the fresh milk, mainly for tea (Cui, personal communication). While in lactation, 172 (1722 x 0.01) g/day of milk would be removed for that purpose, leaving 1551 (1723–172) g/day for butter, *qula* and yoghurt. With a fat content of 6.3% and assuming that butter contains 15% water (Dong et al. 2003; Wiener et al. 2003), then the amount of butter produced would be 115 [(1551 × 0.063)/0.85] g/day. After the butter is removed from the milk, 1436 (1551 − 115) g/day of skimmed milk remains which can be drunk, added to tea or can be used for *qula* and yoghurt. Let us assume that it is used for *qula* and yoghurt.

Fig. 8.6 Yak butter being sold near the Jokhang Temple Monastery in Llasa (Photograph by A. Allan Degen)

Total solids content of yak milk without fat (mainly protein, lactose and ash) is about 11% (Dong et al. 2003) and, therefore, 158 g (1436 g × 0.11) g/day remain in the skimmed milk. The amount of *qula* is typically 7.2% of the skimmed milk; therefore, about 103 (1436 × 0.072) g/day, which would require 936 (103/0.11) g/day of skimmed milk. This would be the case if the *qula* is completely dry and all the whey is removed. However, *qula* can contain up to 40% water (whey), so can weigh up to 172 (103/0.06) g/day, which includes 69 g/day of whey. Therefore 500 (1436–936) g/day is left for yoghurt if the *qula* is completely dry or 431 (1436 − (936 + 69)) g/day if there is 40% whey (Table 8.2).

Milk is vital for Tibetans as it is for many pastoralist societies (Degen 2007; Sadler et al. 2012). Yaks produce particularly nutritious milk (Nikkhah 2011). The antioxidant capacity, vitamin A and vitamin C in yak milk are significantly higher than in commercial cow milk (Cui et al. 2016). Among fatty acids, particular attention has been paid to n-3 PUFA as they are reputed to prevent and treat cancer, coronary artery disease, hypertension, diabetes, and inflammatory and autoimmune disorders (Simopoulos 1991, 1999). The proportion of PUFA in yak milk is significantly higher than in commercial cow milk. Also, the α-linolenic acid of yak milk is higher than in commercial cow milk; conjugated linoleic acids are reputed to possess anti-carcinogenic, anti-atherogenic, anti-inflammatory and anti-lipogenic properties (Wahle and Heys 2002; Nikkhah 2011). The fatty acid γ-linolenic acid (c6c9c12 18:3), reputed to improve eyesight as well as prevent cancer, cardiovascular diseases and hypertension (Wright et al. 1998) was detected in yak milk, but not

Table 8.2 Reported milk yield, lactation period and fat content of different yak breeds from different localities (Dong et al. 2003) and the estimated daily and yearly fresh milk drunk and butter, *qula* and yoghurt produced from these yaks

Localities	Yak breed	Daily Milk yield (g)	Lactation yield (kg) (in days)	Fat (%)	Estimated fresh milk drunk		Estimated butter produced		Estimated qula produced		Estimated yoghurt produced	
					During lactation (g/day)	Whole year (g)	During lactation (g/day)	Whole year (g)	During lactation (g/day)	Whole year (g)	During lactation (g/day)	Whole year (g)
Gansu	Tianzhu White	2300	304 (135)	6.8	230	30,400	166	21,888	149	19,699	549.5	72,628
Gansu	Shandan	2600	464 (180)	5.4	260	46,400	149	26,530	134	23,877	975.0	174,007
Gansu	Gannan	1800	315 (177)	6.3	180	31,500	120	21,012	108	18,911	517.5	90,568
Qinghai	Plateau	1400	214 (153)	5.6	140	21,400	83	12,689	75	11,420	497.8	76,092
Qinghai	Huanhu	3000	487 (153)	6.4	300	48,700	203	33,001	183	29,701	833.4	135,287
Qinghai	Guoluo	1000	162 (153)	6.6	100	16,200	70	11,321	63	10,189	258.4	41,853
Sichuan	Maiwa	1800	365 (150)	6.8	180	36,500	130	26,280	117	23,652	430.0	87,201
Sichuan	Jiulong	2800	414 (150)	5.7	280	41,400	169	24,986	152	22,488	968.4	143,182
Tibet	Plateau	2700	280 (105)	6.3	270	28,000	180	18,678	162	16,810	776.3	80,505
Tibet	Alpine	920		6.4	92		62		56		255.6	
Tibet	Jiali	800	148 (180)	6.8	80	14,800	58	10,656	52	9590	191.1	35,358
Tibet	Pali	1000	200 (180)	5.9	100	20,000	62	12,494	56	11,245	326.4	65,281
Tibet	Sibu	900	180 (180)	7.5	90	18,000	71	14,294	64	12,865	153.8	30,754
Yunnan	Zhongdian	1100	216 (195)	6.2	110	21,600	72	14,180	65	12,762	327.0	64,204
Average ± SD		1723 ± 810	288 ± 115 (149 ± 49)	6.3 ± 0.56	172 ± 81	28,838 ± 11,510	115 ± 51	19,078 ± 7130	103 ± 46	17,170 ± 6417	504 ± 282	84,379 ± 43,021

in commercial cow milk (Ding et al. 2013). The atherogenic index (AI) is significantly lower in yak than in commercial cow milk and yak milk contains significantly higher non-atherogenic fatty acids (NAFA: C4 to C10, and C18:0).

8.7 Dung Production

To estimate dung production by yaks, we used available data on their dry matter intake while free-grazing. Wang et al. (2011) reported that dry matter intake of grazing yaks on the Qinghai-Tibetan Plateau varied greatly among seasons, ranging between 1.9 and 8.5 kg per day or an average of 5.2 kg per day per yak. If we assume that dry matter digestibility of the forage averages 0.55 (Tufarelli et al. 2010) and ash content of the feces averages 0.08, then each yak would produce 2.15 [(5.2 × 0.45) × 0.92] kg dung, in organic dry matter, per day or 786 (2.15 × 365) kg per year.

Worldwide, approximately 2.7 billion people rely on biomass (animal dung, wood, charcoal, and agricultural residues) for cooking and heating. Dung-fuel reduces the number of native shrubs that are harvested and lowers the importation of fossil fuels. Yu (2010) reported that in the alpine region of China, at elevations from 2500 to 4700 m, households burn from 590 to 9200 kg of dung-fuel per person per year. This is roughly equivalent to reducing the burning of coal by approximately 331–5188 kg. If coal, however, were used and imported from outside the alpine region, the carbon sink in the alpine region would increase as dung would be used to a lesser extent.

Most Tibetan families have little choice but to use only yak dung for cooking and heating, as their limited income does not allow them to purchase fossil fuel. Families live in either tents or stone homes and, due to economic reasons, use mainly simple stoves without chimneys (Xiao et al. 2015). Because of the low air temperatures, they heat for an average of 16 h per day (Chen et al. 2011) and, as a result, indoor emissions of black carbon and fine particulate matter rank among the highest in the world. Indoor air pollution has become extreme (Holthaus 2015; Watts 2015), increasing the risk of diseases such as cancer, cardiovascular diseases and respiratory disorders (Pope and Dockery 2006; Holthaus 2015).

It has been calculated that approximately 0.4–1.7 Gg/year of black carbon is emitted by burning yak dung in Tibet (Xiao et al. 2015); black carbon is one of the main causes of both global warming and the melting of snow and ice in the Himalayans (Menon et al. 2002, 2010). Furthermore, it was reported that the mean indoor concentration of fine particulate matter ($PM_{2.5}$; aerodynamic diameter of 2.5 μm or less) in households using a simple stove was 956 μg/m^3, that is, considerably higher than the mean 24-h average of 25 μg/m^3 recommended by the WHO Air Quality Guidelines (Xiao et al. 2015). It was also considerably higher than the air in kitchens in other countries that use biomass as fuel. These countries, such as India and Mexico, are at lower altitudes and have higher air temperatures, and heat less than the homes on the Tibetan plateau (Xiao et al. 2015).

8.8 Conclusions

Yaks provide valuable milk and dung for the inhabitants of the Hindu-Kush Himalayan region and are vital for their livelihoods. On the Qinghai-Tibetan Plateau, a lactating female yak produces an average of 288 kg of milk and 19 kg of butter per year and each yak produces an average of 786 kg of dung organic dry matter per year. The milk is close to a "perfect food" in that it provides most of the vitamins and nutritional needs of the inhabitants in an extremely harsh environment. The dung provides much-needed fuel for heat and for cooking and valuable fertilizer for crop production. Both milk and dung play important roles in the recycling of carbon.

References

Bauer, K. 2015. New homes, new lives—The social and economic effects of resettlement on Tibetan nomads (Yushu prefecture, Qinghai province, PRC). *Nomadic Peoples* 19 (2): 209–220.

Bauer, K., and Y. Nyima. 2011. Laws and regulations impacting the enclosure movement on the Tibetan Plateau of China. *Himalaya* 30 (1): 23–37.

Beall, C.M., and M.C. Goldstein. 1993. Dietary seasonality among Tibetan nomads. *Research and Exploration* 9: 477–479.

Bessho, Y. 2015. Migration for ecological preservation? Tibetan herders' decision making process in the eco-migration policy of Golok Tibetan autonomous prefecture (Qinghai province, PRC). *Nomadic Peoples* 19 (2): 189–208.

Cai, L. 1996. *The yak*. Bangkok: FAO.

Cencetti, E. 2011. Tibetan plateau grassland protection: Tibetan herders' ecological conception versus state policies. *Himalaya* 30 (1): 39–50.

Chen, P.F., C.L. Li, S.C. Kang, et al. 2011. Indoor air pollution in the Nam Co and Ando regions in the Tibetan Plateau. *Huan Jing Kexue* 32: 1231–1236.

Cui, G.X., F. Yuan, A.A. Degen, et al. 2016. Composition of the milk of yaks raised at different altitudes on the Qinghai-Tibetan Plateau. *International Dairy Journal* 59: 29–35.

Degen, A.A. 2007. Sheep and goat milk in pastoral societies. *Small Ruminant Research* 68 (1–2): 7–19.

Degen, A.A., M. Kam, S.B. Pandey, et al. 2007. Transhumant pastoralism in yak herding in the Lower Mustang district of Nepal. *Nomadic Peoples* 11 (2): 57–85.

Devkota, R.C. 2002. *Yak farming in Solukhumbo*. Lailitpur, Nepal: Ministry of Agriculture, Department of Livestock Service.

Ding, L.M., Y.P. Wang, M. Kreuzer, et al. 2013. Seasonal variations in the fatty acid profile of milk from yaks grazing on the Qinghai-Tibetan plateau. *Journal of Dairy Research* 80: 410–417.

Dong, S.K., R.J. Long, and M.Y. Kang. 2003. Milking and milk processing: Traditional technologies in the yak farming system of the Qinghai-Tibetan Plateau, China. *International Journal of Dairy Technology* 56: 86–93.

Fischer, A.M. 2011. The great transformation of Tibet? Rapid labor transitions in times of rapid growth in the Tibet Autonomous Region. *Himalaya* 30 (1): 63–77.

Gao, T., B. Xu, X.C. Yang, et al. 2012. Review of research on biomass carbon stock in grassland ecosystem of Qinghai-Tibetan plateau. *Progress in Geography* 31 (2): 1724–1731.

Goldstein, M.C., C. M. Beall 1987. Anthropological fieldwork in Tibet studying nomadic pastoralists on the Changtang. *Himalaya* 7 (1): Article 4.

Goldstein, M.C., C.M. Beall, and R.P. Cincotta. 1990. Traditional nomadic pastoralism and ecological conservation on Tibet's northern plateau. *National Geographic Research* 6: 139–156.

Guo, X.S., R.J. Long, M. Kreuzer, et al. 2014. Importance of functional ingredients in yak milk-derived food on health of Tibetan nomads living under high-altitude stress: A review. *Critical Reviews in Food Science and Nutrition* 54 (3): 292–302.

Gyal, H. 2015. The politics of standardising and subordinating subjects: The nomadic settlement project in Tibetan areas of Amdo. *Nomadic Peoples* 19 (2): 241–240.

Han, X.T., A.Y. Xie, X.C. Bi, et al. 2002. Effects of high altitude and season on fasting heat production in the yak *Bos grunniens* or *Poephagus grunniens*. *British Journal of Nutrition* 88: 189–197.

Holthaus, E. 2015. *Yak Dung is making climate change worse and there's no easy solution. Future tense: What's to come.* http://www.slate.com/articles/technology/future_tense/2014/12/yak_dung_is_making_climate_change_worse_and_new_cookstoves_don_t_help.html.

IPCC. 2007. *Climate change 2007. The fourth assessment report. The physical science basis. Intergovernmental Panel on Climate Change*. Cambridge, UK: Cambridge University Press.

Iselin, L. 2011. Modern education and changing identity constructions in Amdo. *Himalaya* 30 (1): 91–100.

IUCN. 2002. *International union for the conservation of nature and natural resources*. Gland: The World Conservation Unit.

Joshi, D.D. 2003. Yak in Nepal. In *The yak*, ed. G. Wiener, J. Han, and R. Long, 2nd ed., 316–322. Bangkok: FAO.

Kharel, M., S.P. Neopane, R. Shrestha 2005. *Performance characteristics of the yak in Nepal and its crosses with mountain cattle.* http://agtr.ilri.cigar.org/Casestudy/yak. Yak.htm. Accessed March 26, 2017.

Lal, R. undated. *Soil carbon sequestration: SOLAW background thematic report—TR04B.* Rome: FAO. http://www.fao.org/fileadmin/templates/solaw/files/thematic_reports/TR_04b_web.pdf. Accessed June 1, 2017.

Levine, N.E. 2015. Transforming inequality: Eastern Tibetan pastoralists from 1955 to the present. *Nomadic Peoples* 19 (2): 164–188.

Li, J., and X.A. Yang. 2005. Gender role of females and their social level change of Tibetan population in Gannan. *Lanzhou Xuekan* 6: 110–113.

Lu, H., K. Deng, X. Zi, et al. 2004. 'Yaks' website and its effects. In *Yak Production in Central Highlands, Fourth International Congress on Yak*. Chengdu, China.

Luo, N., J. Gu, Aireti, et al. 1997. Yaks in Xinjiang. In *Conservation and Management of Yak Genetic Diversity, Proceedings of a workshop in Kathmandu, Nepal, 1966*, ed. D.J. Miller, S.R. Craig and G. M. Rana, 115–122. ICOMOD.

Menon, S., J. Hansen, L. Nazarenko, et al. 2002. Climate effects of black carbon aerosols in China and India. *Science* 297: 2250–2253.

Menon, S., D. Koch, G. Beig, et al. 2010. Black carbon aerosols and the third polar ice cap. *Atmospheric Chemistry and Physics* 10: 4559–4571.

Miller, D.J. 2000. Tough time for Tibetan nomads in western China: Snowstorms, settling down, fences and the demise of traditional nomadic pastoralism. *Nomadic Peoples* 4 (1): 83–109.

Miller, D. 2003. *Tibet: Environmental analysis.* Background paper in preparation for USAID's program.

Nikkhah, A. 2011. Science of camel and yak milks: Human nutrition and health perspectives. *Food and Nutrition Sciences* 2: 667–673.

Nyima, T. 2011. Development discourses on the Tibetan Plateau: Urbanization and expropriation of farmland in Dartsedo. *Himalaya* 30 (1): 79–90.

Pope, C.A., and D.W. Dockery. 2006. Health effects of fine particulate air pollution: Lines that connect. *Journal of Air Waste Management* 56: 709–742.

Ptackova, J. 2015. Hor—A sedentarisation success for Tibetan pastoralists in Qinghai? *Nomadic Peoples* 19 (2): 221–240.

Ramanathan, V., and Y. Xu. 2010. *The Copenhagen Accord for limiting global warming: Criteria, constraints and available avenues.* www.pnas.org/cgi/doi/10.1073/pnas.1002293107. Accessed June 1, 2017.

Sadler, K., E. Mitchard, Y.A. Abdi, et al. 2012. *Milk matters: The impact of dry season livestock support on milk supply and child nutrition in Somali Region, Ethiopia.* Addis Ababa: Feinstein International Center, Tufts University and Save the Children.

Shang, Z., A. White, A.A. Degen, et al. 2016. Role of Tibetan women in carbon balance in the alpine grasslands of the Tibetan plateau—A review. *Nomadic Peoples* 20: 106–122.

Sherchand, L., and N.P.S. Karki. 1997. Conservation and management of domestic yak diversity in Nepal. In *Conservation and management of yak genetic diversity, Proceedings of a Workshop in Kathmandu, Nepal, 1996*, ed. D.J. Miller, S.R. Craig and G.M. Rana, 47–56. ICOMOD.

Simopoulos, A.P. 1991. Omega-3 fatty acids in health and disease and in growth and development. *The American Journal of Clinical Nutrition* 54: 438–463.

———. 1999. Essential fatty acids in health and chronic disease. *The American Journal of Clinical Nutrition* 70 (Suppl 3): 560S–569S.

Sulek, E. 2011. Disappearing sheep: The unexpected consequences of the emergence of the caterpillar fungus economy in Golok, Qinghai, China. *Himalaya* 30 (1): 9–22.

Tashi, N., and T. Partap. 2000. Livestock based livelihoods in Tibet, China, and sustainability concerns. In *Contribution of livestock to mountain livelihoods*, ed. P.M. Tulachan, M.A.M. Saleem, J. Maki-Hokkonen, and T. Partap, 171–182. Nepal: ICOMOD.

Tufarelli, V., E. Cazzato, A. Ficco, et al. 2010. Evaluation of chemical composition and *in vitro* digestibility of Appennine pasture plants using yak (*Bos grunniens*) rumen fluid or faecal extract as inoculum source. *Asian-Australian Journal of Animal Science* 23 (12): 1587–1593.

Tulachan, P.M. 2000. Livestock production and management strategies in the mixed farming areas of the Hindu-Kush Himalayas. In *Contribution of livestock to mountain livelihoods*, ed. P.M. Tulachan, M.A.M. Saleem, J. Maki-Hokkonen, and T. Partap, 123–134. Patan, Nepal: ICOMOD.

Vinding, M. 1999. *The thakali: A Himalayan ethnography.* Chicago, IL: Serindia Publications, Inc.

Wahle, K.W., and S.D. Heys. 2002. Cell signal mechanisms, conjugated linoleic acids (CLAs) and anti-tumorigenesis. *Prostaglandins, Leukotrienes, and Essential Fatty Acids* 67: 183–186.

Wang, H.C., R.J. Long, J.B. Liang, et al. 2011. Comparison of nitrogen metabolism in yak (*Bos grunniens*) and indigenous cattle (*Bos taurus*) on the Qinghai-Tibetan Plateau. *Asian-Australian Journal of Animal Science* 24 (6): 766–763.

Wang, Y.X., S.Y. Pan, T. Lu, et al. 2014. Impact of livelihood capital on the livelihood activates of herdsmen on the eastern edge of the Qinghai-Tibet plateau. *Resources Science* 36 (10): 2157–2165.

Watts, A. 2015. *Breaking science news: yak dung burning pollutes indoor air of Tibetan households. WUWT (What's up with that).* http://wattsupwiththat.com/2015/01/16/breaking-science-news-yak-dung-burning-pollutes-indoor-air-of-tibetan-households/.

Wiener, G., J. Han, and R.J. Long, eds. 2003. *The Yak.* 2nd ed. Bangkok: FAO.

Williams, D.M. 1997. The desert discourse of modern China. *Modern China* 23 (3): 328–355.

WMO. 2010. *Greenhouse Gas Bulletin: The state of the greenhouse gases in the atmosphere until December 2009.* Geneva: World Meteorological Organization.

Wright, T., B. McBride, and B. Holub. 1998. Docosahexaenoic acid enriched milk. *World Review of Nutrition and Dietetics* 83: 160–165.

Xiao, Q., E. Saikawa, R.J. Yokelson, et al. 2015. Indoor air pollution from burning yak dung as a fuel in Tibet. *Atmospheric Environment* 102: 406–412.

Yu, X.J. 2010. *The role and mechanism of yak dung on maintenance for Qinghai-Tibet Plateau alpine grassland health.* PhD thesis, Gansu Agricultural University, Gansu, China.

Yu, S.J., Y.M. Huang, and B.X. Chen. 1993. Reproductive patterns of the yak. 1. Reproductive phenomena of the female yak. *British Veterinary Journal* 149: 579–583.

Zhang, R.C. 2000. Effects of environment and management on yak reproduction. In *Recent advances in yak reproduction*, ed. X.X. Zhao and R.C. Zhang. Ithaca, New York: International Veterinary Information Service.

Zhao, X.X. 2000a. Introduction: The yak as a subject of scientific writing. In *Recent advances in yak reproduction*, ed. X.X. Zhao and R.C. Zhang. Ithaca, New York: International Veterinary Information Service.

———. 2000b. An overview of the reproductive performance. In *Recent advances in yak reproduction*, ed. X.X. Zhao and R.C. Zhang. Ithaca, New York: International Veterinary Information Service.

Zheng, R.L., and Z.Y. Huang. 2008. *Free radical biology*. 3rd ed. Lanzhou, China: Higher Education Press.

Zhu, L., D.A. Johnson, W. Wang, et al. 2015. Grazing effects on carbon fluxes in a northern China grassland. *Journal of Arid Environments* 114: 41–48.

Zi, X.D. 2003. Reproduction in female yaks (*Bos grunniens*) and opportunities for improvement. *Theriogenology* 59: 1303–1312.

Chapter 9
The Effect of Ecology, Production and Livelihood on the Alpine Grassland Ecosystem of the Tibetan Plateau

Xingyuan Liu

Abstract The grassland of the Tibetan Plateau is an important base for livestock production and shelter for ecological safety in China. The degradation of alpine grassland in the Tibetan Plateau has not only impacted local livestock production and people's quality of life, but has also damaged the ecosystem security of China and Southeast Asia. By analyzing the function of ecology, production and livelihood (FEPL) of the alpine grassland ecosystem and using the Naqu region of the northern Tibetan Plateau as an example, this study utilizes the annual per-capita income of well-to-do society and the overcome-poverty line for the herdsman to standard livelihood function and establishes an appropriate structure for the FEPL of the alpine grassland ecosystem. Controlling livestock carrying capacity is the key for the development of the FEPL of alpine grassland ecosystem in the Tibetan Plateau.

Keywords Alpine grassland · Ecological service function · Interaction mechanism · Climate change · Titetan Plateau

9.1 Introduction

The Tibetan Plateau is important for animal husbandry and for maintaining grassland culture, Tibetan herdsmen's livelihood, regional economic development and social stability (Hu 2000). The geographical, climatic and natural conditions of the Tibetan Plateau gave birth to the unique grassland ecosystem, which accounts for 50.9% of the total land of the Tibetan Plateau (Yu et al. 2007). The grasslands not only support yak and Tibetan sheep production but also play important roles in

X. Liu (✉)
State Key Laboratory of Grassland Agro-Ecosystems, College of Pastoral Agricultural Science and Technology, Lanzhou University, Lanzhou, Gansu, China
e-mail: liuxingyuan@lzu.edu.cn

carbon sequestration, biodiversity conservation, atmosphere regulation and Tibetan culture preservation (Long et al. 2008). Proper grassland management is closely related to ecological safe shelter zone function for downstream areas and local food security. Due to high altitude, drought and cold, it has a very fragile ecosystem (Wang et al. 2005; Li et al. 2008). About 90% of the total grassland in this area has been degraded to varying extents by misguided human activities (such as overgrazing, reclamation and shrub uprooting) and climate change in the last few decades. Degradation of grassland threatens the subsistence of local herders and the region's economic sustainability.

Global warming and intense human activity, together with poor land management and high population, led to a substantial reduction in the Tibetan Plateau resources (Cheng and Shen 2000; Lin et al. 2010). Excessive population, economic and environmental pressures have led to serious degradation of alpine grasslands and to a decline in ecological services and productivity (Mao et al. 2008; Gao et al. 2010), triggering a series of ecological, economic, environmental and social problems. If the relationship between ecological protection, animal production and herdsman life cannot be coordinated, a further decline of the alpine grassland ecosystem service function is envisioned which will put the Tibetan Plateau region into the "ecological deterioration—economic poverty" vicious circle. Finally, it will lead to the loss of its ecological security barrier function, causing "ecological disaster" (Zhang 2006).

The alpine grassland ecosystem of the Tibetan Plateau is a complex system integrating ecology, economy and social function (Bao 2009). Previous studies dealt mainly with natural properties and the effects on alpine grassland ecosystems from the aspects of soil, plant, animal and microbes, but neglected social attributes of the alpine grassland ecosystem. When population pressure and economic demands are low, the herdsmen's economic activity does not threaten the natural characteristics of the alpine grassland ecosystem, and the system self-regulation can function. However, with continuous growth of the population and economic needs, the impact of herdsmen on alpine grassland ecosystems has been increasing, resulting in the destruction of the ecological and economic harmony between man and nature. Then the social attributes of the alpine grassland have become the decisive factor in its development (Ferraro 2004). Therefore, the herdsmen's activities become the driving force of changes in the ecological and life functions in alpine grassland ecosystem.

9.2 The Connotation of Ecosystem Service Function of Alpine Grassland

The essential importance of the alpine grassland ecosystem is mainly to protect the ecological environment. The grassland ecosystem has ecological, productive and livelihood functions (FEPL), namely ecological protection, animal production and livelihood of herdsmen.

9.2.1 The Ecological Function

The ecological function of the alpine grassland refers to biological properties and ecological processes, which provide natural environmental conditions for the survival of the life system. The system provides a prerequisite for ecological resources and maintains the basis of social and economic development (Daily et al. 2000; Holling 2001). The ecological functions of grasslands include climate regulation, nutrient cycling and storage, fixed CO_2, release of O_2, reduction of SO_2, water conservation, soil formation, erosion control, waste treatment, retention of dust and biodiversity conservation (Costanza et al. 1997). These functions are difficult to control on the grassland ecosystem with the present public funding. Available direct or indirect economic value assessment is required (de Groot et al. 2000). The Tibetan Plateau alpine grassland ecosystem plays an important role in maintaining global CO_2/O_2 balance, absorbing greenhouse gases, regulating downstream water resources, controlling soil erosion and reducing wind, and is an important global biological species gene pool and an important area of biodiversity protection (Lu et al. 2004). It is very important for the alpine grasslands in the Tibetan Plateau to protect the national ecological barrier and the biodiversity (Fig. 9.1).

9.2.2 The Production Function

Alpine grassland production function includes livestock production, grass products and medicinal plants (Fig. 9.2). These can be commercialized and have direct economic values. It plays an important role in supporting the development of unique

Fig. 9.1 The ecological services function (water conservation) of alpine grassland in the Tibetan Plateau (Photography by Xingyuan Liu 2016)

Fig. 9.2 The production function (grazing pasture) of alpine grassland in the Tibetan Plateau (Photography by Xingyuan Liu 2018)

animal husbandry on the plateau (Mao et al. 2008). As the "soil-grass-animal-human" food chain in the alpine grassland ecosystem differs from the "grain-human" food chain in the farmland, the grassland per unit area can raise more livestock and only support a small number of people. To meet the needs of life function, herdsmen paid much attention to the production of grasslands, which led to a weakening of the ecological function (Sala and Paruelo 1997). However, the same area of farmland supports a larger population, but only through agricultural measures to increase the yield per unit area (Kelly et al. 1997). Therefore, the production pressure of grassland is higher than that of farmland because production of food energy from the grassland system is lower than from the farm system. This is the main reason for grassland degradation.

9.2.3 The Livelihood Function

The livelihood function of the alpine grassland ecosystem is a reflection of the social attributes of the grassland and is also an expression of the ecological and production functions (Long et al. 2008). It is an organic carrier of "human-grass-animal-ecology-culture". Alpine grassland ecosystem livelihood function includes mainly economic security, cultural heritage and leisure travel (Straton 2006; Bao 2009). Some of these features can be commercialized functions, with direct economic value, and some cannot be quantified, with indirect economic value. The livelihood life function of grassland ecosystem in the Tibetan Plateau is reflected in two aspects: the first is through the production function of animals and plants and the second is the maintenance of culture, religion and tradition (Fig. 9.3).

Fig. 9.3 The production function (milk products) of alpine grassland in the Tibetan Plateau (Photography by Ruijun Long 2008)

9.3 Interaction Mechanism of "FEPL" in Alpine Grassland Ecosystem

9.3.1 *The Internal Relations of "FEPL"*

Alpine grassland ecological service function is the result of the interaction of physical, chemical, biological and human activities in the ecosystem, which play different service functions for human and ecological environments (de Groot et al. 2000). Ecological function is the foundation, production function is the means and the way, and livelihood function is the ultimate goal (Bao 2009). Production function, as a bridge connecting ecological function and life function, has a leading role in the change of ecological service function in alpine grassland, and the standard of livelihood depends on ecological and production functions, which embody the development and management of the system. When the three factors are in a state of harmony, the grassland system develops steadily and maximizes its potential. When the relationship is in disarray, the grassland ecosystem degrades and the ecosystem service function is weakened (Liu et al. 2011). Alpine grasslands and livestock are the material basis and the main economic source of the herdsmen in the Tibetan Plateau. The quality of life determines the attitude of herders to the production function and ecological function. Production function reflects the level of life function and the strength of ecological function, and the strength of ecological function restricts the size of the production function and livelihood function (Fig. 9.4). Ecological health of the alpine grassland is necessary for economic output to support the population.

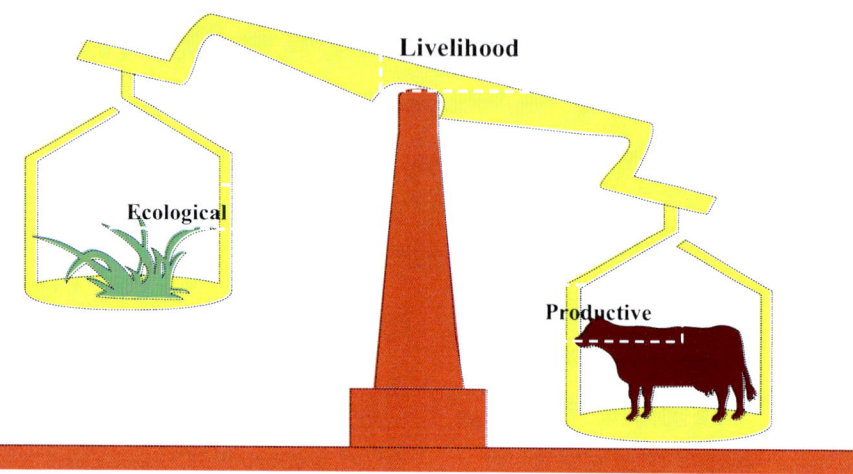

Fig. 9.4 Balance of model the ecological, productive and livelihood functions of alpine grassland ecosystem (Long 2007)

Ecological function protects the ecological security barrier, and its impact reaches beyond its geographical and administrative boundaries, as it has globosity. Production function embodies mainly specific areas of animal husbandry and economic development, it has regionalism. Livelihood function embodies mainly the livelihood of the herdsmen who depend on the grassland for the cultural heritage and social interactions (Liu et al. 2011). FEPL of alpine grassland ecosystem includes the interaction, mutual influence and constraints of complex relationships, expressed as multiple coupling, multi-dimensional chains and multiple feedbacks of the intrinsic relationships (Zhang and Fang 2002). Under the condition of grassland resource constraints, the grassland production function is kept in check, and the "competition" phenomenon of grassland resources is inevitably kept between ecological protection and economic development. With the growth of alpine grassland population and the expansion of the number of livestock, the degradation of alpine grassland is very rapid. Once it exceeds its stable elastic threshold, it affects not only the health of the grassland ecosystem, but also the production of alpine grasslands. Therefore, the production function of the alpine grassland in the Tibetan Plateau can only achieve its ecological value within the elastic threshold of the carrying capacity of the ecosystem.

9.3.2 The Influence of the Structure Change of FEPL on Ecological Service Function

The relationships among the FEPL of the alpine grassland determines the health and sustainable development of the alpine grassland ecosystem. Depending on the proportion of each of the functions, three scenarios can be envisioned.

9 The Effect of Ecology, Production and Livelihood on the Alpine...

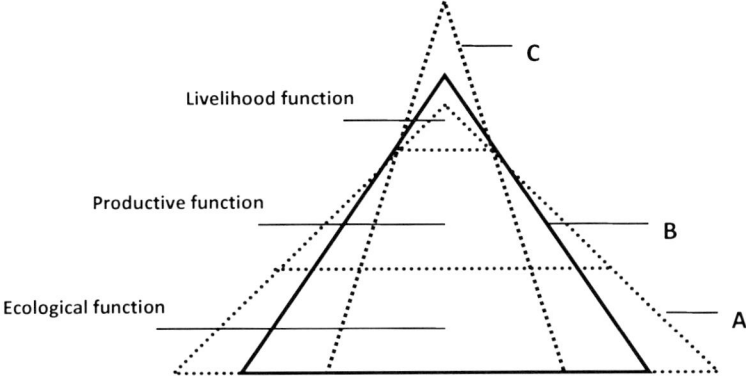

Fig. 9.5 Structure variation of the ecological, productive and livelihood functions of alpine grassland ecosystem (Liu et al. 2012)

1. When the proportion of the ecological function is large and the proportion of livestock production is small, then the support for people's livelihood is reduced (Fig. 9.5A). This state emphasizes that the ecological function of alpine grasslands maintains a modest production function and a higher level of living function. However, this state reduces the external pressure and making grassland eco-system less active and less productive, which not only reduces the economic output efficiency of alpine grassland ecosystem, but also interferes with restoration and renewal of alpine grassland ecosystem.
2. The proportion of ecological function is moderate, according to the level of grassland resources and appropriate animal husbandry production (Fig. 9.5B). This state not only protects the alpine grassland ecological environment, but also meets the requirements of regional economic development and the improvement of the herdsmen's quality of life. It is appropriate for the coordinated development of ecological, production and living functions.
3. The proportion of ecological function is small, but the proportion of livestock production is large, and the of population is growing continuously (Fig. 9.5C). This state emphasizes the grassland production function, increases the carrying pressure of the alpine grassland ecosystem and increases the economic output efficiency of the system, but continuous overuse will lead to the degradation of the alpine grasslands and the ecological service function. When the pressure exceeds the recoverable elastic threshold of the system, it will cause the collapse of the ecosystem, which not only weakens the ecological function, but also reduces the production and life functions.

State A is suitable in areas where the dominant function is the ecological protection, has less stress and has an important role in national and regional ecological security barriers. State B is suitable for load-carrying population size and the appropriate numbers of livestock areas, but needs support of advanced science and technology, management measures, and relevant national policies and regulations. State C is a state in which ecological, production and life functions are not sustainable. As the size of pastoral areas is increasing, and the scale of livestock production increases rapidly, the proportion of State C becomes steeper. If effective measures are not taken, eventually, an "ecological disaster—production collapse—life" will be created.

The change in proportions of the FEPL in the Tibetan Plateau alpine grassland ecosystem plays an important role in improving the national ecological barrier, the regional economic development and the living standards of the herdsmen (Zhong et al. 2006). The state of the three determines the ecological status of the grassland and the number of people who can be supported. At present, the proportions of the production and living functions are large and the proportion of ecological function is small in the Tibetan Plateau alpine grassland ecosystem, so the imbalance leads to alpine grassland ecosystem degradation. Therefore, it is urgent for the state to implement the ecological compensation policy for the grasslands, optimize the utilization structure of the grassland resources, make full use of resource endowments of different grassland types, rationally distribute the production factors, adjust the proportion structure of the FEPL, strengthen the grassland management measures, and build ecological compensation mechanisms to restore degraded and degrading alpine grassland ecosystems. These will promote the coordination and sustainable development of alpine grassland ecology, production and living functions.

9.3.3 The Interaction Mechanism of FEPL

Ecological and production functions are supported by the ecological and economic needs of livelihood functions, and the change in the demand for living function has a negative effect on ecological and production functions. Any change in function will cause other functional changes, which in turn affect the alpine grassland ecosystem (Fig. 9.6). The driving mechanism of these interactions is through the interactions between population, grazing pressure and the carrying capacity of resources and environment, which cause a chain reaction (Zhang and Fang 2002) When the economic needs of the living function increase, then the economic demand forces an increase in the production function and reduction in the ecological function. This reaction force of the living function causes the grazing pressure to exceed the carrying capacity of the grasslands, leading to an imbalance between the production function and the ecological function (Venkatachalam 2007). Therefore, the core of the coordinated development of the FEPL in the Tibetan Plateau alpine grassland ecosystem is based on the change of grassland resources and carrying capacity, which regulates the population and livestock in the system, and the ecological, economic and social benefits. Consequently, ecological, production and living functions in appropriate proportions are required for protection of the alpine grassland ecosystem.

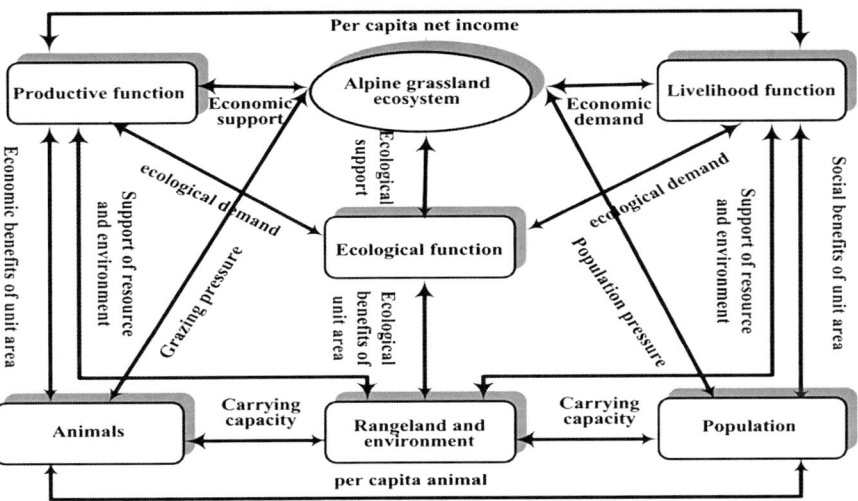

Fig. 9.6 Interaction mechanism for the ecological, productive and livelihood functions of alpine grassland ecosystem (Liu et al. 2011)

9.4 Reasonable Structures and Livestock Carrying Capacity of FEPL in Alpine Grassland Ecosystem

The precondition of alpine grassland ecosystem FEPL requires a reasonable proportion of structure. The life function is the dominant function that forms a reasonable proportion of the structure. But the number of people that can be supported and the livestock carrying capacity of the alpine grassland ecosystem are the basis of the life function. Due to the different types of alpine grasslands, the productivity and ecological functions values differ, so the carrying capacity also differs. Therefore, this study is based on the alpine steppe, alpine desert, alpine desert steppe and alpine meadow in northern Naqu region, and uses the annual per-capita income of well-to-do society and overcome-poverty line for herdsmen proclaimed by the Chinese government in 2008 as the standard of living function.

9.4.1 Determine the Method

9.4.1.1 Proportional Structure Model

The following equation can be used to estimate the minimum health of alpine grasslands for livestock production, and to carry the largest population and its economic output to achieve different living standards.

$$E(x) : P(y) : L(z) = Xmin : Ymid : Zder$$

In the formula, E(x), P(y) and L(z) are ecological function, production function and life function, respectively. Xmin, Ymid and Zder are the ecological function equivalents of ecosystems in maintaining the lowest health level, the number of production functional equivalents under moderately productive conditions and the number of functional equivalents to meet the different economic needs of herders, respectively.

9.4.1.2 Model Conditions

1. The theoretical carrying capacity of alpine grasslands under the non-degraded state is set to moderate carrying capacity, because theoretical carrying capacity is determined according to the lowest level of ecological health and the appropriate carrying capacity of grasslands. (Carrying capacity: sheep unit/ha, one sheep unit is valued at 600 RMB).
2. The annual slaughter rate is 30%.
3. Use the 2008 China's proposed per capita income of 1300 RMB for the poverty line and 8000 RMB for well-off, as standards for measuring life function.
4. Use the "eco-service equivalent" to standardize the ecological, production and life function values. The ecological service equivalent indicates that the ecological, productive and living functions of the alpine grassland ecosystem are defined as an ecological service function equivalent of 1000 RMB per annum.

9.4.1.3 Calculation Method and Procedure

1. According to the theory of carrying capacity calculation method (Ren 1998), evaluation Method of Ecological Service Value (Liu et al. 2011) and model conditions, we can calculate the ecological service value and carrying capacity of the alpine grassland per unit area and convert the carrying capacity into economic value.
2. According to the number of livestock needed for each herder each year, we can achieve the standards of well-off and poverty-free herdsmen, and then determine the amount of livestock under ecological health conditions.
3. Based on the determined number of sheep units, the theoretical stocking capacity is used as the standard to calculate the area of grassland needed to achieve poverty alleviation and well-off herdsmen.
4. According to the grassland area, we can calculate the total ecological service value and economic value, and then convert them into ecological service function equivalent.
5. Determine the number of production and life functions that can be carried by an ecological service function equivalent.
6. Determine the different types of grassland FEPL reasonable structure ratio.
7. For different grassland types per unit area of ecological service value, the total ecological service value of the grassland type is estimated, and the corresponding living function value is estimated by using the proportion of the FEPL of

different types of grassland. Then, divided by the standard, the number of grassland types, and according to the area of the grassland area, the population that can be supported can be determined.

9.4.2 FEPL Reasonable Proportion of the Structure

Using the above method, the theoretical carrying capacity, economic value and ecological service value of a unit grassland area in ungraded states of different types of grasslands in Naqu region were calculated (Table 9.1) (BLMT 1994). It was estimated that the herdsmen in Naqu region need 3 and 44 sheep units of livestock for the standards of overcome poverty line and well-off line, respectively. Under the conditions of poverty alleviation and well-off, the ecological function, production function and living function of alpine steppe, alpine desert, alpine desert steppe and alpine meadow were 1:0.05:0.02, 1:0.09:0.03, 1:0.05:0.02 and 1:0.13:0.04, respectively (Tables 9.2 and 9.3).

Table 9.1 Ecological services value, animal carrying capacity and economic value of the different degraded grasslands in the Naqu region of Tibet (Liu et al. 2012)

	Grassland area ($\times 10^4$ ha)	Utilization rate (%)	Ecological services value (RMB/ha)	Carrying capacity (number/ha)	Economic value (RMB/ha)
Alpine steppe	1890	84	5871	0.49	294
Alpine desert	289	70	1046	0.15	90
Alpine desert steppe	371	82	3491	0.30	180
Alpine meadow	827	97	6234	1.39	834
Total	3378	–	–	–	–

Table 9.2 Proportional relationship for the FEPL based on the condition of overcome poverty in the Naqu region of Tibet (Liu et al. 2012)

	Alpine steppe	Alpine desert	Alpine desert steppe	Alpine meadow
Number of grazing animals (sheep unit)	7.3	7.3	7.3	7.3
Demand grassland (ha)	15.0	49.0	24.3	5.3
Ecological services value (RMB)	86,888	50,921	84,839	33,041
Economic value (RMB)	4380	4380	4380	4380
Ecological function equivalent	87	51	85	33
Productive function equivalent	4.4	4.4	4.4	4.4
Livelihood function equivalent	1.3	1.3	1.3	1.3
$E(x):P(x):L(x)$	1:0.05:0.02	1:0.09:0.03	1:0.05:0.02	1:0.13:0.04

Table 9.3 Proportional relationship for the FEPL based on the condition of well-to-do in the Naqu region of Tibet (Liu et al. 2012)

	Alpine steppe	Alpine desert	Alpine desert steppe	Alpine meadow
Number of grazing animals (sheep unit)	44.4	44.4	44.4	44.4
Demand grassland area (ha)	91.0	296	148.0	32.0
Ecological services value (RMB)	526,611	309,707	516,363	199,491
Economic value (RMB)	26,640	26,640	26,640	26,640
Ecological function equivalent	527	310	516	200
Productive function equivalent	26.6	26.6	26.6	26.6
Livelihood function equivalent	8	8	8	8
$E(x):P(x):L(x)$	1:0.05:0.02	1:0.09:0.03	1:0.05:0.02	1:0.13:0.04

Table 9.4 Farms carrying capacity of alpine grassland in the Naqu region of Tibet (Liu et al. 2012)

	Original vegetation ($\times 10^4$)		Degradation vegetation ($\times 10^4$)	
	Overcome poverty	Well-to-do	Overcome poverty	Well-to-do
Alpine steppe	99.4	16.4	5.4	0.9
Alpine desert	2.4	0.40	1.1	0.2
Alpine desert steppe	12.5	2.06	11.7	1.9
Alpine meadow	151.4	25.1	88.2	14.4
Total	265.8	43.9	106.4	17.5

9.4.3 Livestock Carrying Capacity of Alpine Grassland Ecosystem

In the alpine grasslands, the livestock carrying capacity is a key factor to determine the quality. Therefore, the regulation of livestock carrying capacity is the key to the coordinated development of ecology, production and living functions of the alpine grassland ecosystem in Tibetan Plateau.

The total grassland area of the Naqu region is 33.8 million ha, subtract no man's land area of 2.36 million ha and available grassland area is 29.0 million ha. Based on the appropriate proportion structure and population carrying capacity, the ideal livestock carrying capacity of alpine grassland ecosystem in Naqu region of Tibet was estimated under the condition of Original vegetation status and Degradation vegetation status (Table 9.4). The number of people engaged in animal husbandry was 38.3 million, with an animal production value of 517 million RMB (Statistical Bureau of Tibet 2009). According to the standard of poverty alleviation and well-off, the reasonable livestock carrying capacity of the alpine grassland ecosystem should be 39.8 million and 6.5 million, respectively.

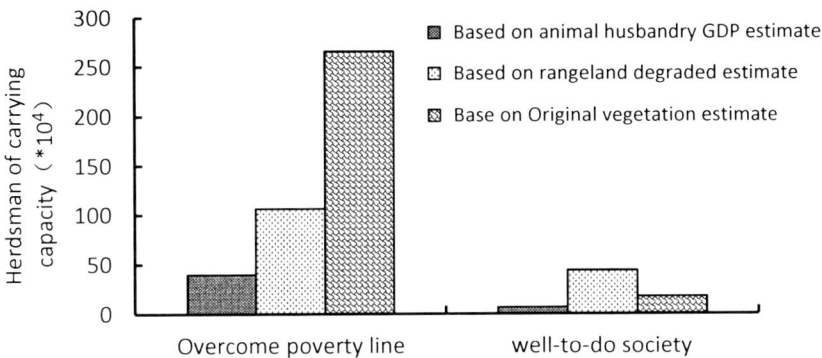

Fig. 9.7 Carrying capacity of pastoral population based on the different economic levels in the Naqu region of Tibet (Liu et al. 2012)

Compared with the estimated livestock carrying capacity of poverty alleviation standards, the actual animal husbandry population accounts for 96.3% of the standard of poverty alleviation, but with the well-off standard, the actual animal husbandry population exceeded the estimate by 489.2%. If the actual population in 2008 corresponded to the appropriate livestock carrying capacity under graded state of the alpine grassland, the actual population accounted for 14.2% and 85.9% of poverty alleviation and well-off standards. However, in the current situation of degradation, the actual livestock population accounted for 36% of the standard of poverty alleviation, while the well-off standard was 118.9% (Fig. 9.7). The results showed that the livestock carrying capacity of degraded alpine grassland was significantly decreased in the Tibet. Compared with the non-degraded status of grasslands, the total livestock carrying capacity of under poverty and well-off standards decreased by 60%.

9.5 The Effect of Climate Change to the FEPL of Alpine Grassland Ecosystem

The alpine grassland ecosystem in the Tibetan Plateau is highly sensitive to climate change. The plateau has experienced significant warming over the past 50 year (Liu et al. 2018), and this warming trend is projected to intensify in the future. The responses of the alpine grassland ecosystem to climate change from ecological function, production function and livelihood function include mainly the aspects of water storage, erosion control, nutrition cycling, waste treatment, resort sand and dust, biological control, release O_2, fixation CO_2, reduce SO_2, recreation, culture and animal production (Fig. 9.8).

Intensive human economic activities influence alpine grassland ecological systems. Fast economic growth leads to increasing environmental stresses in the Tibetan Plateau. In addition, climate change also may have stabilized primary production in

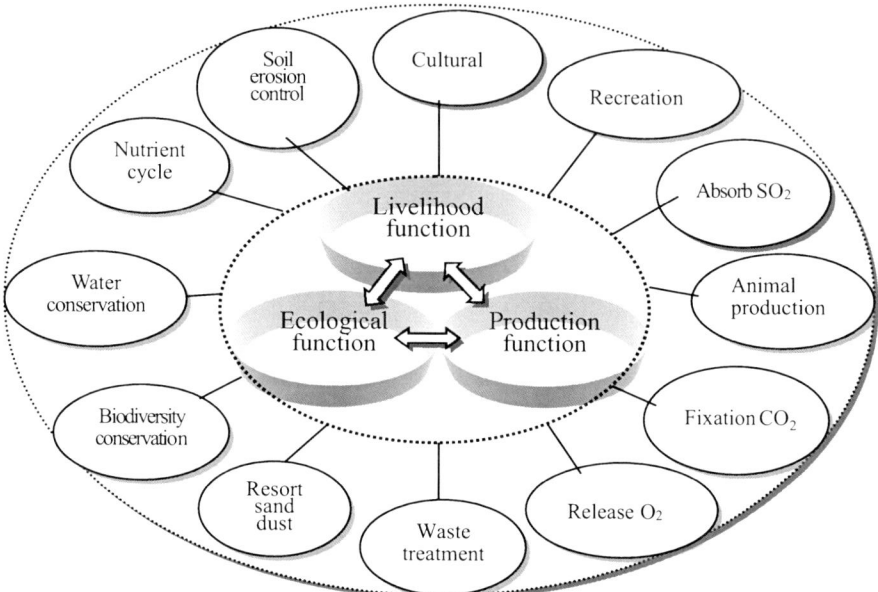

Fig. 9.8 The effect of climate change on the FEPL of alpine grassland ecosystem (Liu et al. 2011)

the high-elevation region, but it also caused a shift from aboveground to belowground productivity. Therefore, the impacts of climate change and human activities on the structure and functioning of ecosystems in Tibetan Plateau are very great.

Grassland ecosystem response to climate change is a great concern. Therefore, it is necessary to strengthen research on the effect of climate change on FEPL of alpine grassland ecosystem. Effective measures should be taken and scientific plans developed to deal with the negative effects of climate change to the alpine grassland ecological system.

9.6 The Theoretical and Practical Significance of Alpine Grassland Ecosystem FEPL

The aim of the sustainable development of the alpine grassland ecosystem in the Tibetan Plateau is to meet the requirements of human livelihood and the environment, to improve the quality of resources and to achieve the coordinated development of ecological, production and living functions. Due to the lack of understanding of the grassland ecosystem, production managers of grassland, under the drive of living function, pay attention only to the production function of grassland. Consequently the carrying capacity of the grassland ecological environment is exceeded, leading to grassland degradation (Foggin 2008). The fundamental reason is that it does not deal with the FEPL of the proportion of structural problems. Under the condition of alpine grassland resources, with the continuous growth of population, there is inevitably contradiction among the FEPL of the alpine grassland ecosystem (Liu et al. 2006).

The alpine grassland ecosystem is different from other ecosystems in terms of their components, ecological processes, and natural environmental conditions and utility. Its ecological function is embodied mainly in the protection of ecological security barrier, which is the basis of system maintenance and development. With the overall situation and deterioration, it will cause alpine grassland ecosystem degradation, water conservation capacity decline, and biodiversity reduction risk. These factors threaten the development of the Tibetan Plateau and its downstream areas not only within its geographical and administrative boundaries but beyond the boundaries (Long 2007). The production function is embodied in the development of animals, which reflects the production attributes of the system. When the number of livestock and grazing pressures of the alpine grassland ecosystem are lower than the environmental carrying capacity, the system is in balance or in a healthy state (Cheng and Shen 2000).

The imbalance between grazing pressure and environmental carrying capacity leads to degradation, and the system is in an unsustainable state of development. The function of life for the herdsmen is dependent on the survival of grasslands and the inheritance of grassland culture, which has the deepening of society and culture. Life function is the ultimate goal of the system and depends on the balance of ecological and production functions, reflecting the comprehensive development of the system and management status. A lack of ecosystem service function will lead to the collapse of the global economic system. Therefore, the FEPL of the alpine grassland ecosystem creates a relationship of global and regional harmony.

The alpine grassland ecosystem FEPL is a nonlinear multi-coupling, multi-dimensional chain and multi-feedback interaction system. Ecological and production functions provide economic needs of life support, while the demand for living function and ecological and production functions have an adverse effect. Any change in function will cause other functional changes, which in turn affect the alpine grassland ecosystem. This occurs mainly through population and grazing pressures and resources and environmental carrying capacities, which is a chain reaction (Cheng and Shen 2000). The increase in population pressure is accompanied by degradation of alpine grasslands, so that the FEPL of the alpine grassland ecosystem are weakened gradually, and if they exceed the stability threshold of the grassland ecosystem, produce negative ecological effects.

Therefore, starting from the FEPL of alpine grassland ecosystem, the management decision of animal husbandry, economic development and ecological protection can be formulated under the multiple standards of ecological, economic and social benefits (Moran et al. 2007). Through the grassland ecological compensation policy of strengthening ecological function, reducing the production function and improving the life function, it is possible to maintain the appropriate level of productivity and herds and promote healthy and sustainable development of the Qinghai-Tibet Plateau grassland ecosystem.

The theory of FEPL of alpine grassland ecosystem provides important theoretical and practical knowledge to understand the ecological barrier status of alpine grassland, the development utility of animal husbandry economy and the living security of herdsmen. It is helpful to guide the producers, managers and policymakers to grasp the changes of the alpine grassland ecosystem and implement the corresponding control measures to understand the relationship among the FEPL (Fisher et al. 2009). The understanding of the interactions among the FEPL are needed to determine the FEPL proportion structure and the suitable livestock carrying capacity for sustainable development, to maintain the ecological security barrier, to plan economic development planning and to reduce herders' poverty.

In practice, according to the FEPL and appropriate proportion structure, the regulation of livestock carrying capacity is the basis for realizing the coordinated development of ecological, production and living functions of the alpine grassland ecosystem. China's grassland pasture areas are distributed mainly in the cold and dry climate, poor environmental conditions in the western and border areas.

Grassland resources are the main source of livelihood for the survival of the herdsmen in these areas. Therefore, the direct victims of grassland degradation are the herdsmen in pastoral areas, which not only affects the ecological safety barrier and herdsmen's standard of living, but also affects national unity and border stability. From the level of national strategy, we should pay greater attention to grassland ecosystem service function, increase the investment of science and technology, improve ecological function, improve production efficiency and plan the coordinated development of the FEPL of the alpine grassland ecosystem.

9.7 Sustainable Development of FEPL in the Tibetan Plateau

How to harmonize the relationship between population, resources, environment and economic development, to rationally develop and utilize grassland resources, and to protect ecological security barriers is related to the overall situation of coordinated and sustainable development of ecology, economy and society in Tibetan plateau. Thus, seeking a rational management pattern of the grassland is an urgent issue for professionals, herders and the Chinese government to achieve sustainable animal production to maintain the health of the alpine grassland ecosystem. To achieve this, the government should take efficient countermeasures as follows:

First, to evaluate the ecological services value and loss value due to grassland degeneration considering the regional differences, spatial heterogeneity, and economic development level of different areas in alpine grassland ecosystem of Tibetan Plateau (Li et al. 2008). According to the results, to develop corresponding protective measures.

Second, in the functional subarea model, to determine ecological compensation based on productivity, seasonal grazing importance, ecological services value and ecological environmental sensitivity of each alpine grassland subtype in the Tibetan Plateau. According to the model, the Tibetan Plateau alpine grassland was classified as the moderate production sector, and needed to restore grassland by reducing livestock number. It is suggested that the ecological compensation mechanism be used based on the Grassland functional subarea.

Third, based on the function subarea and grading of alpine grassland, to present measures of rewards and punishment for different function sectors. To set ecological compensation period and ecological compensation fund for alpine grassland in Tibetan plateau. To safeguard ecological compensation fund for the sustainable development of alpine grassland by establishing institutional commitments, necessary improvements and technology integration in the Tibetan plateau.

Fourth, to combine features of politics, economy and culture in the Tibetan plateau area. The three-dimensional framework of sustainable development to be implemented in the alpine grassland ecosystem of the Tibetan plateau. The laws and regulations of grassland management to be perfected. The concept of ecological civilization to be set up. Dominant function of alpine grassland ecosystem is cleared in ecological barriers and the regional economic development. The structure of alpine grassland resources to be optimized and ecological animal husbandry to be developed. Consequently, to develop a system using function, time, order and space that can promote sustainable development of alpine grasslands resource utilization in the Tibetan plateau.

9.8 Conclusions

Alpine grasslands of the Tibetan Plateau are important for livestock production and for ecological safety in China. The impacts of climate change and human activities on the structure and functioning of ecosystems in Tibetan Plateau are very great. Grassland ecosystem response to climate change is a great concern. Degradation of alpine grasslands in the Tibetan Plateau has not only impacted local livestock production and the people's quality of life but has also seriously damaged the ecosystem security of China and Southeast Asia.

The interaction mechanisms of ecological function, productive function and livelihood function (FEPL) were expounded by driving mechanism among population pressure, grazing pressure and environmental carrying capacity of the northern Tibetan alpine grassland ecosystem. The interaction mechanism of the FEPL indicated a nonlinear internal relation of polybasic coupling, many dimensions linkage and multiple feedback of the northern Tibetan alpine grassland ecosystem. Based on the analysis quantity relative of the FEPL for functional structure effect of alpine grassland ecosystem, quantity relative of the FEPL was determined based equivalent of the ecological services function of alpine grassland. According to this, number of people that can be supported and the livestock carrying capacity of the northern Tibetan alpine grasslands were estimated based on the standard of overcome poverty and well to do of herdsman of established by China government in 2008.

Compared with the state of alpine grassland degradation in previous years, the number of people that can be supported by the northern Tibetan alpine grasslands has decreased by 60%. According to the analysis of gross domestic product of animal husbandry of the Naqu region of northern Tibetan in 2008, the number of people that can be supported at the standard of overcome poverty accounted for 96.3%

of the actual number of people of the Naqu region of northern Tibetan and the number of well to do was at 489.2% of the actual number of people. The percentages were 14.2 and 85.9%, respectively before alpine grassland degradation.

Therefore, controlling population is the key for mutual development of the FEPL of alpine grassland ecosystem in the Tibetan Plateau. It is also necessary to strengthen research on the effect of climate change on FEPL of alpine grassland ecosystem. Effective measures should be taken, and scientific plans developed to deal with the negative effects of climate change to the alpine grassland ecological system.

References

Bao, W. 2009. Development problem and strategy options of Tibetan Plateau rangeland resources. *Research of Agricultural Modernization* 30 (1): 20–23.
BLMT (Bureau of Land Management of Tibet and Bureau of Animal Husbandry of Tibet). 1994. *The Tibet grassland resources*. Beijing: Chinese Science Press.
Cheng, S.K., and L. Shen. 2000. Approach to dynamic relationship between population, resources, environment and development of the Qinghai-Tibet Plateau. *Journal of Natural Resources* 15 (4): 297–304.
Costanza, R., R. d'Arge, R. de Groot, et al. 1997. The value of the world's ecosystem services and natural capital. *Nature* 387 (6630): 253–260.
Daily, G.C., T. Söderqvist, S. Aniyar, et al. 2000. The value of nature and the nature of value. *Science* 289 (5478): 395–396.
de Groot, R.S., J.V.D. Perk, A. Chiesura, et al. 2000. Ecological functions and socio-economic values of critical natural capital as a measure for ecological integrity and environmental health. *Earth and Environmental Sciences* 1: 191–214.
Ferraro, P.J. 2004. Targeting conservation investments in heterogeneous landscapes: A distance-function approach and application to watershed management. *American Journal of Agricultural Economics* 86 (4): 905–918.
Fisher, B., R.K. Turner, and P. Morling. 2009. Defining and classifying ecosystem services for decision making. *Ecological Economics* 68 (3): 643–653.
Foggin, J.M. 2008. Depopulating the Tibetan grasslands: National policies and perspectives for the future of Tibetan herders in Qinghai Province, China. *Mountain Research and Development* 28 (1): 26–31.
Gao, Q.Z., M.J. Duan, Y.F. Wan, et al. 2010. Comprehensive evaluation of eco-environmental sensitivity in Northern Tibet. *Acta Ecologica Sinica* 30 (15): 4129–4136.
Holling, C.S. 2001. Understanding the complexity of economic, ecological, and social systems. *Ecosystems* 4 (5): 390–405.
Hu, Z.Z. 2000. *Grassland development and ecological environment of Qinghai-Tibetan Plateau Tibetan Plateau*. Beijing: China Tibetology Research Press.
Kelly, K., R. Ayling, and G. Elmekki. 1997. Dealing with conflict: Natural resources and dispute resolution. *Commonwealth Forestry Review* 76 (3): 182–185.
Li, D.M., Z.G. Guo, and L.Z. An. 2008. Assessment on vegetation restoration capacity of several grassland ecosystems under destroyed disturbance in permafrost regions of Qinghai-Tibet Plateau. *Chinese Journal of Applied Ecology* 19 (10): 2182–2188.
Lin, L., G.M. Cao, Y.K. Li, et al. 2010. Effects of human activities on organic carbon storage in the *Kobresia hummilis* meadow ecosystem on the Tibetan Plateau. *Acta Ecologica Sinica* 30 (15): 4012–4018.

Liu, X.Y., Q.G. Chen, and Y.N. Wang. 2006. The effect of grassland degeneration for ecological security and economic development in Gannan region of Gansu Province. *Pratacultural Science* 23 (12): 39–41.

Liu, X.Y., R.J. Long, and Z.H. Shang. 2011. Evaluation method of ecological services function and their value for grassland ecosystems. *Acta Prataculturae Sinica* 20 (1): 167–174.

———. 2012. Interactive mechanism of service function of alpine rangeland ecosystems in Qinghai-Tibetan Plateau. *Acta Ecologica Sinica* 32 (24): 7688–7697.

Liu, H.Y., Z.R. Mi, L. Lin, et al. 2018. Shifting plant species composition in response to climate change stabilizes grassland primary production. *Proceedings of the National Academy of Sciences of the United States of America* 115 (16): 4051–4056.

Long, R.J. 2007. Functions of ecosystem in the Tibetan grassland. *Science and Technology Review* 25 (9): 26–28.

Long, R.J., L.M. Ding, Z.H. Shang, et al. 2008. The yak grazing system on the Qinghai-Tibetan plateau and its status. *The Rangeland Journal* 30 (2): 241–246.

Lu, C.X., G.D. Xie, Y. Xiao, et al. 2004. Ecosystem diversity and economic valuation of Qinghai-Tibet Plateau. *Acta Ecologica Sinica* 24 (12): 2749–2756.

Mao, F., Y.H. Zhang, Y.Y. Hou, et al. 2008. Dynamic assessment of grassland degradation in Naqu of northern Tibet. *Chinese Journal of Applied Ecology* 19 (2): 278–284.

Moran, D., A. McVittie, D.J. Allcroft, et al. 2007. Quantifying public preferences for agri-environmental policy in Scotland: A comparison of methods. *Ecological Economics* 63 (1): 42–53.

Ren, J.Z. 1998. *Research methods of grassland science*. Beijing: China Agricultural Press.

Sala, O.E., and J.M. Paruelo. 1997. Ecosystem services in grasslands. In *Nature's services: Societal dependence on natural ecosystems*, ed. G.C. Daily. Washington, DC: Island Press.

Statistical Bureau of Tibet. 2009. *Tibet statistical yearbook in 2009*. Beijing: China Statistics Press.

Straton, A. 2006. A complex systems approach to the value of ecological resources. *Ecological Economics* 56 (3): 402–411.

Venkatachalam, L. 2007. Environmental economics and ecological economics: Where they can converge? *Ecological Economics* 61 (2/3): 550–558.

Wang, Y.B., G.X. Wang, Y.P. Shen, et al. 2005. Degradation of the eco-environmental system in alpine meadow on the Tibetan Plateau. *Journal of Glaciology and Geocryology* 27 (5): 633–640.

Yu, G., C.X. Lu, and G.D. Xie. 2007. Seasonal dynamics of ecosystem services of grassland in Qinghai-Tibetan Plateau. *Chinese Journal of Applied Ecology* 18 (1): 47–51.

Zhang, L.X. 2006. Exploration of ecological pitfalls embedded in the human-land interaction. *Acta Ecologica Sinica* 26 (7): 2167–2173.

Zhang, C.G., and C.L. Fang. 2002. Driving mechanism analysis of ecological-economic-social capacity interactions in oasis systems of arid lands. *Journal of Natural Resources* 17 (2): 181–187.

Zhong, X.H., S.H. Liu, X.D. Wang, et al. 2006. A research on the protection and construction of the state ecological safe shelter zone on the Tibet Plateau. *Journal of Mountain Science* 24 (2): 129–136.

Part III
Impacts of Restoration, REDD and Biochar on Carbon Balance

Chapter 10
Prospects of Biochar for Carbon Sequestration and Livelihood Improvement in the Tibetan Grasslands

Muhammad Khalid Rafiq, Jamila Sharif, Zhanhuan Shang, Yanfu Bai, Fei Li, Ruijun Long, and Ondřej Mašek

Abstract As the key part of HKH, the Qinghai-Tibetan plateau supports the largest population of pastoralists (10 million) in the world. Livestock production on the plateau produces large quantities of dung, but approximately 80% is collected for energy purposes such as cooking and heating needs, which is a link with carbon cycling being a source of carbon to soil and livelihood activity i.e by providing energy and imrpoving grassland productivity. However, inefficient combustion of the dung results in indoor as well as environmental pollution with adverse impact on human health. Heating biomass in oxygen-limited conditions transforms the biomass into bio-oil, syn-gas and a carbon-enriched material known as biochar. Biochar can be used to store carbon in soil and to improve soil quality. This chapter explores the importance of biochar for grasslands restoration and the potential of dung biochar for carbon capture and for increasing grassland productivity. In addition, future biochar research directions to restore grasslands and to improve the livelihood of the pastoralists are discussed.

Keywords Alpine grassland · Land degradation · Biochar · Yak dung · Soil carbon · Grassland productivity · HKH region

---

M. K. Rafiq (✉)
UK Biochar Research Centre, School of GeoSciences, University of Edinburgh, Edinburgh, UK

Rangeland Research Institute, National Agricultural Research Center, Islamabad, Pakistan

J. Sharif
School of Sociology and Philosophy, Lanzhou University, Lanzhou, China

Z. Shang · Y. Bai · F. Li · R. Long
School of Life Sciences, State Key Laboratory of Grassland Agro-Ecosystems, Lanzhou University, Lanzhou, China

O. Mašek
UK Biochar Research Centre, School of GeoSciences, University of Edinburgh, Edinburgh, UK

10.1 Introduction

Grasslands are one of the most widespread terrestrial ecosystems, occupying about 13% of the surface of the earth (Gong et al. 2013) and contain approximately 20% of global organic carbon stock (Scurlock and Hall 1998). Mountain grasslands have been used by grazing livestock for thousands of years (Poschlod and WallisDeVries 2002) and require proper management to be sustainable. They are rich in plant species but are low in productivity due to a short growing season and poor soil nutrients (Hopkins 2009). Grasslands ecosystem services include buffering of extreme weather, preventing floods, purifying water, providing fodder for herbivores, recycling nutrients, conserving biodiversity, improving soil physical and chemical properties and contributing to landscape beautification and recreational benefits (Gibon 2005; Quétier et al. 2010; Lindeman-Matthies et al. 2010; Lavorel et al. 2011; Ocak 2016). Climatic and anthropogenic factors are the basic driving forces changing terrestrial ecosystems and threatening grasslands (Haberl et al. 2007; Vitousek et al. 1997).

The Qinghai Tibetan Plateau (QTP), also known as the third pole and the roof of the world (Fig. 10.1) is important for the livelihood of the local population (Qiu 2008; Zhang et al. 2014) as well as hundreds of million people in China and other countries. However, alpine grasslands are fragile and vulnerable to global change (Chen et al. 2013; Li 2017; Yao et al. 2016) and face a higher degree of global warming than other places on earth (Pepin et al. 2015; Yang et al. 2014). Biochar, produced by thermal decomposition of organic material (Lehmann and Joseph 2009), can improve soil nutrients and sequester carbon in the soil (Lehmann 2006; Sohi et al. 2009; Woolf et al. 2010). Application of biochar to soils have increased crop yields worldwide (Glaser et al. 2002; Jeffery et al. 2011; Kammann et al. 2011; Vaccari et al. 2011; Spokas et al. 2012; Wang et al. 2012) and have reduced nutrient leaching from the soil. This chapter examines the prospects of biochar for grass-

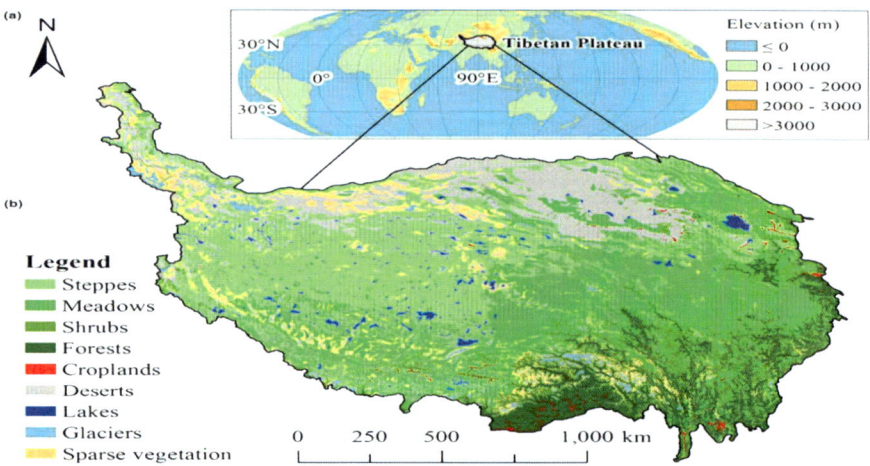

Fig. 10.1 Spatial distribution of topographic features (**a**) and typical vegetation types (**b**) on the Tibetan Plateau (Li et al. 2018)

lands restoration and the potential of yak dung biochar for carbon capture and for increasing grassland productivity in the QTP.

10.1.1 Grassland Ecosystem Services in the Qinghai Tibetan Plateau

The QTP is one of the key ecosystems of the earth, with approximately 85% covered by alpine grasslands. These grasslands offer a number of valuable ecosystem services, such as the maintenance of plant species diversity, storage of carbon, reduction of soil erosion, conservation of water and the maintenance of Tibetan traditions and culture (Lu et al. 2013). The direct and indirect economic and ecological services provided by the QTP could be listed as: (1) playing a major role in the livelihood of the local population as the QTP supports approximately 30 million sheep and goats and 14 million yaks (Lu et al. 2013); (2) sequestering almost 4.0% (ca. 10.7 Pg C) of the soil carbon of the world's grasslands (Ni, 2002) and 0.7–1.0% (up to 920 Tg N) of total global N (Tian et al. 2006); (3) being the watershed basin of major rivers of Asia (Yangtze, Yellow, Indus, Mekong, and Ganges Rivers), the QTP is considered the world's largest river runoff area from a single location. It is a source of water for nearly 40% of the world's population, including China and India (United Nations Environment Program 2007); and (4) being a hub of endangered and endemic species of global importance, including more than 1500 genera of vascular flora with 12,000 species; 700 genera of epiphytes with more than 5000 species; 29 families of mammals with more than 210 species; 57 families of birds with 532 species; and 115 species of fish (Wu and Feng 1992).

10.1.2 Grassland Degradation in the Tibetan Plateau

Despite their globally recognized ecological services, economic significance and cultural values, approximately 90% of the grasslands of northern China have been degraded to some degree during the past 50 years (Nan 2005). Degradation occurs at a yearly rate of approximately 6700 km² (Yang 2002) due to global change (Chase et al. 2000), intensified land uses (Kang et al. 2007), population increase, and socioeconomic improvements. Tibetan pastures are well adapted to the harsh conditions of low mean annual air temperature (below 0 °C, Frauenfeld et al. 2005) and precipitation, (~437 mm, Xu et al. 2008), high solar radiation (Liu et al. 2012), short growing season (~3.5 months, Leonard and Crawford 2002), high wind and water erosion (48 t ha^{-1} year^{-1}, Yan et al. 2000), low soil nutrient levels (e.g. N and phosphorus (P); Li et al. 2014), very shallow soil profiles (~30–50 cm, Chang et al. 2014), and low air O_2 concentration. In addition, warming across the entire plateau is greater and quicker than the overall global mean (Kuang and Jiao 2016). The harsh environment subsequently lowers the water table, increases soil erosion and reduces soil productivity (Chen et al. 2013) and also depletes the shallow soil profile, organic carbon and nutrients of the Tibetan pasture soils. The degradation of grasslands in the Tibetan Plateau can cause huge economic losses. Table 10.1 compares the economic benefits and losses of non-degraded, lightly degraded, moderately degraded, heavily degraded and severely degraded grasslands.

Environmental and socio-economic factors and their interactions have accelerated pasture degradation resulting in losses of soil organic carbon (SOC) and nutrients in Tibetan pastures. Recent economic growth has led to a rapid escalation in infrastructure network development (Liu et al. 2018), which further increased soil deterioration (Fig. 10.2), in addition to the direct adverse effects of increased population and high grazing intensity. There are a number of causes of grassland degradation of the Tibetan Plateau. (1) Overgrazing is a major cause of pasture degradation as livestock numbers are above the carrying capacity in many areas. (2) The imbalance between the spring–winter and summer–autumn grasslands leads to overgrazing of the winter–spring grasslands, especially near the watering points and the gathering areas for animals. (3) Tibetan grasslands have been experiencing increased atmospheric nitrogen deposition that is causing acidification of the grasslands. This occurred especially during the 1980s to the 2000s across northern China. (4) Livestock dung is a source of soil nutrients that benefits Tibetan grasslands. Collection of up to 80% of yak dung from the

Table 10.1 Total economic balance of the alpine grasslands caused by degradation (Wen et al. 2013)

Degradation intensity	Area ($\times 10^7$ ha)	Total economic balance in total area ($\times 10^7$ \$)			
		NPP	Carbon sequestration	Nitrogen sequestration	Biodiversity maintenance
ND	0.19	0	0	0	0
LD	0.49	−54	550	262	−49
MD and HD	0.97	−134	−4785	−5837	−97
SD	0.28	−56	−2249	−3728	−56
Total	1.93	−244	−8485	−6939	−202

Note: ND, LD, MD, HD, SD represent non-degradation, light degradation, moderate degradation, heavy degradation and severe degradation, respectively; NPP represents net primary production

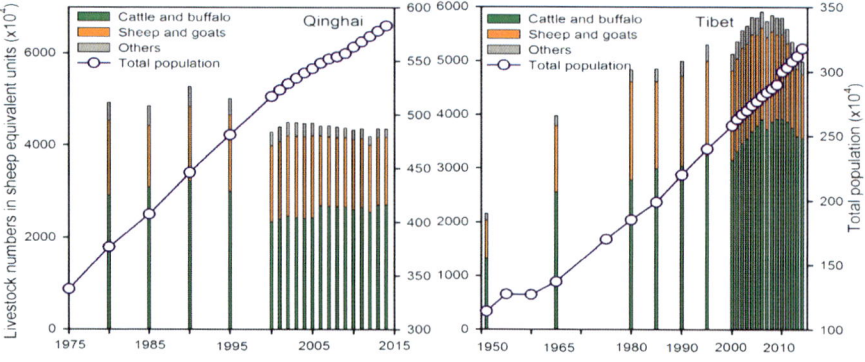

Fig. 10.2 Livestock numbers in Qinghai [(top left) and Tibet (top right) (Liu et al. 2018)]

grasslands affects recycling of nutrients, reduces soil fertility and reduces primary production (He et al. 2009; Shimizu et al. 2009).

10.2 Yaks, Yak Dung and Pastoralist Livelihood in the Tibetan Plateau

Domestic yaks serve as the major livestock through much of the Tibetan Plateau and as the basis of the livelihood for much of the local population. It is a multi-purpose animal that provides leather, wool, milk, meat, dung and transportation (Gerald et al. 2003).

Yak dung is a valuable commodity being used as the main source of energy on the Tibetan plateau and also as fertilizer in pastoral areas of Tibet (Cai 2003). When left on the grasslands, it is one of the major sources of soil nutrients and SOC in the Tibetan plateau. However, the heavy use of dung by the local herders to satisfy their energy needs has turned the plateau from a net carbon sink to a source of GHG emissions. In addition to SOC and nutrient depletion, pasture degradation has a significant negative effect on ecosystem services. Degradation of pastures increases N and P losses directly (Vitousek et al. 2010), which reduces pasture forage productivity and, ultimately, lowers the income of the pastoralists (Wen et al. 2013). In addition, degradation accelerates the leaching of nutrients, resulting in the pollution of surface and subsurface waters (Zhang et al. 2013). Strong winds and dust storms also occur more often as a result of pasture degradation (Wang et al. 2008).

10.3 Biochar: A win-win strategy for the Tibetan Plateau

Pyrolysis of biomass produces several co-products, namely, bio-oil (can be used as liquid fuel), pyrolysis gas, and charcoal ('biochar'). Biochar is a charcoal-like product with a high carbon content produced by heating biomass without or with limited air. The term 'biochar' is used when the charred organic matter is applied to the soil with the aim of storing carbon and improving soil properties (Lehmann and Joseph 2009) (Fig. 10.3). In this way, it not only satisfies the energy requirements of the local communities, but also sequesters carbon (Lehmann 2006; Laird 2008 and Woolf et al. 2010).

About 40% of the total yak dung is deposited in night enclosures (Wang et al. 2002). This dung is collected, dried and stacked in heaps to be used as household energy (Fig. 10.4). In addition, yak dung is also collected from grasslands. Although some studies have reported that alpine meadows on the QTP act as a weak CH_4 sink during the growing season (Wei et al. 2015), dung on grasslands emit CH_4 and N_2O (Lin et al. 2009). However, traditional yak dung collection and drying could provide cost effective opportunities to produce biochar and energy simultaneously and create a win-win situation where enrgy needs of pastoralists are met by use of bio-oil and syn gas, while biochar is used for environmental applications.

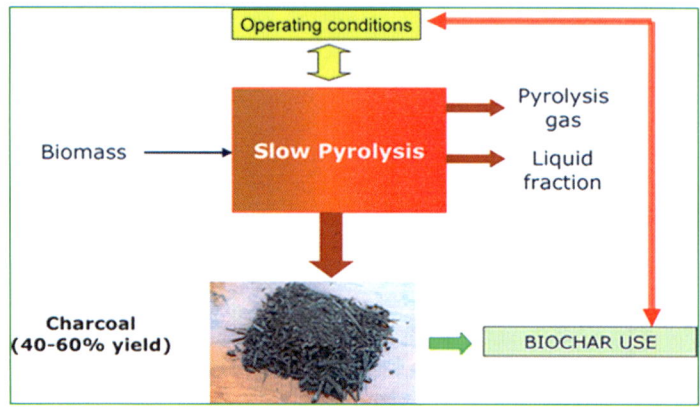

Fig. 10.3 Production of biochar by slow pyrolysis (Manya 2012)

Fig. 10.4 Dung collection and management showing dung heaps and livestock pens (Photograph by Jamila Sharif)

The use of biochar to improve soil quality has two major benefits. Firstly, biochar has the potential to increase soil productivity in terms of plant production, as it increases water -available to plants, soil organic matter (SOM), nutrient cycling, soil bulk density and soil pH, and reduces leaching of pesticides and nutrients to the soil subsurface (Laird 2008; Fig. 10.5). The effect of biochar on soil properties depends on the soil type, as well as on the biochar properties, and, therefore, matching biochar to the soil is very important. Secondly, biochar carbon being more stable than dung, remains in soil for longer periods (Lehmann 2006) and, thus, has the ability to contribute to global climate change mitigation. In addition to the benefits of biochar in soil improvement, agricultural production and carbon sequestration, the pyrolysis process also produces energy-rich co-products in the form of bio-oil

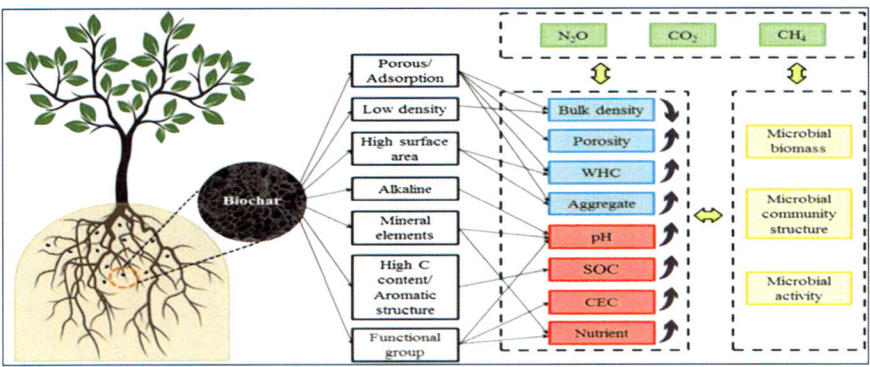

Fig. 10.5 Conceptual scheme of the effects of biochar application on soil properties and greenhouse gases (Li et al. 2018)

and pyrolysis gas (Cong et al. 2018; Crombie and Mašek 2014, 2015), offering a more efficient and environmentally friendly alternative to open fire burning, commonly practiced on the Tibetan plateau.

10.3.1 Yak Dung vs. Yak Dung Biochar for Carbon Capture and Sustainable Livelihood

Yak dung partially decomposes within three months during the growing season, although it can take 3–5 years to decompose fully in grasslands of north China (Jiang and Zhou 2006). The dung emits CH_4 and CO_2, likely being the main source of these two gases in the region. Yak dung is burnt in traditional stoves with a poor overall thermal efficiency ranging from 22 to 33%, which causes indoor as well as atmospheric pollution. The combustion releases many air pollutants, among them, small particulate material which can cause cardiovascular diseases, respiratory disorders, and cancer (Pope and Dockery 2006). Pyrolysis of yak dung could provide biochar and also green energy to meet the cooking needs of the local inhabitants. Table 10.2 presents basic properties of yak dung and biochar from yak dung. Higher pH of the biochar than of yak dung has the potential to reverse the acidification process in the Tibetan plateau and improve grassland productivity. Biochar contains aromatic carbon structures that are highly stable and remains in the soil longer, thus contributing towards carbon sequestration in grasslands. This is the case also for yak dung biochar, as evidenced by the low O/C and H/C ratios (Table 10.2). The porous structure of biochar, higher surface area, and its improved physical and chemical properties could enhance microbial activities, improve soil structure and increase moisture content and, thus, help restore grasslands productivity and improve the livelihood of the pastoralists. Preliminary studies have shown promising results of yak dung biochar applications for forage productivity and soil improvement and for removing fluoride and arsenic from geothermal water (Rafiq et al. 2017; Luo et al. 2018; Zhang et al. 2018).

Table 10.2 Basic properties of yak dung and biochar based on dry matter basis (Rafiq et al. 2017; Wang and Liu 2018)

Items	Yak manure	Yak manure biochar
pH	7.3	10.1
Ash content (%)	27.2	40.9
N (%)	2.71	2.96
C (%)	37.4	40.4
H (%)	5.87	3.55
O (%)	26.8	12.2
Atomic O/C	0.54	0.23
Atomic H/C	1.88	1.06
Ca (%)	4.31	6.48
Fe (%)	0.49	0.73
K (%)	0.54	0.77
Mg (%)	0.95	1.43
S (%)	0.37	0.21

10.4 Recommendations for Future Biochar Research and Application on Grasslands

Applying biochar in mountain grasslands could have a number of benefits, including an increase in organic carbon, in soil pH and in cation exchange capacity (CEC), an improvement in soil structure, and a decrease in soil bulk density and in emissions of greenhouse gases from the soil. The potential benefits of biochar are soil- and crop-specific and their magnitude can differ among systems. Biochar applications are beneficial for polluted soils, particularly with heavy metals and organic compounds (Zhang et al. 2013). However, the effects of biochar application on soil processes require further examination. Here we outlined some specific recommendations to guide future research on the effects of biochar applications to grassland soils.

Biochar helps to improve soil physical, chemical, and biological properties. However, biochar alone is inadequate to meet the nutrient needs for plant growth and productivity. Research on the application of biochar in combination with fertilizers could help to restore grasslands and improve nutrient availability in grasslands. In addition, research on the effects of biochar on microbial functions could help us understand the mechanisms by which biochar affects the biogeochemical processing of soil nutrients in grassland ecosystems.

Biochar application has direct effects on plant growth, but the modification of soil properties by biochar can have indirect benefits. Different grassland types may respond differently to various types of biochar. Thus, it is important to examine the effects of biochar on the growth of different grass species.

The effects of biochar application on soil properties have been studied by short term incubation and laboratory experiments (Nguyen et al. 2017). Biochar carbon is composed of aromatic carbon structures with soil residence times greater than

100 years. Long-term studies on the effect of biochar on soil properties and greenhouse gas emissions are crucial for biochar application in grassland soils to have effective results.

Biochar application has many benefits including the improvement of soil quality and the mitigation of greenhouse gas emissions. In contrast, long-term chemical fertilization has negative effects on soil properties and carbon sequestration. A comprehensive cost-benefit analysis of biochar production and application in Tibetan ecosystems is recommended.

Recently, it has been reported that polycyclic aromatic hydrocarbons (PAHs) and volatile organic compounds (VOCs), which have a negative effect on the soil microbial community and plant productivity, may exist on the surface of biochar particles (Dutta et al. 2017). This can be especially relevant in low-level pyrolysis units with limited processing controls, although proper design and operation of biochar production units can, to a large extent, eliminate these problems (Buss et al. 2016; Buss and Mašek 2016; Ghidotti et al. 2017; Weidemann et al. 2017). Therefore, the screening for toxicity of biochar on the growth of soil microorganisms and grassland ecosystems should be investigated.

Research is required to examine pyrolysis units in terms of biochar and energy production, as well as the energy efficiency of dung burned in the traditional stoves. In addition, studies should be done on the mixing of locally available low cost attapulgite clay, fly ash and other clays with dung to produce biochar that has been proved not only to enhance biochar characteristics but also impove its stability and carbon sequestartion potential (Xu et al. 2017; Buss et al. 2018).

Discussions among the local inhabitants, policy makers, grassland management officials and scientists should be encouraged to generate policies for biochar applications on the Tibetan plateau.

References

Buss, W., and O. Mašek. 2016. High-VOC biochar-effectiveness of post-treatment measures and potential health risks related to handling and storage. *Environmental Science and Pollution Research* 23 (19): 19580–19589.
Buss, W., M.C. Graham, G. MacKinnon, et al. 2016. Strategies for producing biochars with minimum PAH contamination. *Journal of Analytical and Applied Pyrolysis* 119: 24–30.
Buss, W., S. Jansson, and O. Mašek. 2018. Unexplored potential of novel biochar-ash composites for use as organo-mineral fertilizers. *Journal of Cleaner Production* 208: 960–967.
Cai, X. 2003. Characteristics of soil degradation of the "three river" region in Tibet. *Soils Fertility* 3: 4–7.
Chang, X.F., X.X. Zhu, S.P. Wang, et al. 2014. Impacts of management practices on soil organic carbon in degraded alpine meadows on the Tibetan plateau. *Biogeosciences* 11: 3495–3503.
Chase, T.N., R.A. Pielke Sr., J.A. Knaff, et al. 2000. A comparison of regional trends in 1979–1997 depth averaged tropospheric temperatures. *International Journal of Climatology* 20: 503–518.
Chen, H., Q. Zhu, C. Peng, et al. 2013. The impacts of climate change and human activities on biogeochemical cycles on the Qinghai-Tibetan Plateau. *Global Change Biology* 19: 2940–2955.
Cong, H., O. Masek, L. Zhao, et al. 2018. Slow pyrolysis performance and energy balance of corn stover in continuous pyrolysis-based poly-generation system. *Energy & Fuels* 32 (3): 3743–3750.

Crombie, K., and O. Mašek. 2014. Investigating the potential for a self-sustaining slow pyrolysis system under varying operating conditions. *Bioresource Technology* 162: 148–156.

———. 2015. Pyrolysis biochar systems, balance between bioenergy and carbon sequestration. *GCB Bioenergy* 7: 349–361.

Dutta, T., E. Kwon, S.S. Bhattacharya, et al. 2017. Polycyclic aromatic hydrocarbons and volatile organic compounds in biochar and biochar-amended soil: A review. *GCB Bioenergy* 9: 990–1004.

Frauenfeld, O.W., T.J. Zhang, and M.C. Serreze. 2005. Climate change and variability using European Centre for Weather Forecasts reanalysis (ERA-40) temperatures on the Tibetan Plateau. *Journal of Geophysical Research* 110: D02101. https://doi.org/10.1029/2004JD005230.

Gerald, W.N., J.L. Han, and R.J. Long. 2003. *The Yak*. 2nd ed. RAP Publication: Bangkok.

Ghidotti, M., D. Fabbri, O. Masek, et al. 2017. Source and biological response of biochar organic compounds released into water; relationships with bio-oil composition and carbonization degree. *Environmental Science and Technology* 51 (11): 6580–6589.

Gibon, A. 2005. Managing grassland for production, the environment and the landscape. Challenges at the farm and landscape level. *Livestock Production Science* 96: 11–31.

Glaser, B., J. Lehmann, and W. Zech. 2002. Ameliorating physical and chemical properties of highly weathered soils in the tropics with charcoal—a review. *Biology and Fertility of Soils* 35: 219–230.

Gong, P., J. Wang, L. Yu, et al. 2013. Finer resolution observation and monitoring of global land cover: first mapping results with Landsat TM and ETM+ data. *International Journal of Remote Sensing* 34: 2607–2654.

Haberl, H., K.H. Erb, F. Krausmann, et al. 2007. Quantifying and mapping the human appropriation of net primary production in earth's terrestrial ecosystems. *Proceedings of the National Academy of Sciences of the United States of America* 104: 12942–12945.

He, Z.Q., D.M. Endale, H.H. Schomberg, et al. 2009. Total phosphorus, zinc, copper, and manganese concentrations in cecil soil through 10 years of poultry litter application. *Soil Science* 174: 687–695.

Hopkins, A. 2009. Relevance and functionality of semi-natural grassland in Europe? status quo and future prospective. In *Proceedings of the International Workshop of the SALVERE-Project*, 6–11. Raumberg-Gumpenstein: Agricultural Research and Education Centre.

Jeffery, S., F.G.A. Verheijen, M.V.D. Velde, et al. 2011. A quantitative review of the effects of biochar application to soils on crop productivity using meta-analysis. *Agriculture, Ecosystems and Environment* 144: 175–187.

Jiang, S., and D. Zhou. 2006. The impact of cattle dung deposition on grasslands in the Songnen Grassland. *Acta Prataculturae Sinica* 15: 30–35.

Kammann, C.I., S. Linsel, J.W. Gößling, et al. 2011. Influence of biochar on drought tolerance of *Chenopodium quinoa* Willd and on soil–plant relations. *Plant and Soil* 345: 195–210.

Kang, L., X. Han, Z. Zhang, et al. 2007. Grassland ecosystems in China: Review of current knowledge and research advancement. *Philosophical Transactions of the Royal Society B: Biological Sciences* 362: 997–1008.

Kuang, X.X., and J.J. Jiao. 2016. Review on climate change on the Tibetan Plateau during the last half century. *Journal of Geophysical Research: Atmospheres* 121: 3979–4007.

Laird, D.A. 2008. The charcoal vision: A win-win-win scenario for simultaneously producing bioenergy, permanently sequestering carbon, while improving soil and water quality. *Agronomy Journal* 100: 178–181.

Lavorel, S., K. Grigulis, P. Lamarque, et al. 2011. Using plant functional traits to understand the landscape distribution of multiple ecosystem services. *Journal of Ecology* 99: 135–147.

Lehmann, J. 2006. Black is the new green. *Nature* 442: 624–626.

Lehmann, J., and S. Joseph. 2009. Biochar for environmental management: an introduction. In *Biochar for environmental management: Science and technology*, ed. J. Lehmann and S. Joseph, 1–10. Earthscan: London.

Leonard, W.R., and M.H. Crawford. 2002. *The human biology of pastoral populations*, 133. Cambridge, UK: Cambridge University Press.

Li, W. 2017. An overview of ecological research conducted on the Qinghai-Tibetan Plateau. *Journal of Resources and Ecology* 8: 1–4.

Li, J.H., Y.J. Yang, B.W. Li, et al. 2014. Effects of nitrogen and phosphorus fertilization on soil carbon fractions in alpine meadows on the Qinghai-Tibetan Plateau. *PLoS One* 9: e103266. https://doi.org/10.1371/journal.pone.0103266.

Li, L., Y. Zhang, L. Liu, et al. 2018. Spatiotemporal patterns of vegetation greenness change and associated climatic and anthropogenic drivers on the Tibetan Plateau during 2000–2015. *Remote Sensing* 10: 1525.

Lin, X.W., S.P. Wang, X.Z. Ma, et al. 2009. Fluxes of CO_2, CH_4, and N_2O in an alpine meadow affected by yak excreta on the Qinghai-Tibetan plateau during summer grazing periods. *Soil Biology and Biochemistry* 41: 718–725.

Lindeman-Matthies, P., R. Briegel, B. Schüpbach, et al. 2010. Aesthetic preference for a Swiss alpine landscape: The impact of different agricultural land-use with different biodiversity. *Landscape and Urban Planning* 98: 99–109.

Liu, J.D., J.M. Liu, H.W. Linderholm, et al. 2012. Observation and calculation of the solar radiation on the Tibetan Plateau. *Energy Conversion and Management* 57: 23–32.

Liu, S., K. Zamanian, P.M. Schleuss, et al. 2018. Degrdation of Tibetan grasslands: Consquences for carbon and nutrient cycles. *Agriculture, Ecosytems and Environment* 252: 93–104.

Lu, X.Y., J.H. Fan, Y. Yan, et al. 2013. Responses of soil CO_2 fluxes to short-term experimental warming in alpine steppe ecosy, Northern Tibet. *PLoS One* 8: e59054.

Luo, C., J. Tian, P. Zhu, et al. 2018. Simultaneous removal of fluoride and arsenic in geothermal water in Tibet using modified yak dung biochar as an adsorbent. *Royal Society Open Science* 5: 181266. https://doi.org/10.1098/rsos.181266.

Manya, J.J. 2012. Pyrolysis for biochar purposes: A review to establish current knowledge gaps and research needs. *Environmental Science & Technology* 46: 7939–7954.

Nan, Z. 2005. The grassland farming system and sustainable agricultural development in China. *Grassland Science* 51: 15–19.

Nguyen, T.T., C.Y. Xu, I. Tahmasbian, et al. 2017. Effects of biochar on soil available inorganic nitrogen: a review and meta-analysis. *Geoderma* 288: 79–96.

Ni, J. 2002. Carbon storage in grasslands of China. *Journal of Arid Environments* 50: 205–218.

Ocak, S. 2016. Transhumance in Central Anatolia: A resilient interdependence between biological and cultural diversity. *Journal of Agricultural and Environmental Ethics* 29: 439–453.

Pepin, N., R.S. Bradley, H.F. Diaz, et al. 2015. Elevation-dependent warming in mountain regions of the world. *Nature Climate Change* 5: 424–430.

Pope, C.A., and D.W.H. Dockery. 2006. Effects of fine particulate air pollution: Lines that connect. *Journal of the Air & Waste Management Association* 56: 709–742.

Poschlod, P., and M. WallisDeVries. 2002. The historical and socioeconomic perspective of calcareous grasslands-lessons from the distant and recent past. *Biological Conservation* 104: 361–376.

Qiu, J. 2008. The third pole. *Nature* 454: 393–396.

Quétier, F., F. Rivoal, P. Marty, et al. 2010. Social representations of an alpine grassland landscape and socio-political discourses on rural development. *Regional Environmental Change* 10: 119–130.

Rafiq, M.K., S.D. Joseph, F. Li, et al. 2017. Pyrolysis of attapulgite clay blended with yak dung enhances pasture growth and soil health; characterization and initial field trials. *Science of the Total Environment* 607 (16): 184–194.

Scurlock, J.M.O., and D.O. Hall. 1998. The global carbon sink: A grassland perspective. *Global Change Biology* 4: 229–233.

Shimizu, M., S. Marutani, A.R. Desyatkin, et al. 2009. The effect of manure application on carbon dynamics and budgets in a managed grassland of Southern Hokkaido, Japan. *Agriculture, Ecosystem & Environment* 130: 31–40.

Sohi, S., E. Lopez-Capel, E. Krull, et al. 2009. Biochar, climate change and soil: A review to guide future research. In: *CSIRO Land and Water Science Report*. 05/09: 1–56.

Spokas, K.A., K.B. Cantrell, J.M. Novak, et al. 2012. Biochar: A synthesis of its agronomic impact beyond carbon sequestration. *Journal of Environmental Quality* 41: 973–989.

Tian, H.Q., S.Q. Wang, J.Y. Liu, et al. 2006. Patterns of soil nitrogen storage in China. *Global Biogeochemical Cycles* 20: GB1001. https://doi.org/10.1029/2005GB002464.

United Nations Environment Programme (UNEP). 2007. In *Global outlook for ice and snow*, ed. J. Eamer, A. Hugo, and P. Prestrud. Nairobi: United Nations Environment Programme. http://www.unep.org/geo/ice_snow.

Vaccari, F.P., S. Baronti, E. Lugato, et al. 2011. Biochar as a strategy to sequester carbon and increase yield in durum wheat. *European Journal of Agronomy* 34: 231–238.

Vitousek, P.M., H.A. Mooney, J. Lubchenco, et al. 1997. Human domination of earth's ecosystems. *Science* 277: 494–499.

Vitousek, P.M., S. Porder, B.Z. Houlton, et al. 2010. Terrestrial phosphorus limitation: Mechanisms, implications, and nitrogen-phosphorus interactions. *Ecological Applications* 20 (1): 5–15.

Wang, Y., and R. Liu. 2018. H_2O_2 treatment enhanced the heavy metals removal by manure biochar in aqueous solutions. *Science of the Total Environment* 628–629: 1139–1148.

Wang, S., Y. Wang, E. Schnug, et al. 2002. Effects of nitrogen and sulphur fertilization on oats yield, quality and digestibility and nitrogen and sulphur metabolism of sheep in the Inner Mongolia Steppes of China. *Nutrient Cycling in Agroecosystems* 62: 195–202.

Wang, X.D., X.H. Zhong, S.Z. Liu, et al. 2008. Regional assessment of environmental vulnerability in the Tibetan plateau: Development and application of a new method. *Journal of Arid Environment* 72: 1929–1939.

Wang, J., X. Pan, Y. Liu, et al. 2012. Effects of biochar amendment in two soils on greenhouse gas emissions and crop production. *Plant and Soil* 360: 287–298.

Wei, D., X. Ri, and T. Tarchen. 2015. Considerable methane uptake by alpine grasslands despite the cold climate: In situ measurements on the central Tibetan Plateau, 2008–2013. *Global Change Biology* 21: 777–788.

Weidemann, E., W. Buss, M. Edo, et al. 2017. Influence of pyrolysis temperature and production unit on formation of selected PAHs, oxy-PAHs, N-PACs, PCDDs, and PCDFs in biochar—a screening study. *Environmental Science and Pollution Research* 25 (4): 3933–3940.

Wen, L., S. Dong, Y. Li, et al. 2013. Effect of degradation intensity on grassland ecosystem services in the alpine region of Qinghai-Tibetan Plateau, China. *PLoS One* 8: e58432.

Woolf, D., J.E. Amonette, F.A. Street-Perrot, et al. 2010. Sustainable biochar to mitigate global climate change. *Nature Communications* 1: 56. https://doi.org/10.1038/ncomms1053.

Wu, S.G., and J.X. Feng. 1992. Characteristics, exploitation and protection of biological resources in the Tibetan Plateau. In *Proceedings of the First Symposium of the Qinghai-Tibet Plateau Association of China*. Beijing: Science Press.

Xu, X.D., C.G. Lu, and X.H. Shi. 2008. World water tower: An atmospheric perspective. *Geophysical Research Letters* 35: L20815. https://doi.org/10.1029/2008GL035867.

Xu, X.Y., Y.H. Zhao, J. Sima, et al. 2017. Indispensable role of biochar-inherent mineral constituents in its environmental applications: A review. *Bioresource Technology* 241: 887–899.

Yan, P., G.R. Dong, X.B. Zhang, et al. 2000. Preliminary results of the study on wind erosion in the Qinghai-Tibetan Plateau using ^{137}Cs technique. *Chinese Science Bulletin* 45: 1019–1024.

Yang, R.Y. 2002. Studies on current situation of grassland degradation and sustainable development in western China. *Pratacultural Science* 19: 23–27.

Yang, K., H. Wu, J. Qin, et al. 2014. Recent climate changes over the Tibetan Plateau and their impacts on energy and water cycle: A review. *Global and Planetary Change* 112: 79–91.

Yao, T., F. Wu, L. Ding, et al. 2016. Multispherical interactions and their effects on the Tibetan Plateau's earth system: A review of the recent researches. *National Science Review* 2: 468–488.

Zhang, X., H. Wang, L. He, et al. 2013. Using biochar for remediation of soils contaminated with heavy metals and organic pollutants. *Environmental Science and Pollution Research* 20: 8472–8483.

Zhang, Y., B. Li, and D. Zheng. 2014. Datasets of the boundary and area of the Tibetan Plateau. *Acta Geographica Sinica (Supplment)* 69: 164–168.

Zhang, J., B. Huang, L. Chen, et al. 2018. Characteristics of biochar produced from yak manure at different pyrolysis temperatures and its effects on the yield and growth of highland barley. *Chemical Speciation & Bioavailability* 30 (1): 57–67.

Chapter 11
Optimizing the Alpine Grazing System to Improve Carbon Management and Livelihood for Tibetan Herders

Xiaoxia Yang, Quanmin Dong, and Chunping Zhang

Abstract The grazing system is the largest human-ecosystem to link to the indigenous livelihood on the Qinghai-Tibetan Plateau where livestock has been the basis for local and regional economies. However, the grassland on the plateau has been undergoing degradation to various extents mainly due to overgrazing. Grazing exclusion is a widely-used restoration approach and has proven to be effective on stimulating plant growth and soil carbon accumulation, however response of soil carbon accumulation lagged behind. It emerged that the optimum stocking density was 1.77 head ha^{-1} in the warm-season pasture and 0.72 head ha^{-1} in the cold-season pasture. Almost 100% of the cold-season pasture was overgrazed, compared to about 37% in the warm-season. Adjustment of the proportion of seasonal grazing area, accelerating livestock turnover, and supplementary feeding during winters would be effective approaches to reach sustainable balance of husbandry development and ecosystem functions on the Qinghai-Tibetan Plateau.

Keywords Grassland degradation · Grazing exclusion · Optimized grazing management · Carbon and livelihood balance · Qinghai-Tibetan Plateau

11.1 Introduction

The Qinghai-Tibetan Plateau, between the Kunlun Mountain and the Himalayas, consists of several terranes accreted successively to Eurasia (Dewey et al. 1988). It is commonly referred to as "the third pole of the earth" and "the roof of the world", having the highest and largest alpine region in the world and an average altitude of

X. Yang (✉) · Q. Dong · C. Zhang
Qinghai Academy of Animal and Veterinary Science, State Key Laboratory of Plateau Ecology and Agriculture, Qinghai University, Xining, China

State Key Laboratory of Plateau Ecology and Agriculture in the Three River Head Waters Region, Academy of Animal Science and Veterinary Medicine, Qinghai University, Xining, China
e-mail: yxx@qhmky.com

over 4000 m (Huang et al. 2017). The plateau is important for its ecological functions and scientific values due to its unique climate and vegetation and contains the main sources of many major rivers (such as Yangtze river, Yellow river and Lancang river). Xie et al. (2003) evaluated the ecological assets of the Qinghai-Tibetan Plateau noting six categories: forest, grassland, farmland, wetland, water body and desert. The total ecosystem services of the plateau were valued at 936.3 billion *yuan* per year, accounting for 17.7% of the total in China and 0.61% of the world. Grassland is the dominant ecosystem type, occupying over 70% of the total land area of the plateau, and contributes most of the ecological services (48.3%, Xie et al. 2003). A highly various and harsh climate is typical (Wang et al. 2016), with cool and short summers (i.e. growing seasons), and severe, cold and long winters (i.e. non-growing seasons). Snowstorms are frequent and destructive and can cause a large loss of livestock (Gao and Qiu 2011; Wang et al. 2013, 2016).

Although the climate is extremely harsh for livestock grazing, the alpine grassland is one of the most favorable grazing lands for livestock in Asia (Miller 1990). Therefore, livestock grazing plays a crucial role in the livelihoods of the pastoralists on the Qinghai-Tibetan Plateau and neighboring highland countries such as Nepal, India and Bhutan in the HKH region (Rhode et al. 2007; Wang et al. 2016). The people in the region started to raise yaks, sheep and goats in the ancient times. Livestock species provide daily sustenance in the form of milk products, meat and occasionally blood, daily necessities and ornaments in the form of hair, furs, horns and skulls, and dung as fuel (mainly from yaks). The pastoralists also trade livestock products with neighboring farmers, merchants and lamaseries for essentials and luxuries not available locally, such as tea, barley, spices and so on (Rhode et al. 2007). In the spiritual realm, livestock, especially yaks, are central to pastoral and cultural rituals and religious festivals (Long et al. 2008) (Fig. 11.1).

Fig. 11.1 Livestock grazing is still the basis for livelihood of the herdsmen today on the Qinghai-Tibetan Plateau (Left: photography courtesy of Yanlong Wang; Right photography courtesy of Quanmin Dong)

11.2 The Livestock Species on the Qinghai-Tibetan Plateau

Yaks (*Bos grunniens* or *Poephagus grunniens*) and Tibetan sheep (*Ovis aries*), the two most important livestock species, and Tibetan goats (*Capra hircus*) are the main domestic livestock on the plateau. Horses (*Equus caballus*) are raised for working purposes.

Yaks, domesticated by the ancient Qiang people, play an important role in the culture and life of the local residents (Miller 1990; Wiener et al. 2006). Cai (1992) classified the Chinese yak into the Qinghai Tibetan Plateau and the Hengduan Alpine types, which are based on geographic and topographic parameters, as well as morphological and physiological characteristics. There are 12 officially recognized breeds of domestic yak in China according to the Animal Genetic Resources in China, Bovines (2011). The yak is multiple purpose, providing milk, meat, hair and wool, leather, working as a draught animal (packing, riding, and ploughing) and dung–as an important fuel and building material. Yak milk and meat are both of high quality but of low production compared to cattle; the milk is rich in protein, fat, lactose, mineral elements and essential amino acids, and the meat contains high contents of protein and essential amino acids, low content of fat and cholesterol (Long et al. 2008). There are 14–15 million domestic yaks worldwide, of which more than 90% are in China, and the rest are in the bordering highland countries including Nepal, India, Bhutan, Mongolia and Russia (Wiener et al. 2006; Rhode et al. 2007).

Tibetan sheep, with a population of over 50 million, also provide meat, milk and economic benefits for local residents on the plateau (Xin et al. 2011). However, Tibetan sheep have always been "trapped" in a vicious cycle which is described by the locals as "strong in summer, fat in autumn, emaciated in winter, weak in spring" (Dong et al. 2003) due to the very short growing season of forage, and harsh climate (Fig. 11.2).

Fig. 11.2 Yak and Tibetan sheep are the main domestic livestock on the Qinghai-Tibetan plateau (Photography courtesy of Yanlong Wang)

11.3 Grazing Systems on the Qinghai-Tibetan Plateau

It is estimated that the pastoralists have been raising livestock for at least 4000 years on the Qinghai-Tibetan Plateau. The traditional transhumant herding (Huang et al. 2017), the main grazing management in the area, uses several systems depending on weather, forage availability and productivity, topography and cultures (Long et al. 2008). A common system is dividing the grassland into summer–autumn pasture from June to October, and winter-spring pasture from November to the next May (Huang et al. 2017). Besides the two-season system, there are also three types of three-season system: (1) summer pasture, winter pasture and autumn–spring pasture; (2) summer pasture, autumn pasture and winter–spring pasture; and (3) summer–autumn pasture, winter pasture and spring pasture (Long et al. 2008) (Fig. 11.3).

11.4 Historical Vicissitude of Grassland Property Rights and Ownership on the Qinghai-Tibetan Plateau

The institutional framework of livestock husbandry on the Qinghai-Tibetan Plateau has been undergoing fundamental changes since the middle of the last century (Huang et al. 2017). Before the founding of the People's Republic of China, the local feudal landlords, including monasteries, aristocracy, and government officials owned the grassland. The boundaries of grassland among "landlords" were usually natural features, such as mountain ridges and streams. There was no common pasture and the landlords were responsible for their own gains and losses (Ma 2007; Li 2012).

The first huge change in grassland property rights occurred in the 1950s when land reform policy was implemented throughout the China. The grassland was nationalized by transferring ownership form the feudal lords to the collective or the states. However, this grassland property rights and ownership transformation was

Fig. 11.3 The herdsmen usually reside by the water and grass, along with moving their livestock (Photography courtesy of Quanmin Dong)

not completely realized until the late 1950s and 1960s when the people's communes were established. Within these political and economic reforms, all livestock and grassland were shared by all the households. The new system undermined the herdsmen's work enthusiasm and reduced the grassland productivity in the long term, although it had positive impacts on the national stability and economics in the early stage, as did the land reform policies in other parts of China (Guo and Ma 2005; Li and Huntsinger 2011).

In 1984, the Household Responsibility System (HRS), reasserted the household as the basic unit of production and decision-making. It was introduced from the Chinese agriculture sector to the grasslands, and it was covered in the China's Grassland Law which was promulgated in 1985 (Ma 2007). The livestock was divided equally among households based on household size, but the state still maintained the ownership of the grassland. Though this new policy stimulated the herdsmen's work incentives and increased productivity in husbandry, it brought serious grassland degradation (Foggin 2000; Ma 2007), which could be identified as a "tragedy of the commons" (Hardin 1968).

To prevent grassland degradation and improve its functions and productivity, the second round of the grassland HRS was implemented from 1994. The grassland was inventoried and classified according to the forage quality, and then divided and contracted to households based on livestock size. The contractual duration of grassland use rights of the households can be as long as 30 years, and in some special circumstances even up to 50 years. Furthermore, two new options of grassland management emerged within the grassland HRS, which were grounded in voluntarily actions and in which the members were usually relatives or neighbors. In the first option, the "household cooperative groups", the members use their summer–autumn pastures together and fence them as a whole. In the second option the members transfer or lease grassland use rights. The grassland use rights are rented by oral or written contract and the negotiated price is based on grassland quality (Huang et al. 2017).

The grassland property rights and livestock ownership have evolved along with the political regimes change, the technical advances and the development requirements of the state and the local herdsmen. It reflects the continuous exploration of achieving the balance between husbandry development and stability of ecosystems by the local governments and herdsmen on the Qinghai-Tibetan Plateau.

11.5 The Current Situation of Livestock Grazing on the Qinghai-Tibetan Plateau

As a global concern, "grassland degradation" causes serious consequences to both natural ecosystems and human society (Harris 2010). The grassland on the Qinghai-Tibetan Plateau has been undergoing degradation to various degrees (Akiyama and Kawamura 2007; Dong et al. 2013; Zhang 2008; Liu et al. 2018), with climatic changes and overgrazing considered the two main causes (Huang et al. 2017). Overgrazing causes decreased species richness and diversity, and reduced biomass of plant species preferred by livestock (Zhu et al. 2006), and adversely affects soil physics, soil water content (Dong et al. 2006), soil chemistry (Dong et al. 2012b) and ecosystem function at various spatial-temporal scales. Some researchers also theorized that the breakdown of traditional nomadism may be a reason (Foggin and Torrance-Foggin 2011; Huang et al. 2017). For centuries, the local herdsmen practiced transhumant pastoralism on the Qinghai-Tibetan Plateau (Miller 1990), and usually resided by the water and grass (*zhushuicaoerju*), along with moving their livestock. However, they now live in settlements, or semi-settlements, since the 1990s (Long et al. 2008; Cao et al. 2018).

11.6 The Influence of Grazing Exclusion on Plant Community and the Carbon Cycle

To earn higher incomes, the local herdsmen increased livestock numbers, although this can bring economic benefits in the short term, it does not meet the requirements for sustainable grassland-based animal production in the long term (Ren et al. 2008). To restore the degraded grasslands and to protect grassland resources, the Chinese government started the national ecological program of "Returning grazing land to grassland" in temperate and alpine grasslands throughout northern China in 2003 (Xiong et al. 2014).

Grazing exclusion is one of the simple and effective restoration approaches that has been applied widely on the Qinghai-Tibetan Plateau, and has proven to be effective for grassland recovery (Ma et al. 2016; Zhao et al. 2016a; Liu et al. 2018). The positive impact of grazing exclusion on the plant community total coverage, plant height and biomass (both aboveground and belowground) was reported in different locations of the plateau (Wu et al. 2010; Xiong et al. 2014; Zhao et al. 2016b). However, the influence of grazing exclusion on the soil carbon pools and the carbon fluxes differed among studies (Wu et al. 2010; Lu et al. 2015a, b; Zhao et al. 2016a, b).

Ecosystem respiration, which is controlled by plant biomass, soil microbial biomass and abiotic factors such as temperature and moisture (Wan and Luo 2003; Saito et al. 2009; Lin et al. 2011), is a major CO_2 efflux from terrestrial ecosystems to the atmosphere (Davidson et al. 2006). Grazing exclusion's influence on ecosystem respiration is affected by vegetation type, grazing history, climatic factors and other environmental factors, since livestock grazing has comprehensive effects on these biotic and abiotic variables of the grassland. Zhao et al. (2016b) observed that 7-years grazing exclusion promoted ecosystem respiration in the alpine meadow and alpine steppe but depressed it in the swamp meadow on the central Tibetan Plateau. They also reported that even within the same vegetation, the stimulation of grazing exclusion on ecosystem respiration decreased along increasing altitudes (Zhao et al. 2016a), and that changes in plant biomass rather than soil organic carbon content made the major contribution. As the largest CO_2 efflux from the soil to the atmosphere, changes of soil respiration under disturbances could have a substantial contribution to atmospheric CO_2 concentration.

Though grazing exclusion depressed soil respiration by decreasing soil carbon release, which was due to low soil temperate, it also increased the temperature sensitivity of soil respiration. Therefore, grazing exclusion may have the potential to accelerate soil carbon release under future climate warming despite the reduction in soil CO_2 efflux at the early-stage (Ma et al. 2016). Lu et al. (2015a) investigated the carbon storage in plant and soil of three alpine grassland types after 6–8 years of grazing exclusion in Tibet. The impact of grazing exclusion on carbon storage depended on vegetation type and differed between plant and soil. Grazing exclusion increased the aboveground biomass carbon storage in the alpine steppe and alpine desert steppe, had no effect in alpine meadow, and it decreased surface soil carbon storage (0–10 cm) in the alpine meadow and alpine desert steppe.

An analysis of the effect of grazing exclusion on grasslands across northern China (Hu et al. 2016) concluded that carbon dynamics in both plants and soil depended greatly on the duration of exclusion, and resulted in an increase in 1–3 years, moderate increase in 4–15 years, but marginal increase when longer than 15 years. In addition, the accumulation of soil carbon lagged behind the enhancement of plant biomass.

11.7 Optimizing Livestock Grazing Intensity on the Qinghai-Tibetan Plateau

Jones and Sandland (1974) generated a simple linear regression model for the relationship between animal liveweight gain (Y_a) and grazing intensity (i.e. stocking rate) (x).

$$Y_a = a - bx \qquad \text{(Eq. 10.1)}$$

where a, the intercept, indicates the forage quality of a pasture or the maximum potential production per animal on a pasture, and b, the slope, indicates the spatial stability and the resilience of a pasture or the change of liveweight along with grazing intensity; and a quadratic model to describe the relation between liveweight gain per hectare (Y_h) and grazing intensity (x):

$$Y_h = ax - bx^2, \qquad \text{(Eq. 10.2)}$$

from which an optimum grazing intensity (x = $a/2b$) can be calculated (Fig. 11.4).

Jones and Sandland (1974) tested the models through 33 grazing intensity experiments across different vegetation types from tropical to temperate, and confirmed that the negative linear relationship between liveweight gain per animal and grazing

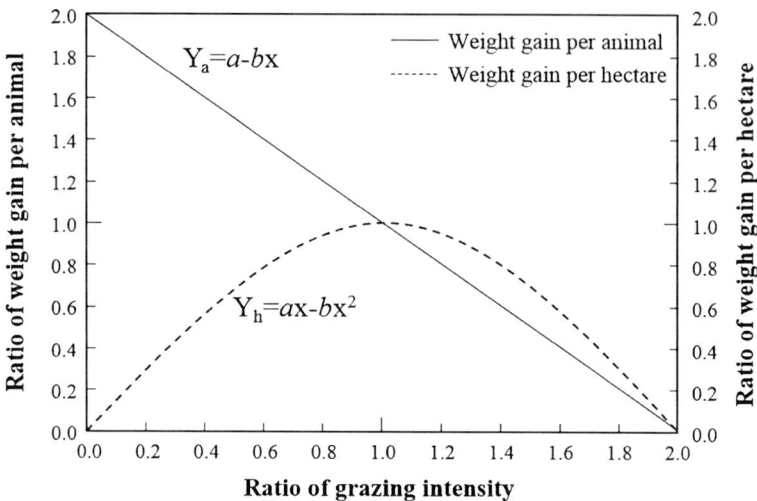

Fig. 11.4 The relation of grazing intensity and livestock weight gain per animal ($Y_a = a - bx$) and per hectare ($Y_h = ax - bx^2$). Ratio of grazing intensity means the ratio of grazing intensity to the optimum grazing intensity; Ratio of weight gain per animal means the ratio of weight gain per animal to weight gain per animal at the optimum grazing intensity; and Ratio of weight gain per hectare means the ratio of weight gain ha^{-1} to weight gain ha^{-1} at the optimum grazing intensity (Redrawn based on Jones and Sandland 1974)

intensity held over a wide range of grazing intensities, and the maximum gain per hectare (Y_{hmax}) occurred when grazing intensity (x) was $a/2b$. Several subsequent studies confirmed the relations between livestock gain and grazing intensity through grazing experiments across a wide range of vegetation types (Wilson 1986; Zhou et al. 1995; Dong et al. 2015).

As local herdsmen depend on livestock production on the Qinghai-Tibetan Plateau, an optimum livestock grazing management is urgently required to balance the livestock production and the sustainability of grassland resources.

To explore the optimum grazing intensity in the Three-River-Source region of the Qinghai-Tibetan Plateau, a yak grazing experiment was carried out by the authors (Dong et al. 2015) from 1998 to 2000 in Wosai Township of Dari County, Tibetan Autonomous Prefecture of Golog in Qinghai Province. The study site is a typical alpine meadow, with an average altitude of 4000 m, and is characterized by a continental monsoon-type climate, with long and cold winter (referred to as cold-season) and short and cool summer (referred to as warm-season). The pastures were grazed traditionally on the two-season system (see text 3.)

The grazing experiment included three grazing intensities, and the grazing intensity used by the local herdsman near the experiment site (referred as native grazing intensity based on Dong et al. 2015), to establish the models (Table 11.1). Both the intercept a_1 (71.86) and slope b_1 (20.33) in the warm-season pasture were higher

Table 11.1 Grazing intensities of the yak experiment in an alpine meadow in the Three-River-Source region of Qinghai-Tibetan Plateau

Treatment	Grazing intensity (head yak ha^{-1})	
	Warm-season pasture	Cold-season pasture
Control (no grazing)	0	0
Light grazing	0.89	0.77
Moderate grazing	1.45	1.29
Heavy grazing	2.08	1.81
Native grazing	2.50	2.30

Redrawn based on Dong et al. (2015)

than a_2 (24.53) and b_2 (16.93) of the cold-season pasture (Table 11.2). Based on the theory of Jones (1981), a represents the nutrient level of the pasture, and b represents the spatial stability and resilience of the pasture. The warm-season pasture had a much higher livestock carrying capacity than the cold-season pasture, whilst the two pastures were similar in grassland recovery. The lower value of a in the cold-season pasture could be explained by the grazing activity occurring in the non-growing season when the forage was withering. The optimal grazing intensity (i.e. the maximum yak liveweight gain grazing intensity) was 1.77 head ha^{-1} in the warm-season, 0.72 head ha^{-1} in the cold-season pasture, and 0.64 head ha^{-1} for the annual pasture (Table 11.3). Accordingly, the maximum carrying capacity for the three pastures were 3.54 head ha^{-1}, 1.44 head ha^{-1} and 1.28 head ha^{-1} respectively. The ratio of warm-season to cold-season in the optimum grazing intensity was 2.46:1.

We also examined the impact of grazing intensity on plant compensatory growth by measuring the aboveground productivity and plant growth ratio (Dong et al. 2012a). Yak grazing at each intensity stimulated plant growth, and the compensatory growth was higher under light and moderate grazing, in the warm-season pasture. However, plant compensatory growth was not observed in the cold-season pasture. It could be partly explained by the lack of direct stimulation of yak foraging, such as the effect of yak saliva as a growth stimulant (McNaughton 1985), since plant growth and yak grazing did not occur at the same time in the cold-season pasture. Even though the plant compensatory growth could not be examined under the optimum grazing intensity, since this intensity was not included in the experiment, the positive influence of grazing on plant growth could be expected under the optimum grazing intensity (1.77 head ha^{-1}), which was lower than the heavy grazing intensity (2.08 head ha^{-1}) in the warm-season pasture.

The actual grazing intensities on the alpine meadow of 18 counties and six townships across the Three-River-Source region over the same period were examined. Widespread overgrazing existed, but the overgrazing differed between seasons; 100% of the cold-season pastures was overgrazed, but only 37.5% of the warm-season pastures was overgrazed (Fig. 11.5). In addition, 96% of the cold-season pastures were over the pastures' maximum carrying capacity (Fig. 11.5). Therefore, grazing pressure should be reduced during the cold-season pastures by accelerating livestock turnover and supplementary feeding, as well as by adjusting the proportion of seasonal pasture areas.

Table 11.2 Regression models between yak weight gain and grazing intensity of the grazing experiment on an alpine meadow in the Three-River-Source region of Qinghai-Tibetan Plateau

	Yak gain (kg head^{-1})			Yak gain (kg ha^{-1})		
	Regression equation	R^2	P	Quadratic equation	R^2	P
Warm-season pasture	$Y_a = 71.86 - 20.33x$	0.950	<0.01	$Y_h = 71.86x - 20.33x^2$	0.795	<0.05
Cold-season pasture	$Y_a = 24.53 - 16.93x$	0.994	<0.01	$Y_h = 24.53x - 16.93x^2$	0.994	<0.01
The whole year	$Y_a = 96.40 - 75.13x$	0.984	<0.01	$Y_h = 96.40x - 75.13x^2$	0.969	<0.01

Y_a yak weight gain per head, Y_h yak weight gain per hectare, x grazing intensity
Redrawn based on Dong et al. (2015)

Table 11.3 Optimum grazing intensity and pasture maximum carrying capacity for warm-season pasture, cold-season pasture and the whole year during the grazing experiment on an alpine meadow in the Three-River-Source region of Qinghai-Tibetan Plateau

	Optimum grazing intensity (head ha^{-1})	Maximum carrying capacity (head ha^{-1})
Warm-season pasture	1.77	3.54
Cold-season pasture	0.72	1.44
The whole year	0.64	1.28

Redrawn based on Dong et al. (2015)

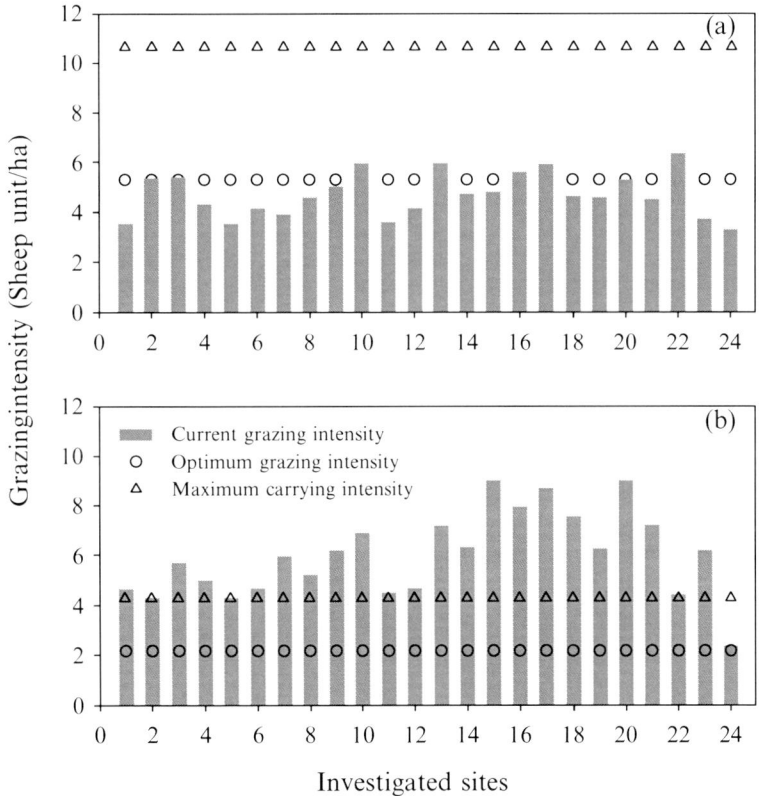

Fig. 11.5 Grazing intensities in alpine meadows of 18 counties and six townships across Three-River-Source region from 1998 to 2000. (**a**) warm-season pasture, (**b**) cold-season pasture (Redrawn based on Dong et al. 2015)

11.8 Conclusion

The alpine grassland on the Qinghai-Tibetan Plateau is one of the most favorable grazing lands in Asia, in spite of the extremely harsh environment. Livestock husbandry has been undergoing fundamental changes since the middle of the twentieth century. The grassland property rights and livestock ownership have evolved with political regime changes, technical advances and development requirements of the state and local herdsmen. Changes have attempted to balance the win-win situation of husbandry development and ecological protection. However, the grasslands have been undergoing degradation to various extents mainly due to overgrazing. On the one hand, to restore the degraded grassland and protect grassland resources, the Chinese government implemented the national ecological program of "Returning grazing land to grassland".

Grazing exclusion with fencing, as the main approach, was used widely and has proven to be very effective for grassland recovery. Grazing exclusion stimulated plant growth at the early stage, however its effect on soil properties (such as soil carbon accumulation) lagged behind. On the other hand, an optimum grazing management option which could achieve the balance between livestock production and grassland sustainability would be beneficial to both economic development and ecological protection in the long run. Based on a yak grazing experiment and on the actual grazing intensity on the alpine meadow in the Three-River-Source region, almost 100% of the cold-season pasture was overgrazed, compared to about 37% in the warm-season. Adjustment of the proportion of seasonal grazing area, accelerating livestock turnover, as well as supplementary feeding during winters could be effective approaches to reach the balance of husbandry development and stability of the ecosystems on the Qinghai-Tibetan Plateau.

References

Akiyama, T., and K. Kawamura. 2007. Grassland degradation in China: Methods of monitoring, management and restoration. *Grassland Science* 53 (1): 1–17.

Cai, L. 1992. *Chinese Yak*. Beijing: Chinese Agriculture Press.

Cao, J.J., M.T. Li, R.C. Deo, et al. 2018. Comparison of social-ecological resilience between two grassland management patterns driven by grassland land contract policy in the Maqu, Qinghai-Tibetan Plateau. *Land Use Policy* 74: 88–96.

China National Commission of Animal Genetic Resources. 2011. *Animal genetic resources in China bovines*. Beijing: Chinese Agriculture Press.

Davidson, E.A., I.A. Jassens, and Y.Q. Luo. 2006. On the variability of respiration in terrestrial ecosystems: moving beyond Q_{10}. *Global Change Biology* 12: 154–164.

Dewey, J.F., R.M. Shackleton, C.F. Chang, et al. 1988. The tectonic evolution of the Tibetan Plateau. *Philosophical Transactions of the Royal Society of London A: Mathematical Physical & Engineering Sciences* 327: 379–413.

Dong, S.K., R.J. Long, M.X. Kang, et al. 2003. Effect of urea multinutritional molasses block supplementation on liveweight change of yak valves and productive and reproductive performances of yak cows. *Canadian Journal of Animal Science* 83 (1): 141–145.

Dong, Q.M., X.Q. Zhao, Y.S. Ma, et al. 2006. Effect of stocking rate on water content of soil and standing crop in *Elymus nutans/Puccinellia tenuflora* mixed-sown pastures in Yangtze and Yellow River headwater region. *Journal of Anhui Agricultural Sciences* 34 (21): 5611–5614.

———. 2012a. Influence of grazing on biomass, growth ratio and compensatory effect of different plant groups in *Kobresia parva* meadow. *Acta Ecological Sinica* 32: 2640–2650.

Dong, Q.M., X.Q. Zhao, G.L. Wu, et al. 2012b. Responses of soil properties to yak grazing intensity in a *Kobresia parva-* meadow on the Qinghai-Tibetan Plateau, China. *Journal of Soil Science and Plant Nutrition* 12 (3): 535–546.

———. 2013. A review of formation mechanism and restoration measures of "black-soil-type" degraded grassland in the Qinghai-Tibetan Plateau. *Environmental Earth Sciences* 70: 2359–2370.

———. 2015. Optimization yak grazing stocking rate in an alpine grassland of Qinghai-Tibetan Plateau, China. *Environmental Earth Sciences* 73 (5): 2497–2503.

Foggin, J.M. 2000. *Biodiversity protection and the search for sustainability in Tibetan Plateau grasslands (Qinghai, China)*. Ph.D. Dissertation in Arizona State University.

Foggin, J.M., and M.E. Torrance-Foggin. 2011. How can social and environmental services be provided for mobile Tibetan herders? Collaborative examples from Qinghai Province, China. *Pastoralism: Research, Policy and Practice* 1: 21.

Gao, M.F., and J.J. Qiu. 2011. Characteristics and distribution law of major natural disasters in Tibetan Plateau. *Journal of Arid Land Resources & Environment* 25 (8): 101–106.

Guo, Z.Y., and H.B. Ma. 2005. Institutional choice of the sustainable development in the Sanjiangyuan District of Qinghai Province. *Reformation* 10: 46–50.

Hardin, G. 1968. The tragedy of the commons. *Science* 162: 1242–1248.

Harris, R.B. 2010. Rangeland degradation on the Qinghai-Tibetan plateau: A review of the evidence of its magnitude and causes. *Journal of Arid Environments* 74 (1): 1–12.

Hu, Z.M., S.G. Li, Q. Guo, et al. 2016. A synthesis of the effect of grazing exclusion on carbon dynamics in grasslands in China. *Global Change Biology* 22 (4): 1385–1393.

Huang, W., B. Bruemmer, and L. Huntsinger. 2017. Technical efficiency and the impact of grassland use right leasing on livestock grazing on the Qinghai-Tibetan Plateau. *Land Use Policy* 64: 342–352.

Jones, R.J. 1981. Interpreting fixed stocking rate experiments. In *Forage evaluation: concepts and techniques*, ed. J.L. Wheeler and R.D. Mochrie. Melbourne: CSIRO.

Jones, R.J., and R.L. Sandland. 1974. The relation between animal gain and stocking rate: derivation of the relation from the results of grazing trials. *Journal of Agricultural Science* 83 (2): 335–342.

Li, J. 2012. Land tenure change and sustainable management of alpine grasslands on the Tibetan Plateau: A case from Hongyuan County, Sichuan Province, China. *Nomadic Peoples* 16 (1): 36–49.

Li, W.J., and L. Huntsinger. 2011. China's grassland contract policy and its impacts on herder ability to benefit in Inner Mongolia: Tragic feedbacks. *Ecology and Society* 16 (2): 1. http://www.ecologyandsociety.org/vol16/iss2/art1/.

Lin, X.W., Z.H. Zhang, S.P. Wang, et al. 2011. Response of ecosystem respiration to warming and grazing during the growing seasons in the alpine meadow on the Tibetan Plateau. *Agricultural & Forest Meteorology* 151 (7): 792–802.

Liu, J., R.I. Milne, M.W. Cadotte, et al. 2018. Protect third pole's fragile ecosystem. *Science* 362: 1368.

Long, R.J., L.M. Ding, Z.H. Shang, et al. 2008. The yak grazing system on the Qinghai-Tibetan Plateau and its status. *The Rangeland Journal* 30 (2): 241–246.

Lu, X.Y., Y. Yan, J. Sun, et al. 2015a. Carbon, nitrogen, and phosphorus storage in alpine grassland ecosystems of Tibet: Effects of grazing exclusion. *Ecology & Evolution* 5 (19): 4492–4504.

———. 2015b. Short-term grazing exclusion has no impact on soil properties and nutrients of degraded alpine grassland in Tibet, China. *Solid Earth* 6 (4): 1195–1205.

Ma, H.B. 2007. Explanation about the causes of ecological degradation in the "Sanjiangyuan" district under the perspectives of new institutional economics. *Tibetan Studies* 3: 88–96.

Ma, W.M., K.Y. Ding, and Z.W. Li. 2016. Comparison of soil carbon and nitrogen stocks at grazing-excluded and yak grazed alpine meadow sites in Qinghai–Tibetan Plateau, China. *Ecological Engineering* 87: 203–211.

McNaughton, S.J. 1985. Ecology of a grazing ecosystem: The Serengeti. *Ecological Monographs* 55: 259–294.

Miller, D.J. 1990. Grasslands of the Tibetan Plateau. *Rangelands* 12 (3): 159–163.

Ren, J.Z., Z.Z. Hu, J. Zhao, et al. 2008. A grassland classification system and its application in China. *Rangeland Journal* 30 (2): 199–209.

Rhode, D., D.B. Madsen, P.J. Brantingham, et al. 2007. Yaks, yak dung, and prehistoric human habitation of the Tibetan Plateau. *Developments in Quaternary Science* 9 (07): 205–224.

Saito, M., T. Kato, and Y.H. Tang. 2009. Temperature controls ecosystem CO_2 exchange of an alpine meadow on the northeastern Tibetan Plateau. *Global Change Biology* 15: 221–228.

Wan, S.Q., and Y.Q. Luo. 2003. Substrate regulation of soil respiration in a tallgrass prairie: results of a clipping and shading experiment. *Global Biogeochemical Cycles* 17 (2): 1054. https://doi.org/10.1029/2002gb001971.

Wang, W., T.G. Liang, X.D. Huang, et al. 2013. Early warning of snow-caused disasters in pastoral areas on the Tibetan Plateau. *Natural Hazards & Earth System Sciences* 13 (6): 1411–1425.

Wang, J., Y. Wang, S.C. Li, et al. 2016. Climate adaptation, institutional change, and sustainable livelihoods of herder communities in northern Tibet. *Ecology and Society* 21 (1). https://doi.org/10.5751/es-08170-210105.

Wiener, G., J.L. Han, and R.J. Long. 2006. *The yak*. 2nd ed. Bangkok: FAO.

Wilson, A.D. 1986. Principles of grazing management systems. In *Rangeland: A resource under siege*, ed. P.J. Joss, P.W. Lynch, and O.B. Williams. New York: Cambridge University Press.

Wu, G.L., Z.H. Liu, L. Zhang, et al. 2010. Long-term fencing improved soil properties and soil organic carbon storage in an alpine swamp meadow of western China. *Plant & Soil* 332 (1/2): 331–337.

Xie, G.D., C.X. Lu, Y.F. Leng, et al. 2003. Ecological assets valuation of the Tibetan Plateau. *Journal of Natural Resources* 18 (2): 189–195.

Xin, G.S., R.J. Long, X.S. Guo, et al. 2011. Blood mineral status of grazing Tibetan sheep in the northeast of the Qinghai–Tibetan Plateau. *Livestock Science* 136 (2–3): 102–107.

Xiong, D.O., P.L. Shi, Y.L. Sun, et al. 2014. Effects of grazing exclusion on plant productivity and soil carbon, nitrogen storage in alpine meadows in northern Tibet, China. *Chinese Geographical Science* 24 (4): 488–498.

Zhang, J.F. 2008. Ecological current situation of grassland and its counter measures in the Three Rivers District of Qinghai. *Prataculture & Animal Husbandry* 1: 30–32.

Zhao, J.X., T.X. Luo, R.C. Li, et al. 2016a. Grazing effect on growing season ecosystem respiration and its temperature sensitivity in alpine grasslands along a large altitudinal gradient on the central Tibetan Plateau. *Agricultural & Forest Meteorology* 218–219: 114–121.

Zhao, J.X., X. Li, R.C. Li, et al. 2016b. Effect of grazing exclusion on ecosystem respiration among three different alpine grasslands on the central Tibetan Plateau. *Ecological Engineering* 94: 599–607.

Zhou, L., Q.J. Wang, J. Zhao, et al. 1995. Studies on optimum stocking intensity in pasturelands of alpine meadow stocking intensity to maximize production of Tibetan sheep. In *Alpine meadow ecosystem*, vol. 4, 365–418. Beijing: Science Press.

Zhu, S.H., C.L. Xu, Q.E. Fang, et al. 2006. Effect of white yak grazing intensity on species diversity of plant communities in alpine grassland. *Journal of Gansu Agricultural University* 4: 71–75.

Chapter 12
Promoting Artificial Grasslands to Improve Carbon Sequestration and Livelihood of Herders

Huakun Zhou, Dangjun Wang, Meiling Guo, Buqing Yao, and Zhanhuan Shang

Abstract Under the pressure of human population growth, resource exploitation and over grazing, the pasture in the Hindu Kush-Himalayan region (HKH) is deteriorating fast, which seriously affects the social and economic development and carbon fixation. Artificial grasslands a powerful and effective way to achieve it. It not only increases the carbon sink of the ecosystem by increasing the leaf area and the light energy utilization rate of the plant and solves the problem of carbon leakage to increase the carbon storage quantity of the grassland ecosystem, but also increases the grassland biomass for livestock's forage supply, that improves the living standards of pastoralists and promotes social and economic development. Therefore, reasonable pasture management, such as appropriate mix of plant species and livestock breeds, fertilization, irrigation, poison weed control and rats and pests control are equally important in keeping balance between ecological function and livelihood for pastoral system in HKH region.

Keywords Hindu Kush-Himalayan region · Artificial grassland · Carbon sequestration · Herder livelihood

---

H. Zhou (✉) · D. Wang · M. Guo · B. Yao
Key Laboratory of Restoration Ecology of Cold Area in Qinghai Province, Northwest Institute of Plateau Biology, Chinese Academy of Sciences, Xining, China
e-mail: hkzhou@nwipb.cas.cn

Z. Shang
Key Laboratory of Restoration Ecology of Cold Area in Qinghai Province, Northwest Institute of Plateau Biology, Chinese Academy of Sciences, Xining, China

School of Life Sciences, State Key Laboratory of Grassland Agro-Ecosystems, Lanzhou University, Lanzhou, China

© Springer Nature Switzerland AG 2020
Z. Shang et al. (eds.), *Carbon Management for Promoting Local Livelihood in the Hindu Kush Himalayan (HKH) Region*,
https://doi.org/10.1007/978-3-030-20591-1_12

12.1 Introduction

Grassland is the most important land cover type in the Hindu Kush-Himalayan region (HKH). The area of natural alpine grassland is 1.28×10^8 hm^2, 51% of the HKH. The ecosystem service value accounts for 48% of the total service value of the HKH ecosystem, and accounts for 17.7% of China's grassland ecosystem service value (Xie et al. 2003a, b). In terms of ecosystem service value, gas regulation and climate regulation together account for 11.0%, with food production, raw materials, and entertainment culture related to human production and living accounting directly for about 14.6% of the total (Xie et al. 2003a, b). In the value of the ecological services of gas and climate regulation, the holding effect of carbon is the most important. Grassland soil organic matter on the HKH contains 33.5 pg carbon, 23.4% of China's soil organic carbon reserves, and about 2.5% of the global soil carbon pool (Wang et al. 2002). The pastoral population's economic income is almost entirely from livestock products. The development of the grassland ecosystem in the HKH is crucial to maintaining the global carbon cycle, promoting regional economic development, and improving the people's livelihood. Most grassland ecosystem of the HKH is located in China and, therefore, China's grassland policy is crucial for the development of grassland in the HKH. The following discussion on grassland related research and development strategies in the HKH will also focus on the Chinese territory.

Subject to the harsh natural environment, grassland ecosystems in the HKH are relatively fragile and sensitive to various types of disturbances. In recent years, due to population growth, overgrazing, resource development, climate change, and rodent damage (Li et al. 2013), grassland ecosystems in the HKH have been degraded extensively. With the intensification of grassland degradation, the aboveground biomass and under-ground biomass have decreased, and soil organic carbon content has been reduced (Wu et al. 2013). From the perspective of economic development and improvement of people's livelihood, the productivity of degraded grassland was reduced, and the carrying capacity of livestock decline. This directly leads to a reduction in income for the people in pastoral areas. From the perspective of the carbon cycle, the degradation of grassland lead to the release of carbon, which lead to the conversion of grassland from a carbon sink to a carbon source (Wang et al. 2005a, b; Abdalla et al. 2018). Due to the unreasonable development and utilization, the grassland total carbon stocks in the HKH decreased by 11.5% from 1990 to 2000 (Li et al. 2015). The improvement of the herdsmen's living standards and the increase in grassland carbon require the restoration of degraded grassland ecosystems to guarantee the long-term healthy development of grassland ecosystems.

The establishment of artificial or semi-artificial grassland (planting local forage seed to build sown grassland) is the most effective way to restore severely degraded grassland (Wu et al. 2010; Dong et al. 2013). The establishment of artificial grassland can increase grassland above-ground and under-ground biomass (Li et al. 2014), as well as the productivity of heavily degraded grassland. In addition, although the soil carbon content of artificial grassland is far less than that of natural non-degraded grassland (Wang et al. 2010b; Zhu et al. 2015), the establishment of artificial grassland on heavily degraded grassland can steadily increase soil carbon content (Wang et al. 2005a, b; Wu et al. 2010; Feng et al. 2010). The restoration of degraded natural grassland relying on artificial grassland is not limited to the artificial grassland itself, but more important on the restoration of other natural grasslands. The high productivity of artificial grassland provides forage for livestock, which eases the pressure on the natural grassland and is extremely helpful for the recovery of moderately mild natural grassland. The restoration of natural grassland also increases its carbon sequestration capacity while increasing grassland productivity. This has an important role in the production of grassland ecosystems and the improvement of ecological functions.

The efficient, long-term and healthy development of the grassland ecosystem is also inseparable from the establishment of artificial grassland. With the rapid development of China's social economy, the demand for meat, eggs, and dairy products has also gradually increased. The increase in demand for animal husbandry products has stimulated the development of animal husbandry. However, traditional nomadic pastoralism can only increase production by increasing the number of livestock. This creates an imbalance between supply and demand between grassland and storage, which leads to grassland degradation and an unsustainable situation. The solution to the problem of grass-storage imbalance is to increase grassland productivity. Artificial grassland can produce more than 10 times of biomass per unit area than natural grassland (Fang et al. 2016a, b). A certain area of artificial grassland can be established to supplement the needs of excess livestock, and at the same time, the reduce pressure on natural grassland carrying animals, and protect natural grassland while improving the income of herdsmen. The scale and level of production of artificial grassland is a measure of the degree of development of animal husbandry in a country (Li et al. 2016a).

Artificial grassland in livestock husbandry in developed countries or regions has received attention, and of the total grassland area in New Zealand, Australia, Europe, Canada, and the United States reached 73%, 58%, 50%, 22%, and 9.5%, respectively, while the area of artificial grassland in China accounts for only 3.2% of the national grassland area. Artificial grassland in the HKH is far below the level in China. The artificial grassland in Qinghai Province, the main province of the HKH, is 43.8×10^4 hm^2, only 1.04% of the total grassland area of the province (Li et al. 2016b). Zhang et al. (2016) proposed five development periods of grassland animal husbandry, and the grass-roots nutrition pyramid structure is shown in Fig. 12.1. Grassland animal husbandry in the HKH is in stage (d), and there is still quite distant from stage (e) of the ideal animal husbandry development model. The key to the transition from (d) to (e) is the establishment of artificial grassland: establish a small area of intensive, high-yield artificial grassland to provide forage grass for the development of animal husbandry, and rational grazing of large areas of natural grassland to maintain natural grassland ecology and production functions.

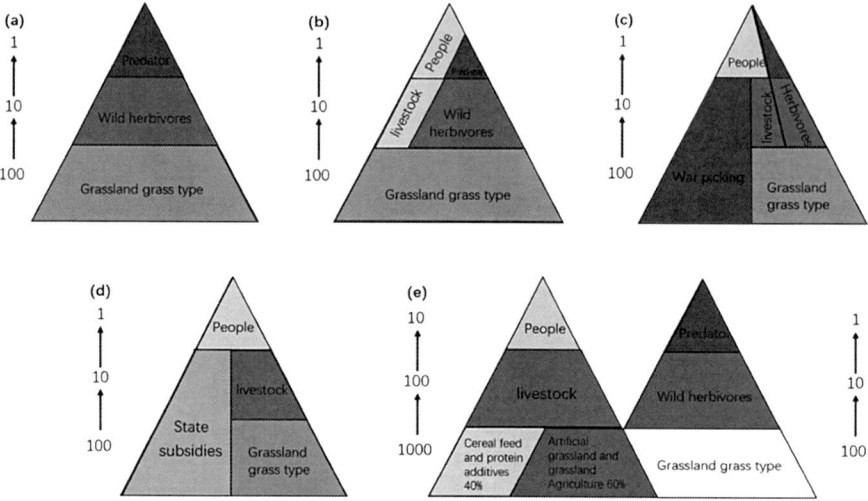

Fig. 12.1 The five developmental periods of grassland husbandry and structural schema for their trophic pyramid. (**a**) Prehistoric period of primitive grassland; (**b**) neolithic period of initial grazing grassland; (**c**) early middle age of nomadism in natural grassland; (**d**) modern age of over-grazed and degraded grassland; and (**e**) contemporary age of combined system of natural grassland with industrial livestock husbandry. (Translated from Zhang et al. 2016)

12.2 Development Status of Artificial Grassland in HKH

The establishment of artificial grassland in the HKH began in the 1950s. At that time, because of the instability of natural grassland production, lack of fodder in winter and spring, and the shortage of pastoral animal husbandry productivity, some state-owned farms in Qinghai Province introduced perennial pasture (Dong 2013). As of 2008, the area of artificial grassland in Qinghai Province reached 702,100 hm^2, an increase of 21% compared with 1978 and the area of artificial grassland in the Tibet Autonomous Region reached 784,800 hm^2, an increase of about 15 times compared with 1978. In addition to Qinghai and Tibet, HKH edge areas such as southern Gansu (Gannan region) and Qilian Mountains, western Sichuan (Aba and Ganzi regions), and southwestern Xinjiang (Pamirs mountain area) are also actively developing artificial grassland (Table 12.1).

According to the "General Plan for the Conservation, Utilization, and Utilization of Grasslands in China", artificial grassland will reach 1 million hm^2 and the area for pasturing seed breeding will reach 90,000 hm^2 by 2010 (Ministry of Agriculture of the People's Republic of China, 2007). The development of artificial grassland and the implementation of ecological replacement (the replacement of natural grassland with the benefits of artificial grassland) have become the basic strategies for grassland ecological environment and sustainable development of grassland and animal husbandry in western China and HKH. Agriculture on artificial grassland, has developed into the main form of grassland agriculture on the HKH (Dong 2013).

Table 12.1 Change of artificial grassland area in Tibetan plateau from 1978 to 2008 (Dong 2013) (unit: million hm^2)

Province (autonomous region)	Area (State)	1978	2006	2007	2008
Qinghai	Haibei state	0.40	1.08	1.08	1.08
	Huangnan state	0.00	1.10	1.60	1.60
	Hainan state	0.00	10.56	8.96	8.96
	Guoluo state	0.00	0.04	0.05	0.50
	Haixi state	57.37	58.07	58.07	58.07
Tibet	Lhasa	0.00	4.52	4.52	3.14
	Naqu area	0.00	36.00	36.21	38.09
	Changdu area	0.00	0.53	0.53	0.70
	Shannan area	0.00	0.07	0.07	0.19
	Shigatse area	4.81	27.42	28.83	33.57
	Ali area	0.00	2.38	2.38	2.45
	Nyingchi area	0.00	0.01	0.01	0.01
Xinjiang	Bavaria area	0.03	0.11	0.11	1.70
Gansu	Wuwei area	0.00	0.80	0.93	1.23
	Jiuquan area	0.03	0.09	0.10	0.67
	Gannan state	0.00	5.53	5.06	1.87
Sichuan	Aba state	0.00	0.52	0.86	1.14
	Ganzi state	0.00	0.00	0.00	4.95
	Liangshan state	0.00	1.93	1.85	1.44

The types of artificial grassland are diverse. They can be divided into perennial artificial grassland and annual artificial grassland. As to the composition of sowing plants, they can be divided into mixed artificial grassland and unicast artificial grassland. Unicast artificial grassland is divided mainly into legume grass pasture and gramineous pasture (Bai et al. 2016). At present, the annual artificial grassland in the HKH is dominated by oat unicast (Xu 2005). Compared with natural grassland, oat unicast grass is higher in biomass but shorter in growing period. Besides, the land of oat unicast grass is exposed for a long period of time and the soil water loss seriously, which is not conducive to the protection of the ecological environment. Therefore, the establishment of perennial artificial grassland is an important approach to solve the sustainable development of alpine grassland on the HKH (Gu 2008).

From the aspects of yield, quality and stability, legumes and grasses are the ideal combination of perennial artificial grasslands (Wang et al. 2007). However, the low air temperature of the HKH has limited the cultivation of perennial forage legumes. Grasses are more adapted to the low-temperature and are the main type of artificial grassland planted in the HKH. Although perennial unicast grassland can achieve high yields, it has poor stability, degrades quickly and the utilization period is short. In contrast, the mixed-sowing grassland can avoid the above-mentioned problems faced by unicast artificial grasslands while providing a high yield and is more conducive to the sustainable development of animal husbandry and ecological environment in high-cold areas (Dong 2001).

In recent years, Chinese scholars have systematically studied the establishment and management of perennial grasses mixed-sowing pastures, including screening of mixed-sowing combinations (Dong et al. 2003a, b; Li et al. 2013), mixed-sown grassland cleaning (Kou and Hu 2003), cultivation management, and the use of management (Dong et al. 2007), which provides technical support for the sustainable development of mixed perennial artificial grassland. Artificial grassland in the HKH is moving towards scale, industrialization, and scientific development.

12.3 Promotion of Artificial Grassland to Strengthen Carbon Fixation in the HKH

Grassland accounts for about one-third of the global land area and is an important global carbon sink (Fang et al. 2010). It stores more than 34% of the carbon in the global terrestrial ecosystem and plays a key role in maintaining the balance of the global carbon cycle (Scurlock and Hall 1998; Cheng et al. 2011). The total carbon storage in the global grassland ecosystem is approximately 308 pg. Plants, litter, and soil constitute the three major carbon pools of grassland ecosystems (Qi et al. 2003), of which about 92% is stored in soil (Schuman et al. 2002), and in the alpine grassland it is as much as 95% (Zhang and Liu 2010). Small changes in soil carbon pools have an important impact on atmospheric carbon dioxide concentration. Grassland soil carbon pool storage and its changes and regulatory mechanisms are the core of grassland carbon cycle research (Qi et al. 2003).

Because of high altitude, low temperature, and weak microbial activity, decomposition of plant litter in alpine meadows is slow. Soil organic matter accumulates year by year and becomes a major carbon pool (Wang et al. 2006). The total carbon stock in the grassland of China about 44.1 pg, which accounting for 54.5% in the high-cold grassland area. And More than 40% occurs in the alpine meadow and alpine grassland ecosystem (Jian 2002). Ecosystems at high altitudes are very sensitive to global climate change. Therefore, changes in carbon storage in the HKH have become a hot topic in the study of the earth's carbon cycle.

However, the current trend of changes in grassland carbon storage in the HKH is still controversial. Li et al. (2015) estimated that total carbon stocks in the HKH fell by 11.5% from 1990 to 2000, while Piao et al. (2017) showed that carbon stocks increased by 13.0% between 1982 and 1999. However, regardless of the overall development of grassland carbon stocks in the HKH, grassland degradation and the resulting decline in grassland carbon stocks are unquestionable. According to existing research, artificial grassland is significant for rehabilitating severely degraded grasslands and even of mild to moderately degraded grasslands. The contribution of artificial grassland in the HKH to carbon sequestration is vital as is the increase in carbon sequestration brought about by the restoration of degraded grassland.

12.3.1 Carbon Sequestration of Artificial Grassland

Carbon sequestration of artificial grassland can be divided into two parts. The first is the absorption of CO_2 during the plant's growth and the carbon transport from the plant to the soil. The source of soil organic carbon is mainly the decomposition of plant residual roots and litter layers. Regarding carbon sequestration of artificial grasslands overall, many current studies consider artificial grasslands as CO_2 sinks (Su et al. 2013; Jones et al. 2005). Zhao et al.'s research on the *Elymus nutans* artificial grassland in the Three-River headwaters region showed that it was a carbon sink with an annual carbon fixation capacity of approximately 49.4 $g \cdot m^{-2}$ (Zhao et al. 2008). Separately, due to human management, artificial grassland plants are less subject to natural stress, and their productivity is 2.7–12.1 times higher than that of natural grassland (Shen et al. 2016).

The strong carbon sequestration capacity of artificial grassland vegetation is obvious. However, artificial grassland has a good recovery effect on above-ground vegetation, but the recovery of under-ground biomass is not ideal (Li et al. 2016b). The transfer of carbon from the root system to the soil is limited. In addition, the greater the harvest intensity of artificial grassland, the greater the cumulative effect of soil carbon transport through plant litter pathways. The key to the conversion of carbon source and sink function of artificial grassland is the change of soil carbon storage after artificial grassland establishment, which depends on the contribution of artificial grassland to soil carbon.

Soil carbon sequestration capacity of artificial grassland is mainly related to years of grass planting (Li et al. 2013) and grassland vegetation types (Zhang et al. 2013). In general, long-term farming will lead to a continuous reduction of soil organic carbon, which will change from a carbon pool to a carbon source (Li et al. 2006). This is mainly due to the increased soil permeability and water holding capacity by reclamation, which intensifies microbial activity in the soil and accelerates the decomposition of plant residues in the soil. However, artificial grassland cultivation, especially of perennial herbs, does not require annual plowing. With the increase in the number of years of planting, the soil becomes tight and microbial activities are gradually limited. Therefore, the loss of soil carbon is limited mostly to the initial stage of establishment. While soil respiration intensity decreases, the artificial grassland plants gradually transfer carbon to the soil through the roots, and the soil carbon content gradually accumulates.

Some studies have pointed out that the cultivation period of *Medicago sativa* artificial grassland has a significant positive correlation with soil organic carbon content (Li et al. 2013; Potter et al. 1999). It should be noted that for mowing artificial grassland, high-strength mowing reduces the transport of litter to soil carbon, leading to a decrease in soil organic carbon (Wang 2007). The relationship between forage grass carbon sequestration and grassland vegetation types can be summarized as follows: legume and grass mixed > perennial legume > perennial grass > annual grass (Zhang et al. 2013). Leguminous plants have strong nitrogen-fixing capacity and relatively less nutrient stress, resulting in stronger carbon sequestration capacity. However, from the perspective of the root system, the fibrous root system of Gramineae dominates the output of soil organic matter compared with the root system of leguminous plants. Therefore, mutual assistance between them can produce a stronger carbon fixation effect. Species diversity or richness can characterize mutual assistance among plants.

Wang et al. (2015) found that there was a positive correlation between plant diversity and primary soil productivity and carbon storage in alpine grasslands. Zhao et al. (2015), in a study of natural alpine meadows, concluded that carbon sequestration capacity of the soil depends, in part, on species richness. In research on artificial grassland, Xu et al. stated that species richness determines ecosystem carbon storage (Xu et al. 2017). Compared with unicast artificial grassland, the mixed artificial grassland has higher species diversity and a relatively more stable community structure, which enhances grassland community stability and stress resistance (Tilman and Downing 1994), so that the carbon fixation capacity can also be strengthened.

The conversion of soil carbon source and sink function after artificial grassland establishment is not only affected by the carbon sequestration capacity of artificial grassland, but also depends on the soil carbon stocks in the established land. Some studies showed that, compared with different degrees of degraded natural grassland, artificial grassland had the lowest soil organic carbon storage (Dong et al. 2012) while some studies showed that the soil organic carbon stock of artificial grassland was higher than that of seriously degraded grassland (Wang et al. 2005a, b; Li et al. 2013). Zhu et al. (2015) reported that soil organic carbon decreased by 14.5% from alpine meadows to artificial grassland, and by 53% from degraded to bare land. As the degree of degraded grassland increases, soil carbon stocks decrease (Li et al. 2013; Yang et al. 2014), so the carbon release by the establishment of artificial grassland on heavily degraded grassland is undoubtedly the lowest.

Artificial grassland is a carbon source or carbon sink that depends strongly on grassland management. Improvement of grassland management is a vital consideration when computing the national carbon budget for Australia. Many management techniques sequester atmosphere carbon. Conant et al. (2001) reviewed management method of improving grassland carbon stock, that including fertilization, improved grazing management, cultivation transformation and native vegetation, planting leguminous plants and grass, earthworm introduction and irrigation. Soil carbon content and concentration increased 74% by management totally, which suggested that grassland soil organic carbon is strongly influenced by management, improve grassland management and increase primary production and soil carbon stocks by eliminating soil disturbance (Conant et al. 2001).

12.3.2 The Effect of Carbon Fixation on Natural Grassland Restoration

The contribution of artificial grassland to carbon sequestration capacity cannot be limited only to artificial grassland as we cannot neglect the spatial compensation of grassland carbon sequestration. The level of carbon sequestration in artificial soils and severely degraded natural grasslands are controversial (Zhang et al. 2013), but there is a huge gap between the two compared to non-degraded natural grassland. Wang et al. (2005a, b) found that the organic carbon content of the heavily degraded grassland in the 0–20 cm soil layer was only 57% of the non-degraded grassland on the alpine meadows of the Tibetan plateau. Lin et al.'s study on the degradation succession series of alpine meadow grass found that in the 0-40 cm soil layer, soil organic carbon content in the non-degraded grassland was about twice that of the heavily degraded grassland (Li et al. 2013). The grassland degradation rate on the HKH is as high as 41%.

The contribution of degraded grassland's restoration to soil carbon accumulation is large. The significance Carbon fixation in artificial grassland is important in relieving and reversing the declining trend of natural grassland carbon stocks. But to use this function of artificial grassland cannot be done without appropriate management.

The main factors affecting soil organic carbon content generally include climate, plant species composition, soil microbial activity intensity, and human activities such as reclamation and raising livestock. Human management and grazing intensity are important factors affecting the carbon content of grassland vegetation (Yang et al. 2014). Consistent with Yang et al. (2014) observations, Schuman et al. (2002) also believe that land use change and land degradation have important effects on the carbon balance between the atmosphere and soil, and land use patterns and intensity are the key factors for the conversion of system carbon sources and sinks. Wang et al. (2011) estimated that by the year 2020, with the implementation of approximately 150 million hm^2 of fenced grassland and 30 million hm^2 of artificial grassland, the Chinese grassland will fix 0.24 Pg carbon per year. Activities such as grazing and enclosure will have a strong impact on grassland biomass, soil carbon pools and their dynamic changes (Fang et al. 2010).

The transformation of grassland carbon sources and sinks is ultimately due to human management. Over-grazing will greatly reduce grassland vegetation and soil-fixed carbon. In alpine meadows in the Aba pastoral area of Sichuan, soil organic carbon reserves decrease as grassland grazing intensity increases (Yang et al. 2014). However, due to the compensatory or over-compensated growth characteristics of plants after they have been damaged (Dong et al. 2006), mild and moderate grazing can stimulate pasture growth. In addition, plant species composition has an important influence on the spatial and temporal distribution of above-ground biomass (Durante et al. 2017). Moderate disturbances produce the highest biodiversity and over-grazing leads to a decrease in the diversity of grassland vegetation species.

The greater number of plant species, the more stable the above-ground biomass production of ecosystem plants (Tilman and Downing 1994). Choosing an appropriate grazing intensity and time of fencing has important implications for the improvement of plant carbon sequestration capacity of the entire system (Qi et al. 2003). Without prejudice to the steady development of animal husbandry, grassland enclosure and rational grazing cannot be separated from the establishment of artificial grassland.

12.4 Promotion of Artificial Grassland to Improve People's Livelihood in the HKH

Livestock size and carrying capacity of the pasture affect the livelihood of local herders. For centuries, grazing has been the main form of land use on the HKH and animal husbandry has been the backbone of the regional. Any prerequisite to improve pastoralist income or pasture productivity requires a good understanding of the linkages between existing systems and components (Squires et al. 2010). The original grassland system can be composed of three parts: pastoralists, livestock and grassland. Due to religious beliefs, the herdsmen's focus on maintaining the balance between the three, and their willingness to improve material life is not strong. Nowadays, with the development of a market-oriented economy, the consumer is no longer limited to herdsmen but extends to the entire consumer market.

With the rapid development of social economy in recent decades, the demand for material culture has also increased, in particular, for foods such as meat and dairy products and for clothing such as leather and wool products. Yak, Tibetan sheep and Tibetan goats are the main livestock in the HKH and their products such as meat, milk and wool are the main sources of income (Long et al. 2008). The increase in the demand for animal products has stimulated an increase in livestock carrying capacity on grassland in the HKH. The imbalance between grass and livestock leads to grassland degradation and declining productivity of the grassland. The most prominent manifestation of grass-storage imbalances is in the winter and spring. Due to lack of forage, livestock die in large numbers in the face of cold weather, resulting in economic losses to pastoralists. The performance of livestock in the four seasons is: strong in summer, fat in autumn, thin in winter and weak or dead in spring.

The mitigation of land degradation and poverty alleviation in China's pastoral areas (two closely linked issues) depend on the interaction between socio-economic and biophysical factors. Poor people do not destroy the environment intentionally, but often lack resources to avoid environmental degradation (Squires et al. 2010). The key to the lack of resources can be solved through the establishment of artificial grassland. Grass cultivation and livestock raising, establishment of stable and high-yielding artificial grassland, and insufficient forage supplement are important ways to solve the imbalance of grass-animal season imbalance and to guarantee the nutritional needs of grazing livestock in the cold season (Wang et al. 2010a). Fang et al. estimated that if 10% of the grassland with good hydrothermal conditions is used to establish intensive artificial grassland and its productivity is raised to the average level of arable land, then China's grassland can increase the biological output annually by 3.7×10^8 t, which is more than grass production in all existing grasslands (Fang et al. 2016b). This can greatly increase the livestock population and living standards in pastoral areas.

In addition to the direct income from the increase in animal production, artificial grassland can also improve people's living standards in pastoral areas through other indirect ways. For example, the reduction of grazing pressure in natural grassland reduces the degree of grassland degradation, increases the ornamental nature of natural grasslands and is conducive to the development of regional tourism. Intensive grassland animal husbandry can also liberate part of the labor force, allowing some herdsmen to seek work elsewhere and increase their sources of income.

12.5 Problems Facing the Establishment of Artificial Grassland in the HKH

Artificial grassland is the fundamental way to increase the carbon fixation of the overall grassland ecosystem and improve the people's living standards in the region. However, there are still many problems in the promotion and development of artificial grassland in the HKH. The key to promoting artificial grassland to larger regions lies in the transformation of herdsmen's ideas and land management models. The problem of grassland degradation is a major issue facing the establishment of artificial grassland.

12.5.1 The Change of Herdsmen's Ideas and Land Management Patterns

Due to the adverse climate and geographical environment, the level of overall cultural education in the HKH is relatively low, especially in pastoral areas. The inhabitants are more inclined to follow traditional experience to solve the problems encountered. However, in recent decades, the HKH has undergone rapid development and has encountered new problems. Traditional experience is no longer sufficient to solve many of the problems. There is a lack of ideas for the management of grasslands. When addressing the problem of grassland degradation, China has followed the "packaged-to-household" model that has achieved remarkable results in rural areas and has allocated grasslands to households.

Therefore, in the promotion of artificial grassland, the problems of non-support and non-cooperation among some herdsmen are often encountered. This can affect the establishment of artificial grassland in the entire region, which constitutes an obstacle to the construction of large-scale artificial grassland and, ultimately, increases costs. This requires the government to increase information on artificial grassland establishment while carrying out pilot work to enable herders to recognize the advantages of artificial grassland. It is also necessary to promote the transformation of the land management model and provide institutional support for collective decision-making. Fundamentally, it is necessary to increase investment in educational resources in pastoral areas so that pastoralists can effectively share the educational dividend brought about by social development.

12.5.2 Degradation of Artificial Grassland

Livestock-artificial grassland interaction is divided into grazing artificial grassland, mowing artificial grassland, and grazing and pasturing artificial grassland (Zhang and Liu 2004). The three different forms of grassland use have a common feature, that is, they continuously provide nutrients and propagation components and remove nutrients. After completion, there is often a decline in productivity within a few years (Shang et al. 2008). The reason for artificial grassland degradation is essentially the loss of effective nutrients. When artificial grass is planted, a large amount of chemical fertilizer is generally applied at the same time to provide sufficient nutrients for growth. However, with time after planting, harvesting of grasses and leaching of soil gradually reduce soil nutrients.

There is usually a single plant species in artificial grassland and the competition for and utilization of common resources are high, resulting in the lack of certain nutrients in the soil pool (Wang et al. 2005b). A survey by Dong et al. (2012) on soil nutrient content of alpine grassland found that total soil nitrogen, total phosphorus, and total potassium showed a downward trend as the time after establishment increased. Li et al. (2017) found that artificial planting resulted in the leaching of soil nitrate. Loss of nutrients causes weakening of the grass's competitiveness and the invasion of miscellaneous grasses. Zhang et al. (2014) found that during the degradation of artificial grassland, the main species were gradually replaced by poisonous weeds, which further exacerbated the degradation of artificial grassland.

Li et al. theorized that the "degeneracy" of artificial grassland after establishment is an important stage of grassland ecosystem self-regulation and it is a key stage of transition from simple artificial grassland communities to complex and stable primary communities (Li et al. 2013). Zhou et al. (2007) stated that the artificial community of grassland transforms from "productive stability" to "ecological stability" within 4 years after completion of the artificial grassland and thus presents a deteriorating trend. For the restoration of heavily degraded grassland, the degradation of artificial grassland is normal and reasonable. However, if it is to promote the development of pastoral animal husbandry, a continuous high yield of artificial grassland is necessary.

The amelioration of artificial grassland degradation can be achieved through two approaches: one is to supplement the nutrients and propagules needed by degraded artificial grassland and the other is to increase the artificial grassland's ability to resist environmental stress. The main reason for degradation of artificial grassland is the loss of soil nutrient elements. The absence of nutrients can lead to the invasion of miscellaneous grasses and increase the nutritional competition with pasture (Zhou et al. 2007). Regular extermination of weeds is necessary for the continued stability of artificial grassland. In addition, artificial grassland propagules with more serious degradation are also missing. Therefore, apart from fertilization and removal of miscellaneous grasses, appropriate exogenous propagules should be added to expand the grassland seed bank to achieve gradual restoration of artificial grassland (Wang et al. 2009).

The stress tolerance of grassland is closely related to species diversity. Higher species diversities have the stronger community's resistance to stress (Tilman and Downing 1994). The decline of available soil nutrients in artificial grassland under unicast mode is the main reason for the unsustainable productivity of artificial grassland. By introducing pastures with different pathways to obtain restricted nutrients, not only can the nutrients needed by each of them be obtained from the nutrient pools, but the total nutrient uptake can also increase with the increase of species diversity (Wang et al. 2016).

In the single species seed planting method, the intraspecific competition is often strong, and the species combination is an important way to adjust the intraspecific competition of artificial grassland. The alpine meadows established through the combination of upper and lower grasses, clumps and rhizomes, and dense clumps and sparse clumps not only increase the community structure complexity and capacity but also weaken the interspecific and intraspecific competition (Shi et al. 2009). A community composed of multiple populations can use environmental resources more effectively than a single population and maintain long-term higher productivity and greater stability. Mixed sowing can also extend the pasture grass period, increase soil fertility and contribute to soil carbon stability (Yang et al. 2011). Soil organic carbon plays a key role in maintaining the sustainability of pasture grassland ecosystems (Xu et al. 2018), which affects soil water conservation and erosion potential and is a key factor affecting soil structure (Tisdall and Oades 1982).

In addition, as the richness of species increases, the resistance of artificial grassland communities to weed invasion is enhanced (Li et al. 2009). Increasing grass diversity maintains the stability of artificial grassland. However, we should realize that the establishment of artificial grasslands has changes the status of various biological and abiotic factors in the entire grassland ecosystem and changes the energy and material circulation patterns (Zhou et al. 2007), thus the production function of artificial grassland cannot be separated from human's intervention. Increasing the diversity, fertilization, propagating bodies and eradication of weeds are effective methods to prevent degradation of artificial grassland.

12.6 The Development Direction of Artificial Grassland in the HKH

Grass production and animal husbandry in the HKH are undergoing a transition from backward and traditional to intensive. In this process, the restoration of degraded grassland and the rational use of natural grassland are the basis. The rational allocation of grassland production functions and ecological functions are signs of success or failure and the establishment of artificial grassland is the key to the success or failure of the transformation (Bai et al. 2016). Artificial grassland in heavily degraded areas not only restores the vegetation, but more importantly, reduces the pressure on grazing grasslands and ensures the continued use of existing natural grasslands (Ma and Lang 1998). It should be noted that the idea of restoration of any degraded ecosystem cannot separate the degraded system from other healthy or surrounding areas, nor can it separate the internal relations of the ecological landscape, ecological economy, or regional economy (Bao et al. 2001). So far, little progress has been made on the integration of farmland societies and ecosystems and their interdependence.

The establishment of artificial grassland and basic pastureland is also the key to achieving a rational allocation of production functions and ecological functions (Bai and Wang 2017). Fang et al. (2016a, b) proposed the grassland protection and utilization model of "taking care of small farmers": use less than 10% (or even 5%) of the grasslands to establish intensive, high-yield artificial grasslands to provide forage grass for animal husbandry and fundamentally solve the contradiction between livestock and environment; to protect, restore and utilize more than 90% of the natural grassland to achieve higher production and ecological functions. This model has achieved remarkable results in Inner Mongolia and is worthy of promotion in the HKH.

Intensive pastoral animal production is inseparable from the ecological grass production. The ecological turf grass technology system is based on the rational utilization of natural grassland, artificial grassland and grassland. . It integrates the existing advanced technologies with newly developed technologies, making the natural grasslands based mainly on ecological functions, supplemented by production functions and artificial grasslands undertake major production functions, so that the production functions and ecological functions of pastoral areas are coordinated, which makes the grass industry efficient and sustainable (Fig. 12.2) (Bai

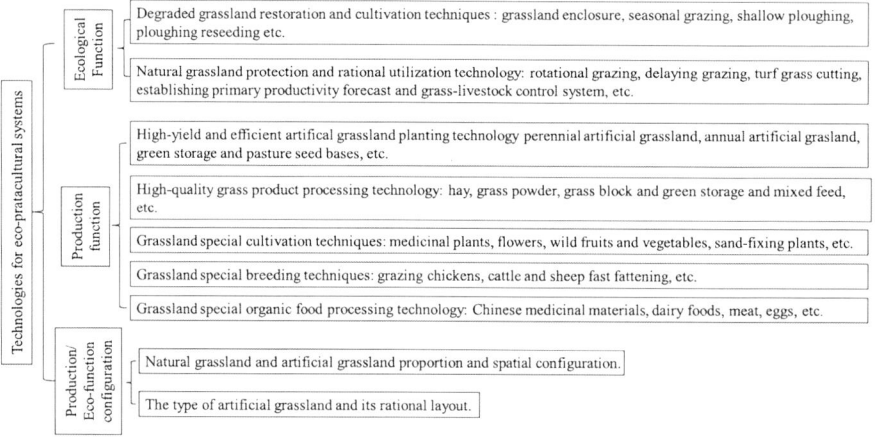

Fig. 12.2 Framework of major technologies for eco-pratacultural systems. Translated form (Bai et al. 2016)

et al. 2016). Fang's research group proposed the concept of building an "ecological grass and animal husbandry pilot area" in the pastoral areas of the grasslands, that is, the allocation and control of various elements of the natural-economic-social complex system in the larger areas and the formation of a multi-industry structure with artificial grassland and modern animal husbandry as the main features, bio-industries and cultural industries as supplements, and greatly enhanced ecological functions of grassland (Fang et al. 2016b).

At present, grassland ecosystems in the HKH are facing problems such as low productivity and diminished soil carbon sequestration capacity. This has seriously hindered regional economic development and threatened regional ecological security. The promotion of artificial grassland plays a key role in improving grassland carbon fixation and living standards in the HKH. Artificial grassland replaces the natural grassland production function and gradually restores the natural grassland ecological function, so as to achieve a rational allocation of grassland production functions and ecological functions. The artificial grassland should be taken as a starting point to build an ecological grass industry and a regional natural-economic-social complex system to promote the long-term development of animal husbandry in the HKH.

Acknowledgement This work was supported by the National Natural Science Foundation of China (31672475), National Key Research and Development Program of China (2016YFC0501901), and Qinghai Innovation Platform Construction Project (2017-ZJ-Y20).

References

Abdalla, K., M. Mutema, P. Chivenge, et al. 2018. Grassland degradation significantly enhances soil CO_2 emission. *Catena* 167: 284–292.
Bai, Y.F., and Y. Wang. 2017. Long-term ecological research and demonstrations support protection and sustainable management of grassland ecosystems. *Bulletin of the Chinese Academy of Sciences* 32 (8): 910–916.
Bai, Y.F., Q.M. Pan, and Q. Xing. 2016. Fundamental theories and technologies for optimizing the production functions and ecological functions in grassland ecosystems. *Chinese Science Bulletin* 61 (2): 201–212.
Bao, W.K., Z.G. Liu, and Q. Liu. 2001. Ecological restoration and rehabilitation: development, researching features and existing major problems. *World Review of Science, Technology and Sustainable Development* 23: 44–48.
Cheng, J., G.L. Wu, L.P. Zhao, et al. 2011. Cumulative effects of 20-year exclusion of livestock grazing on above- and belowground biomass of typical steppe communities in arid areas of the Loess Plateau, China. *Plant, Soil and Environment* 57 (1): 40–44.
Conant, R.T., K. Paustian, and E.T. Elliott. 2001. Grassland management and conversion into grassland: effects on soil carbon. *Ecological Applications* 11 (2): 343–355.
Dong, S.K. 2001. *The stability of mixture grassland of cultivated perennial grass and its regulation in alpine region of Qinghai-Tibetan Plateau of China*. Lanzhou: Gansu Agricultural University.
———. 2013. *Productive and ecological paradigm of alpine cultivated grasslands in the region of Qinghai-Tibet Plateau*. Beijing: Science Press.
Dong, S.K., Z.Z. Hu, R.J. Long, et al. 2003a. Community characteristics of mixed grassland with perennial grasses in alpine region of Tibetan Plateau. *Chinese Journal of Ecology* 22 (5): 20–25.
Dong, S.K., J.X. Ma, X.P. Pu, et al. 2003b. Study on the ecological adaptability of introduced perennial grasses and the selection of combinations in alpine region. *Grassland and Turf* 1: 38–48.

Dong, Q.M., X.Q. Zhao, Y.S. Ma, et al. 2006. Effects of yaks stocking rates on primary productivity and its dynamic changes for mixed-sown sward of *Elymus nutans/Puccinellia tenuiflora* in Yangtze and Yellow river headwater region. *Chinese Journal of Grassland* 28 (3): 5–15.

Dong, Q.M., X.Q. Zhao, and Y.S. Ma. 2007. Situations and strategy of sustained development on alpine grassland-livestock industry in headwater region of Yangtze and Yellow rivers. *Research of Agricultural Modernization* 28 (4): 438–442.

Dong, S.K., L. Wen, Y.Y. Li, et al. 2012. Soil-quality effects of grassland degradation and restoration on the Qinghai-Tibetan Plateau. *Soil Science Society of America Journal* 76 (6): 2256–2264.

Dong, Q.M., X.Q. Zhao, G.L. Wu, et al. 2013. A review of formation mechanism and restoration measures of 'black soil land' degraded grassland in the Qinghai-Tibetan Plateau. *Environmental Earth Sciences* 70 (5): 2359–2370.

Durante, M., G. Piñeiro, J.G.N. Irisarri, et al. 2017. Primary production of lowland natural grasslands and upland sown pastures across a narrow climatic gradient. *Ecosystems* 20 (3): 543–552.

Fang, J.Y., Y.H. Yang, W.H. Ma, et al. 2010. Ecosystem carbon stocks and their changes in China's grasslands. *Science China Life Sciences* 40 (7): 566–576.

Fang, J.Y., Y.F. Bai, L.H. Li, et al. 2016a. Scientific basis and practical ways for sustainable development of China's pasture regions. *Chinese Science Bulletin* 61 (2): 155–164.

Fang, J.Y., Q.M. Pan, S.Q. Gao, et al. 2016b. 'Small vs large area' principle: Protecting and restoring a large area of natural grassland by establishing a small area of cultivated pasture. *Pratacultural Science* 33 (10): 1913–1916.

Feng, R.Z., R.J. Long, Z.H. Shang, et al. 2010. Establishment of *Elymus natans* improves soil quality of a heavily degraded alpine meadow in Qinghai-Tibetan Plateau, China. *Plant and Soil* 327 (1–2): 403–411.

Gu, M. 2008. *Study on the productivity and stability of perennial cultivated grassland in Qinghai-Tibetan alpine meadow*. Lanzhou: Lanzhou University.

Jian, N. 2002. Carbon storage in grasslands of China. *Journal of Arid Environments* 50 (2): 205–218.

Jones, S.K., R.M. Rees, U.M. Skiba, et al. 2005. Greenhouse gas emissions from a managed grassland. *Global and Planetary Change* 47 (2–4): 201–211.

Kou, J.C., and Z.Z. Hu. 2003. Weeds control of mixture perennial grassland in alpine region. Grassland and Turf 4: 33–36.

Li, Y.M., G.M. Cao, and Y.S. Wang. 2006. Effects of reclamation on soil organic carbon in Haibei alpine meadow. *Chinese Journal of Ecology* 25 (8): 911–915.

Li, A., M.H. Gu, S.T. Zhang, et al. 2009. Effects of species richness on weed invasion in an artificial grassland ecosystem in eastern Tibetan Plateau. *Chinese Journal of Ecology* 28 (2): 177–181.

Li, X.L., J. Gao, G. Brierley, et al. 2013. Rangeland degradation on the Qinghai-Tibet Plateau: Implications for rehabilitation. *Land Degradation & Development* 24 (1): 72–80.

Li, Y.Y., S.K. Dong, L. Wen, et al. 2014. Soil carbon and nitrogen pools and their relationship to plant and soil dynamics of degraded and artificially restored grasslands of the Qinghai-Tibetan Plateau. *Geoderma* 213: 178–184.

Li, X.W., M.D. Li, S.K. Dong, et al. 2015. Temporal-spatial changes in ecosystem services and implications for the conservation of alpine rangelands on the Qinghai-Tibetan Plateau. *The Rangeland Journal* 37 (1): 31–43.

Li, L.H., P. Lu, X.Y. Gu, et al. 2016a. Principles and paradigms for developing artificial pastures. *Chinese Science Bulletin* 61 (2): 193–200.

Li, Y.J., C.C. Sun, G.M. Cao, et al. 2016b. Study on plant biomass and soil nutrients under different land use patterns in Three-River headwater region. *Grassland and Turf* 36 (4): 48–53.

Li, Y.J., Y.Y. Wang, G.M. Cao, et al. 2017. Effect of land use patterns on soil nitrogen characteristics in Three-River headwater region. *Agricultural Research in the Arid Areas* 35 (3): 272–277.

Long, R.J., L.M. Ding, Z.H. Shang, et al. 2008. The yak grazing system on the Qinghai-Tibetan Plateau and its status. *The Rangeland Journal* 30 (2): 241–246.

Ma, Y.S., and B.N. Lang. 1998. Establishing pratacutural system—a strategy for rehabilitation of 'black soil type' deteriorated grassland on Qinghai-Tibetan Plateau. *Pratacutural Science* 15 (1): 5–9.

Piao, S.L., J.Y. Fang, L.M. Zhou, et al. 2017. Changes in biomass carbon stocks in China's grasslands between 1982 and 1999. *Global Biogeochemical Cycles* 21 (2): 1–10.

Potter, K.N., H.A. Tobert, H.B. Johnson, et al. 1999. Carbon storage after long-term grass establishment on degraded soils. *Soil Science* 164 (10): 718–725.

Qi, Y.C., Y.S. Dong, Y.B. Geng, et al. 2003. The progress in the carbon cycle researches in grassland ecosystem in China. *Progress in Geography* 22 (4): 342–352.

Schuman, G.E., H.H. Janzen, and J.E. Herrick. 2002. Soil carbon dynamics and potential carbon sequestration by rangelands. *Environmental Pollution* 116 (3): 391–396.

Scurlock, J.M.O., and D.O. Hall. 1998. The global carbon sink: A grassland perspective. *Global Change Biology* 4 (2): 229–233.

Shang, Z.H., Y.S. Ma, R.J. Long, et al. 2008. Effect of fencing, artificial seeding and abandonment on vegetation composition and dynamics of 'black soil land' in the headwaters of the Yangtze and the Yellow Rivers of the Qinghai-Tibetan Plateau. *Land Degradation & Development* 19 (5): 554–563.

Shen, M.H., Y.K. Zhu, X. Zhao, et al. 2016. Analysis of current grassland resources in China. *Chinese Science Bulletin* 61 (2): 139–154.

Shi, J.J., F.Z. Hong, Y.S. Ma, et al. 2009. Artificial control impact on community character of gramineous mixed grassland. *Acta Agrestia Sinica* 17 (6): 745–751.

Squires, V., H. Limin, L. Guolin, et al. 2010. Exploring the options in north-west China's pastoral lands. In *Towards sustainable use of rangelands in north-west China*, ed. R.V. Squire et al., 41–59. Berlin: Springer, Dordrecht.

Su, C.X., Y.T. Mi, D. Wang, et al. 2013. Research progress on exchanging fluxes of greenhouse gases from artificial grassland. *Advanced Materials Research* 726–731: 4131–4134.

Tilman, D., and J.A. Downing. 1994. Biodiversity and stability in grasslands. *Nature* 367 (6461): 363–365.

Tisdall, J.M., and J.M. Oades. 1982. Organic matter and water-stable aggregates in soils. *European Journal of Soil Science* 33 (2): 141–163.

Wang, G.L. 2007. *Study on sustainable utilization of meadow steppe at edge of south Mongolia Altiplano*. China: Chinese Academy of Agricultural Sciences.

Wang, G.X., J. Qian, G.D. Cheng, et al. 2002. Soil organic carbon pool of grassland soils on the Qinghai-Tibetan Plateau and its global implication. *Science of the Total Environment* 291 (1–3): 207–217.

Wang, W.Y., Q.J. Wang, C.Y. Wang, et al. 2005a. The effect of land management on carbon and nitrogen status in plants and soils of alpine meadows on the Tibetan plateau. *Land Degradation & Development* 16 (5): 405–415.

Wang, W.Y., Q.J. Wang, and H.C. Wang. 2005b. The effect of land management on plant community composition, species diversity, and productivity of alpine *Kobersia* steppe meadow. *Ecological Research* 21 (2): 181–187.

Wang, C.T., R.J. Long, and G.M. Cao. 2006. Soil carbon and nitrogen contents along elevation gradients in the source region of Yangtze, Yellow and Lantsang rivers. *Journal of Plant Ecology* 30 (3): 441–449.

Wang, X., Z.H. Zeng, Y.G. Hu, et al. 2007. Progress and prospect on mixture of gramineae herbage and leguminosae herbage. *Chinese Journal of Grassland* 29 (4): 92–98.

Wang, C.T., R.J. Long, Q.L. Wang, et al. 2009. Community succession of differently aged artificial grasslands and their soil nutrient changes in Three-River headwaters region in Qinghai, China. *Chinese Journal of Applied and Environmental Biology* 15 (6): 737–744.

Wang, Q.J., F.G. Wang, H.K. Zhou, et al. 2010a. The status of eco-environment of northeast regions of Yangtze, Yellow and YaluTsangpo rivers and protecting strategies. *Pratacultural Science* 27 (2): 59–65.

Wang, X.J., W.D. Willms, X.Y. Hao, et al. 2010b. Cultivation and reseeding effects on soil organic matter in the mixed prairie. *Soil Science Society of America Journal* 74 (4): 1348–1355.

Wang, S., A. Wilkes, Z. Zhang, et al. 2011. Management and land use change effects on soil carbon in northern China's grasslands: a synthesis. *Agriculture, Ecosystems & Environment* 142 (3–4): 329–340.

Wang, X.X., S.K. Dong, R. Sherman, et al. 2015. A comparison of biodiversity–ecosystem function relationships in alpine grasslands across a degradation gradient on the Qinghai–Tibetan Plateau. *The Rangeland Journal* 37 (1): 45–55.

Wang, W.Y., W.Q. Li, H.K. Zhou, et al. 2016. Dynamics of soil dissolved organic nitrogen and inorganic nitrogen pool in alpine artificial grasslands. *Ecology and Environmental Sciences* 25 (1): 30–35.

Wu, G.L., Z.H. Liu, L. Zhang, et al. 2010. Effects of artificial grassland establishment on soil nutrients and carbon properties in 'black soil land' degraded grassland. *Plant and Soil* 333 (1–2): 469–479.

Wu, X., H.X. Li, B.J. Fu, et al. 2013. Study on soil characteristics of alpine grassland in different degradation levels in Three-River headwaters regions in China. *Chinese Journal of Grassland* 35 (3): 77–84.

Xie, G.D., C.X. Lu, Y.F. Leng, et al. 2003a. Ecological assets valuation of the Tibetan Plateau. *Journal of Natural Resources* 18 (2): 189–196.

Xie, G.D., C.X. Lu, Y. Xiao, et al. 2003b. The economic evaluation of grassland ecosystem services in Qinghai-Tibet Plateau. *Journal of Mountain Science* 21 (1): 50–55.

Xu, C.L. 2005. The study on production potential of nutritive agriculture of *Oat* artificial Grassland in Qinghai-Tibetan Plateau. *Grassland of China* 27 (6): 64–66.

Xu, L.H., B.Q. Yao, W.Y. Wang, et al. 2017. Effects of plant species richness on C-13 assimilate partitioning in artificial grasslands of different established ages. *Scientific Reports* 7: 1–11.

Xu, S., M.L. Silveira, L.E. Sollenberger, et al. 2018. Conversion of native rangelands into cultivated pasturelands in subtropical ecosystems: Impacts on aggregate-associated carbon and nitrogen. *Journal of Soil and Water Conservation* 73 (2): 156–163.

Yang, H.S., J.C. Tai, and F. Fan. 2011. Effects of sowing methods on the oxidation stability and the chemical bound forms of soil organic carbon in artificial grassland. *Acta Prataculturae Sinica* 20 (3): 36–42.

Yang, S.J., T. Li, Y.M. Gan, et al. 2014. Response of soil organic carbon storage in alpine meadow in Aba pastoral areas to different ways of using and degree. *Chinese Journal of Grassland* 6 (6): 12–17.

Zhang, Z., and T.M. Liu. 2004. Study on types of artificial grassland in China. *Grassland of China* 26 (5): 32–36.

Zhang, L.J., and S. Liu. 2010. Protecting grassland and strengthening the function of carbon sequestration of grassland. *Chinese Journal of Grassland* 32 (2): 1–5.

Zhang, Y.J., G.M. Yang, N. Liu, et al. 2013. Review of grassland management practices for carbon sequestration. *Acta Prataculturae Sinica* 22 (2): 290–299.

Zhang, R., Y. Wang, L.N. Ma, et al. 2014. Species diversities of plant communities of degraded artificial grassland, 'black soil land' and natural grassland in the Three-River headwaters region. *Acta Agrestia Sinica* 22 (6): 1171–1178.

Zhang, X.S., H.P. Tang, X.B. Dong, et al. 2016. The dilemma of steppe and it's transformation in China. *Chinese Science Bulletin* 61 (2): 165–177.

Zhao, L., S. Gu, H.K. Zhou, et al. 2008. CO_2 fluxes of artificial grassland in the source region of the Three-Rivers on the Qinghai-Tibetan plateau, China. *Acta Phytoecologica Sinica* 32 (3): 544–554.

Zhao, L., D.D. Chen, N. Zhao, et al. 2015. Responses of carbon transfer, partitioning, and residence time to land use in the plant–soil system of an alpine meadow on the Qinghai-Tibetan Plateau. *Biology and Fertility of Soils* 51 (7): 781–790.

Zhou, H.K., X.Q. Zhao, L. Zhao, et al. 2007. The community characteristics and stability of the *Elymus nutans* artificial grassland in alpine meadow. *Chinese Journal of Grassland* 29 (2): 13–25.

Zhu, P., R.S. Chen, Y.X. Song, et al. 2015. Effects of land cover conversion on soil properties and soil microbial activity in an alpine meadow on the Tibetan Plateau. *Environmental Earth Sciences* 74 (5): 4523–4533.

Chapter 13
Prospects for REDD+ Financing in Promoting Forest Sustainable Management in HKH

Shambhavi Basnet, Jagriti Chand, Shuvani Thapa, and Bhaskar Singh Karky

Abstract Results based payment is the main instrument of REDD+ through which emission reduction activities are rewarded in HKH region. The principle of "additionality" is used in this process which means the incentives are rewarded to the forest managers for conserving forest areas, and, thus, mitigating climate change. Incentives for conserving forests are provided in an ex-post payment model, which is a challenge to countries receiving the payments. The establishment of community forest user groups (CFUGs) for the management and conservation of forests have made locals more accountable towards forests. The involvement of the private sector is equally vital in REDD+ by contributing to sustainable forest management and rural development. Hence, this chapter focuses on the prospects of REDD+ financing by adopting various mechanisms, analyzing the challenges in REDD+, encouraging the involvement of CFUGs and private sector through improved forest management system and addressing problems with the support of national policies.

Keywords REDD+ · Community forestry management system · Private sector · Results-based payment · Benefits

Abbreviations

CFM	Community Forest Management
CFUG	Community Forest User Group
COP	Convention of the Parties
DANAR	Dalit Alliance for Natural Resources, Nepal

S. Basnet (✉) · J. Chand · S. Thapa · B. S. Karky
International Centre for Integrated Mountain Development (ICIMOD), Lalitpur, Kathmandu, Nepal

FECOFUN	Federation of Community Forest Users Nepal
FREL	Forest Reference Emission Level
GHG	Greenhouse Gas
ICIMOD	International Centre for Integrated Mountain Development
MRV	Monitoring, Reporting and Verification
NEFIN	Nepal Federation of Indigenous Nationalities
NRs	Nepalese Rupees
NTFP	Non-timber Forest Product
REDD+	Reducing Emissions from Deforestation and Forest Degradation, and the role of Conservation, Sustainable Management of Forests and Enhancement of Forest Carbon Stocks
SBSTA	Subsidiary Body for Scientific and Technological Advice
tC/ha	tons of Carbon per Hectare
tCO_2	tons of Carbon dioxide
UNFCCC	United Nations Framework Convention on Climate Change
USD	United States Dollar
VDC	Village Development Committee

13.1 Introduction

The 13th Conference of Parties in Bali, December 2017, decided on "Policy approaches and positive incentives on issues relating to reducing emissions from deforestation and forest degradation; and the role of conservation, sustainable management of forests and enhancement of forest carbon stocks in developing countries" (The Red Desk 2016). The aims were to halt forest cover loss in developing countries by 2030 and to reduce gross deforestation in developing countries by at least 50% by 2020. The conference further acknowledged the contributions of emissions from deforestation, global anthropogenic greenhouse gas (GHG) emissions and forest degradation that lead to emissions and, thus, need to be addressed. The conference also recognized that reducing emissions from deforestation and forest degradation in developing countries can promote benefits, but these developing countries require stable and predictable availability of resources as well (UNFCCC 2014).

The concept of REDD+ was introduced in Poznan during the Subsidiary Body for Scientific and Technological Advice (SBSTA) in December, 2008, in which REDD+ was defined as "reducing emissions from deforestation and forest degradation, and the role of conservation, sustainable management of forests and enhancement of forest carbon stocks in developing countries" (Holloway and Giandomenico 2009). REDD+, in addition, must also consider the social issues of forest management, which can be achieved through transparency and accountability, informed consent as well as equitable benefit sharing and social inclusion (Poudyal et al. 2013).

13.2 Results Based Payment

REDD+ has been portrayed as a win-win approach to deal with climate change, promote sustainable forestry, reduce poverty and provide large-scale carbon emission reductions at comparatively low abatement costs, while also enhancing rural livelihoods and protecting biodiversity (Poudel et al. 2014). The main highlight of REDD+ is the method through which emission reductions are rewarded. It is based on the principle of "additionality", meaning payments will be provided as an incentive to forest managers for the conservation of forest area, which mitigates climate change (Rosenbach et al. 2013).

As outlined in the Paris Agreement, results-based finance is the basis in the approach to REDD+, which follows the concept of transferring funds from developed countries to developing countries in exchange for verifiable emission reductions (Rosenbach et al. 2013; Wong et al. 2016), that is, result-based finance provides ex-post payments. Developing countries can implement any one, or a combination, of the REDD+ interventions to address the drivers of deforestation and unlock carbon enhancement potentials. The main difference between REDD+ and previous attempts to promote sustainable forest management and conservation is that countries must prove successful outcomes through measurable reductions in the level of atmospheric GHGs to be eligible for financial rewards. When emission reduction results are demonstrated, monitored, and verified, developing countries can claim financial payment from the developed countries in the form of incentives.

For result-based payments to work effectively, the following conditions must be met: (1) a clear agreement on the definition of the results; (2) a robust measurement, reporting, and verification system that shows with reasonable confidence that results have been achieved; and (3) an appropriate institutional arrangement that manages and oversees the implementation of the actions while also complying with safeguards.

For meeting these conditions, countries that receive payments are required to have appropriate capacities and systems in place, supported by a regulatory framework that demonstrates the effectiveness of result-based payments, as mentioned in COP 13. To access the payments from REDD+, the developing countries undergo a three-phased approach, which includes the following three points. (1) Readiness: In this initial phase, countries prepare a National REDD+ Strategy and initiate mechanisms to ensure social and environmental safeguards are met. Countries must also prepare a National Forest Monitoring System and Safeguard Information System. (2) Implementation: REDD+ countries start the implementation of strategies and enabling processes, as well as undertake policy and legal reforms for the execution of demonstration activities. (3) Results: Emission reductions resulting from the implementation of activities in Phase 2 are submitted to UNFCCC in tCO_2 units. By complying with the conditions, the country is eligible for receiving result-based payments (Kipalu 2011).

Many countries lack appropriate human resources, and both activity and emission factor data. Methodological guidance on Monitoring, Reporting and Verification

(MRV) and Forest Reference Emission Levels (FRELs) is required to be interpreted by the countries according to their national prerequisites. Substantial investment is required to build the necessary capacity for interpreting the REDD+ objectives, formulating intervention strategies and implementing policy measures that result in emission reductions. According to Article 5 of the Paris Agreement, the three major challenges of result-based finance of REDD+ are:

1. Achieving adequate, predictable and effective REDD+ finance: For this, variations in costs and deforestation levels must be understood clearly. It is also important to know who bears the burden of costs in order to design an effective, efficient and equitable results-based payment system. Another vital aspect is the timing and sequencing of payments. Ex-post payments may lead to high rates of non-participation, particularly for lower-income countries.
2. Identifying and measuring 'performance' in REDD+: The major issue here is that results-based payments may itself be insufficient to fund the needed policy and governance reforms for REDD+. Performance must be clearly defined, and its metrics should be appropriate and measurable. Reference levels that are the comparable base figures should be strictly unbiased and also ensure additionality. Stable and sufficient funding is crucial as it is known that if the funding of REDD+ is insufficient, then converting forest to other land uses will be more profitable than conserving it.
3. Integrating and safeguarding non-carbon benefits in a results-based mechanism: For safeguards to be effective in monitoring issues such as social equity and environmental concerns, the use of mixed methods at various scales and indicators that are relevant to local realities will be required. Safeguards that are important in supporting more equitable outcomes can also be used to manage risks such as non-legitimacy of participatory processes (Wong et al. 2016).

The success of any REDD+ policy is highly dependent on generating sufficient and sustainable finance, especially when the funds are used for creating incentives for local forest users as well as for establishing payment mechanisms for government actions and projects to reduce emissions originated from various drivers of deforestation and forest degradation (Logan-Hines et al. 2012).

13.3 Nepal's Result-Based Payment for Landscape REDD+ Programme

The annual cost of reducing deforestation in Nepal is between USD 654/ha and USD 3663/ha, and the associated opportunity cost of carbon sequestration ranges from USD 1.11 to USD 3.56 per tCO_2 in forests managed by the communities (Rai et al. 2017). With an increase in degradation the cost of reducing deforestation decreases but the cost of carbon sequestration increases. This means that forest dependent communities in Nepal receive more benefits from felling trees rather than preserving them. To encourage communities who use forest resources to reduce deforestation, REDD+ must provide funds to offset these gains and to stop deforestation (Rai et al. 2017).

The government of Nepal has developed an Emission Reduction Programme covering 12 contiguous districts in the Terai Arc Landscape. This area of 2.2 million hectares includes nearly 15% of the country's land area, 20% of the total forest area and 25% of the country's population. Half the area is covered by forest (1.17 million ha) and of this forested area, nearly 29% (0.34 million ha) is managed by local communities (FCPF 2018).

Under this proposed landscape REDD+ programme, the country intends to reduce emissions by 35.6 million tCO_2 over a period of 10 years. Here, payments for reducing emissions are provided, on the basis of the carbon stored in the forests, to the government and local authorities so that they can utilize it on improving the management of forests and provide technical support through trainings, etc. Thus, payments based on results are not given to the communities/user groups directly per se. This programme is estimated to cost USD 177.1 million and so sufficient co-investment will be required from the federal government, private sector and local communities to finance the implementation of REDD+. The government is negotiating a cost sharing basis through co-finance. It is envisioned that the federal government may contribute up to USD 70 million. Additional contributions from CFUGs could total USD 25 million and from rural energy programme could total USD 26 million. This leaves Nepal to negotiate around USD 70 million from result-based payments. Nepal has recently managed to leverage USD 35 million as concessional finance from the Forest Invest Programme and International Development Association (FCPF 2018).

13.4 Community Forestry in Nepal

Forestry sectors all around the world have shown increased involvement of local communities in forest management activities. Community forestry has been regarded as a promising approach to achieve sustainable management of forests and improve livelihoods in developing countries having the potential for carbon sequestration. Estimates by various development agencies point out that community-managed forests provide livelihood benefits to more than half a billion poor people (Wong et al. 2016).

Community forestry in Nepal started in the 1980s and has improved forest protection and management (Binod 2016). The policy had originally intended to meet the basic forest products requirements of communities through active participation in management and development of forests. Later, it was expanded to include the mobilization and empowerment of the CFUGs in the development of their local communities (Binod 2016). More than three decades later, 1.8 million ha of forest land in Nepal is managed by approximately 19,300 forest user groups, involving nearly 1.45 million households (35% of the total population of Nepal) in the community forestry management (CFM) program and benefiting more than 2.4 million households (MFE 2018). In other words, one-third of the forest area handed over as community forests is managed by nearly one-quarter of the total households in Nepal.

Timber and firewood for fuel, fodder and grass for livestock, and medicinal plants are important forest products that the communities either use for their own household purposes or sell. From the sale of these forest products, the user groups generate income for themselves. Group incomes are also obtained from monthly membership fees, fines and donations from organizations. The income is then invested in various community development activities such as irrigation, canal improvement, community building, and drinking water schemes. This has played a vital role in improving forest conditions by adopting better forest protection and management measures (Joshi 2017). In Nepal, although ownership of the forests remains with the government, under the CFM system, all management decisions are made by individual community forest user groups (CFUGs). Each member in a user group has equal rights as well as access to the forest resources, while non-members are excluded from resource use and management. CFUGs also have the right to develop management rules and benefit sharing mechanisms. The Government of Nepal provides technical assistance to CFUGs when needed, in return for improved forest management. In this way, decentralized forest governance has enabled forest users to develop autonomous organizations and to reclaim traditional forestry practices (Rosenbach et al. 2013).

According to Koirala (2007), there are numerous objectives regarding the proper implementation of the community forestry plan in Nepal. The short-term objectives are, supply timber, firewood, fodder, litter, and grass regularly; manage silviculture operation to supply basic needs of users; plant medicinal plants and other plant species in the forest area; generate income support programs for poor and oppressed users to raise livelihood; control forest encroachment; develop community funds; and encourage forest conservation and initiate participation.

Accordingly, the author mentions the long-term objectives of the community forestry plan to, support for biodiversity conservation through wildlife conservation and protection of various plant species; minimize soil erosion, develop greenery, and ensure environmental balance; make self-reliance of users on forest products; develop a feeling of communal participation; and conserve water sources (Koirala 2007).

Nepal has a strong CFM system that has been able to enhance the resource base and address vital issues such as social exclusion and gender equality (Poudel et al. 2014). Through involvement in CFM, activities such as unregulated livestock grazing have been reduced to a minimum as members are realizing the true benefits of the forestry system. When the forest dependent communities were involved actively in decisions regarding forest uses, it was observed that forest conditions were kept in check. Numerous degraded forest ecosystems have improved due to decentralized development strategies in CFM (Karky 2008).

Community forests have been responsible for creating natural capital in the form of new forests, and improved biodiversity in existing forests, with 86% of the forests showing improvements (Pandey and Paudyall 2015). These forests have shown better environmental quality in mountainous regions where erosion reductions, protection of watersheds and increased agricultural output are evident. Besides creating income generating opportunities for local communities, community forests have contributed to local economic development through on-site added-value processing of raw materials (Pandey and Paudyall 2015). Some of the positive outcomes generated through CFM system are stated below, (1) legally empowered local communities to manage forests; (2) development and establishment of appropriate institutional structures at different levels, from local to national; (3) provision of subsistence income for poor families; (4) local employment and income by establishment of forest-based enterprises; (5) empowerment of rural people through more inclusive governance and provision of technical skills and training; (6) improvement of forest conditions; and (7) other social welfare activities such as building schools and road networks and scholarships for poor children (Poudel et al. 2014).

The CFM policies in Nepal have shown a progressive trend for the past 30 years and through various favorable features towards economic, social and environmental welfare, it has been concluded that CFM could also support carbon trading for REDD+. Studies have also shown that communally managed forests were able to sequester carbon, which is one of the major aims of REDD+ (Karky 2008). A CFM system may be successful in contributing to potential REDD+ advantages such as equitable co-benefits and increased carbon storage. Nepal's decentralized community forestry system has a long history of not only reducing deforestation but also encouraging reforestation, conservation and enhancement of carbon stocks and, thus, has garnered considerable global attention. In this case, CFM system can facilitate REDD+ so that they can better compete for funds that may become available. If CFM institutions, similar to those in Nepal, have relevant and appropriate qualities needed for REDD+ implementation, REDD+ may be able to harness and possibly improve the capital that CFM has already established (Rosenbach et al. 2013). Thus, a strong community forestry network and a supportive legal and policy framework are required to build a sustainable foundation for the implementation of REDD+ activities (Poudel et al. 2014).

13.5 Benefits from Community Forestry

Community forestry in Nepal provides livelihood capitals to the forest dependent communities. These livelihood capitals come in the form of access to forest products, income and employment through forest related activities and improvement of agricultural produce and biodiversity (Kanel and Niraula 2017). On a household level, incomes in the Churia region of Nepal have risen for almost all the middle- and poor-income households due to the decline in dependency on agricultural and livestock production through their involvement in community forest management activities (Maharjan et al. 2009). The annual income of CFUGs in Nepal from sale of forest products, governmental and non-governmental grants, memberships and entrance fees, and fines among others is NRs 913.8 million (=8.4 million USD with the exchange rate as of June, 2018), of which nearly 82% comes from the sale of forest products (Kanel and Niraula 2017). Similarly, the annual income per household based on stumpage price (number of trees harvested or per stump price) is NRs 1346 (=12.37 USD with the exchange rate as of June, 2018) (Kanel and Niraula 2017). This shows that there is a huge potential of community development and poverty reduction through CFM in Nepal (Kanel and Niraula 2017). Regarding benefit flows within a CFM system, the forest management responsibilities are shared among the Government, Village Development Committees (VDCs) and elected community forest committees (Mahanty and Guernier 2008). CFUGs keep 100% of takings from timber as well as non-timber forest products (NTFPs), except for two species of timber (*Shorea robusta* and *Acacia catechu*), for which 15% royalty is paid to the Government ((RECOFTC 2007), (Mahanty and Guernier 2008)).

In the final stage of REDD+ financing, where rewards are provided according to the results (carbon stored in the forests), a CFM system has been deemed to aid in successful benefit sharing. A recent study posits that it is not the CFM system that sequesters higher carbon in the forests, but it is the collective action that has positive effects on forest condition. Thus, it is not the formalization, but the group activities that increase carbon biomass (Bluffstone et al. 2018).

13.6 Private Sector in Forestry

The major aim of any private firm is profit maximization. With this goal in mind, the involvement of the private sector is found mainly in processing, manufacturing and trade, with limited participation in production of forest products and services (Subedi et al. 2014). The most prominent example of a private company in the forestry sector is Plantec Coffee Estate Pvt. Ltd., a company which produces shaded grown coffee (i.e. coffee and agroforestry system).

According to a study done by ICIMOD in 2016, the total carbon sequestered in the plantation, over a 20 year period, was 176.5 tC/ha which at a rate of USD 5 per tCO_2, is valued at approximately USD 124,693 (Timalsina et al. 2017). Such an agroforestry scheme would fulfill both the mitigation and adaptation strategies of climate change. Along with being the largest coffee producer and exporter in Nepal, this private business helps the rural community of Nepal to generate income by providing opportunities of employment through activities such as application of manure, weeding, thinning, picking and processing coffee. A recent analysis of the company's profitability showed that Plantec Coffee Estate is the largest coffee producer and exporter in Nepal, with an annual revenue of USD 600,000 in 2016. The company gains major economic benefits through its exports since 60% of its production is sold outside Nepal. Thus, we can conclude that private enterprise has the potential to contribute to sustainable forest management and rural development. Such partnerships can help communities reduce risk, achieve better returns on land use, diversify income sources, access paid employment, develop new skills, upgrade infrastructure, and enhance ecosystem management (Chipeta and Joshi 2001).

Forest-based industries in Nepal face major hurdles which include, (1) weak governance and lack of law enforcement, bureaucratic hurdles and procedural delays, and insecurity; (2) inadequate and uncertain supply of raw materials, limited effort on scientific forest management, and unpredictable policy decisions including bans and restrictions; (3) difficulty in accessing finances; (4) lack of infrastructure; and (5) inadequate business development services such as inputs and technologies, extension services, and business information (Subedi et al. 2014).

13.7 Discussion

There are a number of unresolved issues about REDD+ and how it delivers multiple outcomes. Uncertainty remains about the implementation, effectiveness and comparability of REDD+ schemes at local, national and international levels. Furthermore, various stakeholders of the REDD+ initiative have expressed concerns about the possible impacts of REDD+ on benefit-distribution, recentralization, additionality, co-benefits and stakeholder engagement (Paudel and Karki 2013). To develop an effective policy framework based on practical experiences, REDD+ has been experimenting through pilot projects in various countries (Poudel et al. 2014), including Nepal.

In the case of Nepal, there is little financial and institutional capacity to conduct the MRV necessary for REDD+ (Rosenbach et al. 2013). Interviews with various stakeholders give contrasting perspectives on REDD+ implementation. Those in favor of REDD+ activities believe it would incentivize the communities and government to carry out sustainable forest management activities, on the condition that the rights of local communities, indigenous people, marginalized communities (e.g. Dalits) and women and poor are not violated and that they are given a fair share of benefits. They conclude that if REDD+ is designed in a way to ensure fair sharing of benefits among the local users, it will benefit the country. Most interviewees also stated that REDD+ will be able to address the issues related to inclusion, participation and capacity building. However some thought the communities would not be able to meet some criteria of REDD+, such as permanence, leakage, and carbon measurements. They believed that REDD+ success is more inclined towards developed than developing countries (Paudel and Karki 2013).

Looking at the impacts of REDD+ pilot projects on CFUGs in Nepal, we see that REDD+ had both positive and negative effects. Forest fires, livestock grazing and harvesting of grass and fodder from the forest decreased, which eventually increased the productivity and sustainability of the forest ecosystem. However, the increase in participation of stakeholders in meetings and discussions, in forest protection and patrolling and in cultural and tending operations need to be kept in check and compensated. This leads to the conclusion that result-based payments from REDD+ can only be effective for community forests implementing REDD+ projects if the additional direct and indirect time and labor costs, as well as the benefits, do not exceed payments. In some of the CFUGs in Nepal, payments based purely on carbon rates at USD 10 per tCO_2 would not even compensate the local users of the costs incurred from the increased number of committee meetings (Maraseni et al. 2014).

Analyzing the dual impacts of REDD+ on forest dependent communities has provided some interesting results. On a positive note, the project has strengthened forest protection by tightening regular access and has regulated harvesting and grazing activities in the REDD+ pilot areas in Nepal. These pilot projects seek carbon enhancement in community forests by developing group management (i.e. social aspects) and forest management (forest protection, utilization and development aspect) activities. REDD+ has also improved local capacities to understand changes in climate and other forestry related issues, including skills to generate income. In addition, forest protection has been encouraged through distribution of subsidized biogas and improved cooking stoves.

REDD+ has been able to enhance the institutional capacity of CFUGs by organizing regular meetings, maintaining transparent record keeping, auditing and reporting systems. Through benefit sharing, REDD+ has prioritized poorer households, women and marginalized communities. We can conclude that REDD+ has the potential to solidify local approaches to community forestry by improving internal governance and by adding financial value to locally managed forests from the sale of carbon credits. On the negative side, REDD+ can possibly threaten the decision-making power at the community level by imposing externally developed terms and conditions and by CFM's diminishing role in community development and local autonomy. Through the introduction of controlled grazing, customized harvesting and limited access to charcoal burning, there is a possibility that REDD+ will destabilize the livelihood of the supporting community forestry as REDD+ is likely to change the traditional managed community forestry approaches.

The role of REDD+ in addressing the social gaps that exist in these communities may potentially hamper the internal harmony. Various authors speculate that conflicts within societies are possible due to differences in fund allocations between the marginalized and wealthier populations. It has been assessed that addressing social inequity based on ethnicity may drive a wedge in the real objective of REDD+, which is to reduce deforestation and forest degradation. Furthermore, the whole concept of social inclusion in terms of ethnicity may be invalid if forest dependency is not associated with ethnicity or caste. Thus far, REDD+ pilots have had little success in bringing disadvantaged groups into key decision-making posts, despite the positive discrimination approach in benefit sharing (Poudel et al. 2014).

Nevertheless, various service providers and policy developers are optimistic that the positive aspects can be strengthened, and negative aspects can be diminished through intensifying efforts at financial, technical and governance levels. REDD+ can take the steps in planning and implementation which include the following points (Poudel et al. 2014). They are, (1) the rights of CFUGs to make autonomous decisions are not neglected; (2) access and use of forest and forest resources to meet subsistence needs for forest dependent population are continued; (3) people who have lost their traditional employment due to implementation of REDD+ are properly compensated; (4) provision of alternative options when resource access and supply are reduced; (5) women and other disadvantaged groups are provided with equitable access to decision-making operations while removing preferential access based on ethnicity or gender; and (6) leadership capacity is built for efficient deliverance of services and that good governance attributes, including equity and fairness, are established.

Scholars have argued that policy makers can improve the success probabilities of REDD+ with the incorporation of biodiversity considerations and livelihood goals as well as with the use of success factors of CFM management. These success factors include sufficient size, clear boundaries of forests, predictability of benefit flows and local autonomy in decision-making (Poudyal et al. 2013).

13.8 Conclusion

Nepal has shown great potential in generating income and improving livelihoods through community level participation in forest related activities. Similarly, the private sector also plays a major role in encouraging employment generation as well as carrying out environment conservation operations. The main aim of REDD+ in Nepal is to boost livelihood options of the rural poor, while at the same time enhancing the carbon potential of the forests. Results based payments, in this regard, turn out to be one of the most efficient methods of achieving these two objectives. Carbon payments have not yet started in Nepal, but doing so would help in achieving global conservation objectives of enhancing the economic status of rural forest-dependent populations by securing the forest resources available to them. The Emissions Reductions Programme initiated in Nepal aims to focus on encouraging local participation through community forestry as well as providing a platform to private entrepreneurs to invest in the forestry sector.

In Nepal, the government, international development partners and civil society such as FECOFUN, NEFIN and DANAR have accepted REDD+ as an important climate change mitigation strategy and are engaged in appropriate policy making and piloting processes. However REDD+ policies have been affected mainly by the government, donor/international NGOs and civil society sectors, but with minimal consultation with the local communities that manage the forests. Despite these deficiencies, new modes of collaborations are emerging around piloting, awareness-raising efforts and advocacy campaigns for the rights of forest-dependent communities and also for transforming existing institutions of forest governance. The ultimate test for REDD+ will be how inclusive it is and what benefits is passed down to the local communities that manage the forest. The real incentive to bring about the change in forest management must be realized at the local level for the REDD+ performance-based payment to meet its objectives and purpose.

References

Binod, B.B. 2016. History of forestry and community forest in Nepal. *Imperial Journal of Interdisciplinary Research* 2 (11): 424–439.

Bluffstone, R.A., E. Somanathan, P. Jha, et al. 2018. Does collective action sequester carbon? Evidence from the Nepal Community Forestry Program. *World Development* 101: 133–141.

Chipeta, M.E., and M. Joshi. 2001. *The Private sector speaks: Investing in sustainable forest management*, 303p. Bogor: Center for International Forestry Research.

FCPF. 2018. *Emission Reductions Program Document (ER–PD). People and forests—A Sustainable Forest Management-based Emission Reduction Program in the Terai Arc Landscape, Nepal.* Paris: Forest Carbon Partnership Facility.

Holloway, V., and E. Giandomenico. 2009. *The history of REDD policy. Carbon Planet White paper.* Adelaide.

Joshi, M.R. 2017. *Community forestry programs in Nepal and their effects on poorer households.* Rome: United Nations Food and Agriculture Organization. http://www.fao.org/docrep/ARTICLE/WFC/XII/0036-A1.HTM.

Kanel, K.R., and D.R. Niraula. 2017. Can rural livelihood be improved in Nepal through community forestry? *Banko Janakari* 14 (1): 19–26.

Karky, B.S. 2008. *The economics of reducing emissions from community managed forests in Nepal Himalaya.* Enschede: University of Twente. https://ris.utwente.nl/ws/portalfiles/portal/6083008.

Kipalu, P. 2011. *Introducing the FCPF readiness package (R-package) and the carbon fund operational.* Washington, DC: Bank Information Centre, The World Bank.

Koirala, P.N. 2007. *Benefit sharing in Community Forests in Nepal. A case study in Makawanpur District of Nepal.* (Doctoral and MSc dissertations). Wageningen: Wageningen University.

Logan-Hines, E., L. Goers, M. Evidente, et al. 2012. *REDD+ policy options: Including forests in an international climate change agreement. Managing forest carbon in a changing climate,* 357–376. Netherlands: Springer.

Mahanty, S., and J. Guernier. 2008. *A fair share: sharing the benefits and costs of community-based forest management. Theme on understanding the benefits of the commons.* Cheltenham: University of Gloucestershire.

Maharjan, M.R., T.R. Dhakal, S.K. Thapa, et al. 2009. Improving the benefits to the poor from community forestry in the Churia region of Nepal. *International Forestry Review* 11 (2): 254–267.

Maraseni, T.N., P.R. Neupane, F. Lopez-Casero, et al. 2014. An assessment of the impacts of the REDD+ pilot project on community forests user groups (CFUGs) and their community forests in Nepal. *Journal of Environmental Management* 136: 37–46.

Ministry of Forests and Environment (MFE). 2018. *Community forestry.* Department of Forests. http://dof.gov.np/dof_community_forest_division/community_forestry_dof.

Pandey, G.S., and B.R. Paudyall. 2015. *Protecting forests, improving livelihoods—Community forestry in Nepal.* FERN. http://fern.org/sites/default/files/news-pdf/fern_community_forestry_nepal.pdf.

Paudel, G., and R. Karki. 2013. REDD+ governance, benefit sharing and the community: Understanding REDD+ from stakeholders perspective in Nepal. *Journal of Forest and Livelihood* 11 (2): 55–64.

Poudel, M., R. Thwaites, D. Race, et al. 2014. REDD+ and community forestry: Implications for local communities and forest management-a case study from Nepal. *International Forestry Review* 16 (1): 39–54.

Poudyal, B.H., G. Paudel, and H. Luintel. 2013. Enhancing REDD+ outcomes through improved governance of community forest user groups. *Journal of Forest and Livelihood* 11 (2): 14–26.

Rai, R.K., M. Nepal, B.S. Karky, et al. 2017. *Costs and benefits of reducing deforestation and forest degradation in Nepal. ICIMOD Working Paper.* Kathmandu: International Centre for Integrated Mountain Development.

RECOFTC. 2007. *Sharing the wealth, improving the distribution of benefits and costs from community forestry: Policy and legal frameworks.* Synthesis of discussions at the Second Community Forestry Forum, 21–22 March 2007, Bangkok, Thailand, RECOFTC.

Rosenbach, D., J. Whittemore, and J. DeBoer. 2013. *Community Forestry and REDD+ in Nepal.* (Thesis). Ann Arbor: University of Michigan.

Subedi, B.P., P.L. Ghimire, A. Koontz, et al. 2014. *Private sector involvement and investment in Nepal's Forestry Sector: status, prospects and ways forward.* Indonesia: Multi Stakeholder Forestry Programme.

The Red Desk. 2016. *What is REDD+? the REDD desk.* https://theredddesk.org/what-redd.

Timalsina, N., N. Bhattarai, B.S. Karky, et al. 2017. *Contributions by the private sector to climate change mitigation: Lessons from the Plantec Coffee Estate in Nepal. ICIMOD working paper.* Kathmandu: International Centre for Integrated Mountain Development.

United Nations Framework Convention on Climate Change (UNFCCC). 2014. *Key decisions relevant for reducing emissions from deforestation and forest degradation in developing countries (REDD+).* Decision booklet REDD+. Framework Convention on Climate Change. UNFCCC Secretariat.

Wong, G., A. Angelsen, M. Brockhaus, et al. 2016. *Results-based payments for REDD+: Lessons on finance, performance, and non-carbon benefits.* Vol. 138. Bogor: CIFOR.

Part IV
Policies and Strategies for Livelihood Improvements and Carbon Management

Chapter 14
Designing Water Resource Use for Poverty Reduction in the HKH Region: Institutional and Policy Perspectives

Madan Koirala, Udhab Raj Khadka, Sudeep Thakuri, and Rashila Deshar

Abstract The Hindu Kush-Himalayan (HKH) region covers over 4 million km^2 (about 2.9%) of the global land area and approximately 18% of the mountain area. These mountains are headwaters of ten major river systems that provide livelihood to 210 million people living there, and indirectly provide goods and services to 1.3 billion people downstream. Despite local people being safeguards of this valuable water resource, many inhabitants are amongst the poorest of the poor, many being marginalized subsistence farmers of diverse ethnic groups and minorities, not benefiting from these vast resources. Available water resources in the HKH could be used for economic prosperity in the region by maximizing the benefits of this renewable resource. This chapter reviews the water resource use for poverty reduction, and pertinent institutional arrangements and policy provisions amidst sustainable development agendas, underlining the needs of international cooperation, institutional framework, and regional thinking.

Keywords HKH region · Water resource · Poverty reduction · Water institution and policy · Water security · International cooperation · Regional thinking

14.1 Introduction

Mountains comprise 24% of the global land surface area and are home to 12% of the world's population. The Hindu Kush-Himalayan (HKH) region covers over 4 million km^2 (about 2.9%) of the global land area and approximately 18% of the mountain area, extending 3500 km over all or part of eight countries including Afghanistan, Bangladesh, Bhutan, China, India, Nepal, Myanmar and Pakistan (Fig. 14.1).

M. Koirala (✉) · U. R. Khadka · S. Thakuri · R. Deshar
Central Department of Environmental Science, Tribhuvan University,
Kirtipur, Kathmandu, Nepal
e-mail: mkoirala@cdes.edu.np

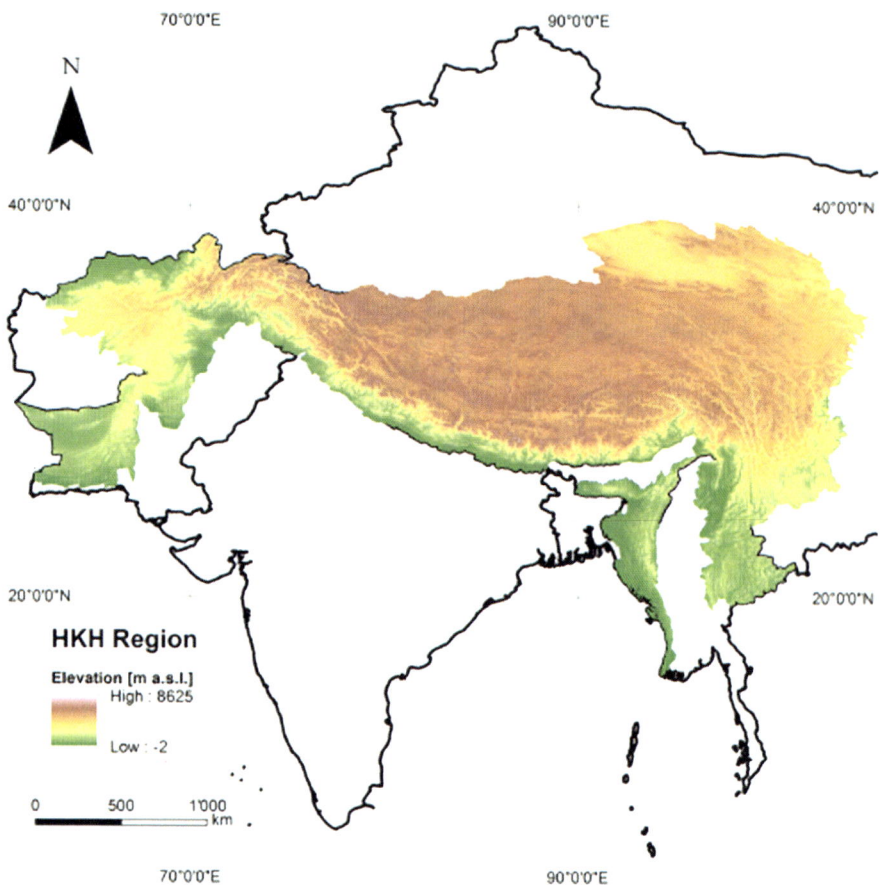

Fig. 14.1 The HKH region as defined by ICIMOD which includes high mountains in Afghanistan, Bhutan, China, India, Myanmar, Nepal, and Pakistan. The SRTM DEM version 4.1 from CGIAR at a spatial resolution of 90 m (Jarvis et al. 2008)

The area is dominated by high mountains, vast glaciers, and large rivers, and has earned many names. As it contains all 14 of the world's highest mountains, those reaching over 8000 m in height, as well as most of the peaks over 7000 m, it has been dubbed the 'roof of the World' and because it is the world's most glaciated place outside the Polar Regions, it is also known as the 'Third Pole'. It is the freshwater tower of South Asia and part of Southeast Asia (ICIMOD 2018). Water originating from snow, glaciers and rain feed the ten largest river systems—the Amu Darya, Indus, Ganges, Brahmaputra (Yarlungtsanpo), Irrawaddy, Salween (Nu), Mekong (Lancang), Yangtse (Jinsha), Yellow River (Huanghe), and Tarim (Dayan)—in Asia that provide water, ecosystem services and the basis of livelihoods to more than 210 million people upstream in the mountains and about 1.3 billion people, a fifth of the world's population, downstream (Karki et al. 2012; Shrestha et al. 2015), the region is also referred to as the 'Water Towers of Asia' (Immerzeel et al. 2010).

14.2 Socio-Economy of the HKH

The Himalayan range alone has the total snow and ice cover of 35,110 km^2 containing 3735 km^3 of eternal snow and ice (Qin 2002). The total for the region is not yet calculated. Hills and mountains, particularly the Hindu Kush Himalaya mountain system, have always constituted places where adaptation, mitigation, and resilience are hallmarks of the people and the landscape they inhabit. Since time immemorial, the people of the Himalaya have maintained a rich cultural identity and have maintained food security and biogenetic diversity within the parameters of their own traditions.

The south Asian countries, including Afghanistan, Bangladesh, Bhutan, India, Maldives, Nepal, Pakistan, and Sri Lanka, are where more than 40% of the world's poor lives and 51% of food and energy deficit exists (Ahmed et al. 2007). The region includes only about 3% of the world's land area but houses 25% of the world population (1.6 billion people). Rice and wheat are the staple foods in this region and they require large amounts of water and energy. Overall, around 3 billion people benefit from food and energy produced in these river basins. Despite being safeguards of a valuable water resource, many inhabitants are amongst the poorest of the poor, many being marginalized subsistence farmers of diverse ethnic groups and minorities, not benefiting from these vast resources. Water for poverty reduction in the HKH region is given relatively little priority (SEI and UNDP 2006; Sharma and Pratap 1994). Economic prosperity is the most powerful way of reducing poverty. Available water resources in the HKH could be used for economic prosperity in the region by maximizing the benefits of this renewable resource. Therefore, poverty reduction and food security by better water management should be major goals.

14.3 Water Resources in the HKH

In the HKH mountains, snow and ice is a dominant feature. Over 54,000 glaciers exist in the HKH region covering a total area of more than 60,000 km^2. Altogether, the glaciers comprise over 6000 km^3 of ice reserves, acting as fresh water reservoirs for the greater region. The Himalayan range alone has a total snow and ice cover of 35,110 km^2 containing 3735 km^3 of eternal snow and ice (Qin 2002). It produces one of the world's largest renewable supplies of freshwater, which is vital for the survival and well-being of billions of people and precious ecosystems in the river basins that stretches beyond the HKH region (Sandhu and Sandhu 2015). In the drier parts of Asia, more than 10% of local river flows come from ice and snow melt (Sivakumar and Stefanski 2011; Shrestha et al. 2015).

The HKH region, being a source of water for the major rivers in Asia, provides water and various ecosystem services for both upstream and downstream communities. The water resource availability of the major rivers in HKH is shown in Table 14.1. The various mountains in the HKH regions are covered by 112,767 km^2 of glaciated areas (Dyurgerov and Meier 2005; Table 14.2).

Table 14.1 Water resources availability of the major rivers in the HKH region

River	River			River basin		
	Annual mean discharge $m^3 s^{-1}$	% of glacier melt in river flow	Basin area (km^2)	Population density (person km^{-2})	Population x 1000	Water availability (m^3 person^{-1} year^{-1})
Amu Darya	1376	Not available	534,739	39	20,855	2081
Brahmaputra	21,261	~12	651,335	182	118,543	5656
Ganges	12,037	~9	1,016,124	401	407,466	932
Indus	5533	Up to 50	1,081,718	165	178,483	978
Irrawaddy	8024	Unknown	413,710	79	32,683	7742
Mekong	9001	~7	805,604	71	57,198	4963
Salween	1494	~9	271,914	22	5982	7876
Tarim	1262	Up to 50	1,152,448	7	8067	4933
Yangtze	28,811	~18	1,722,193	214	368,549	2465
Yellow	1438	~2	944,970	156	147,415	308
Total					**1,345,241**	

Source: IUCN et al. (2003); Mi and Xie (2002); Chalise and Khanal (2001), Merz (2004); Tarar (1982); Kumar et al. (2007); Chen et al. (2007); as cited in Eriksson et al. (2009)

Table 14.2 Glaciated areas in the HKH region

Mountain range	Area (km^2)
Tien Shan	15,417
Pamir	12,260
Qilian Shan	1930
Kunlun Shan	12,260
Karakoram	16,600
Qiantang Plateau	3360
Tanggulla	2210
Gandishi	620
Nianqingtangla	7540
Hengduan	1620
Himalayas	33,050
Hindu Kush	3200
Hinduradsh	2700
Total	112,767

Source: Dyurgerov and Meier (2005)

In addition to its vast water resources, the region is gifted with rich biodiversity and diverse ecosystems, providing a basis for the livelihoods of people inhabiting in the region. Four of the world's 36 biodiversity hotspots are located in the HKH region. Thirty-nine percent of the HKH region is covered with protected areas that harbor a range of ecosystems and provide numerous services in terms of food, water, and climate regulation (Sharma et al. 2010; Molden et al. 2017). However, its physical characteristics make the HKH one of the most hazard-prone regions in the world. Being the youngest mountain chain in the world, the HKH region is also the most fragile.

The features including heavy rains, steep slopes, weak geological formations, accelerated rate of erosion and high seismicity contribute to serious flooding and mass movements of rock and sediment affecting lives and livelihood of millions.

Communities in the HKH region are largely agrarian, relying heavily on local natural resources and subsistence farming on small plots of land. As a result, the communities experience high levels of poverty making them vulnerable to both environmental and socio-economic changes (Shrestha et al. 2015). Moreover, already situated in one of the poorest regions of the world, poverty in the mountains is on average 5% more severe than the national average of the HKH countries, with 31% of the HKH population living below the official poverty line (Gerlitz et al. 2014). In addition, most ecosystems in the region are subject to climatic and non-climatic changes affecting their function and sustainability, thereby affecting livelihoods and community resilience (ICIMOD 2017; Molden et al. 2017).

Further climate related consequences have been projected to occur in the future which can be addressed by policy provisions (Shrestha et al. 2015). Despite having such huge resources and services, due to poverty and limited access to the development services, the people are vulnerable to natural hazards. Therefore, better water resource development and management should be the major goals for poverty reduction and food security (Kayastha 2001).

14.4 Water for Economic Prosperity

14.4.1 Water and Poverty Nexus

The concept of water and poverty is extremely relevant to the rural areas of the HKH region, as the rural poverty in Bangladesh, Pakistan and in Nepal (56%, 39% and 35%, respectively) are higher than urban poverty (36%, 23% and 7%, respectively) (Amarasinghe et al. 2005). Since a major part of the population in this region live in rural areas, rural poverty dominates overall poverty. Agriculture is the dominant source of livelihood for a majority of the rural population in HKH regions and a major contributor to the GDP of countries in the HKH. Thus, reducing rural poverty through improving agriculture income is a major pathway for reducing poverty.

There is adequate evidence of strong linkage of growth in agriculture productivity and reduction in rural poverty. Adequate access to water and quality land resources are crucial for agricultural productivity growth. However, with increasing pressure on limited land resources, enhancing the value of agriculture productivity per unit of water is becoming crucial for rural poverty reduction (Sharma and Pratap 1994). However, the linkage of poverty and the extent of water productivity are not clearly understood. Such knowledge would be valuable in designing appropriate and geographically targeted interventions for increasing water productivity and thereby reducing rural poverty. Groundwater exploitation, water logging, salinity and climate change are threatening agriculture in many parts. To what extent these inhibit the growth of agriculture productivity and poverty alleviation is not adequately understood. Furthermore, there is increasing evidence of climate change causing a gradual depletion of the Himalayan glaciers. Depleting glaciers could have profound effect on the seasonal and spatial water availability. Given these changes, it is extremely important to understand the changing dynamics of the nexus of water, land and poverty.

14.4.2 Water Exploitation and Use in HKH

The Hindu Kush Himalayan (HKH) region depends heavily on water resources for irrigation, food, hydropower, sanitation, and industry, as well as for the functioning of many important ecosystem services. Water thus directly contributes to the national GDP and to livelihoods and income generation at the local level. In the HKH region, hydropower is one of the most promising environmentally friendly sources of energy. With a potential estimated at 500,000 MW, the region has abundant opportunities for hydropower development. Energy security can open up opportunities for development and employment and contribute to the national GDP. However, many countries in the region have been able to tap only a small fraction of their available potential. Out of the 42,000 MW potential reported in Nepal, only about 2% is harnessed so far, whereas Pakistan has harnessed 11% of its total potential. Still, people in both these countries face many hours of scheduled power cuts.

Water and food share a strong nexus, both being essential ingredients for human survival and development. Agriculture is a major contributor to the GDP of countries in the HKH. The Indus River system is a source of irrigation for about 144,900 hectares of land, whereas the Ganges basin provides irrigation for 156,300 hectares of agricultural land. Access to water resources for food production and their sustainable management is a concern from the local to national level. Amid rapid environmental and socio-economic changes, the growing population will require more water and food, and equitable access to vital resources has become a major question (Schreier and Shah 1996). Sustainable solutions to these problems require efficient use of water resources for agricultural use in which technological innovation plays a vital role. The intervention of infrastructure development and the flow regime changes in the downstream areas where communities depend on water resources for livelihoods such as fishing are important. A major concern is how to make sure that a certain minimum flow is maintained so as to sustain freshwater supply and support dependent ecosystems. Although water is the foundation of sustainable development, water management in the HKH region remains fragmented and uncoordinated and does not take relevant regional issues into account (Kayastha 2001).

Due to its physical setting, the HKH region is prone to various water-induced hazards (e.g., landslides, floods, glacial lake outburst floods, and droughts). Every year, during the monsoon season, floods bring havoc to the mountains and the plains downstream. Globally, 10% of all floods are transboundary, and they cause over 30% of all flood casualties and account for close to 60% of all those displaced by floods. The social and economic settings of the region make the people vulnerable to natural hazards. Lack of supportive policy and governance mechanisms at the local, national and regional levels, and the lack of carefully planned structural and non-structural measures of mitigation lead to increased vulnerability.

With poverty affecting larger portions of the burgeoning population, pressure on natural resources will increase, as will the demand for food, water and energy (Rasul 2016), leading to conflict in the region. The potential impacts of climate change are likely to accelerate the retreat of glaciers and change the regional hydrological regime, increase the magnitude and frequency of natural disasters, reduce water availability during low flow periods and deteriorate water quality (Immerzeel et al. 2013).

14.5 Integrated Approach: Water, Energy, and Food Nexus

Issues and challenges in the water-energy-food sectors are interwoven in complex ways in the HKH region (Fig. 14.2; Endo et al. 2017; Rasul 2016). The downstream populations are heavily dependent to the upstream ecosystem services (Biggs et al. 2015).

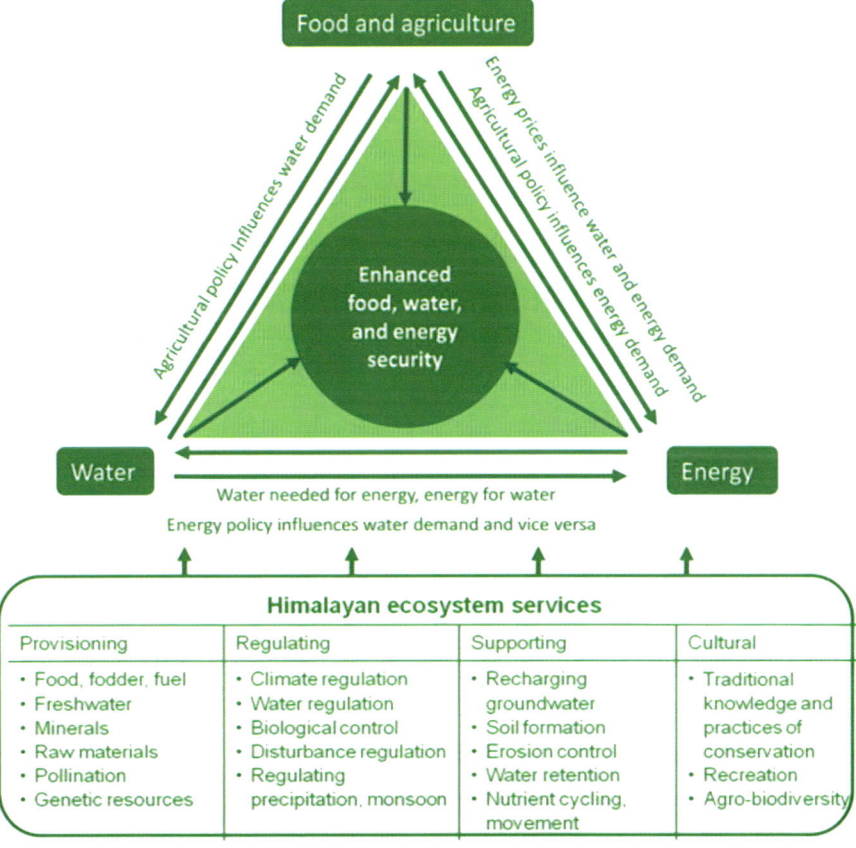

Fig. 14.2 Interdependencies of food, water, energy, and ecosystem services (Adopted from Rasul 2014)

14.6 Trans-Boundary Issues of International Rivers

The river systems in the HKH are of transboundary nature and need regional cooperation for ensuring water availability and use (Mukherji et al. 2015; Rasul 2014). The river system has important implications for social, economic and political aspects. The transboundary issues create interdependencies between the countries (Choudhury 2015). For example, most of the rivers flowing to and from Nepal are transboundary rivers. Large part of the Koshi River basin falls in the Tibetan part of China and flows south into the Ganges in India. Flood and sedimentation in the rivers, water sharing, and upstream-downstream linkages are some of the potential areas of international water issues. The adjoining countries must work in collaboration and cooperation to address these issues.

Several pertinent transboundary issues exist (Choudhury 2015; Rasul 2016; Molden et al. 2017) that need to be addressed to meet the water and poverty objectives of the region: (1) solving boundary issues, inadequate knowledge of the hydrological regime (e.g., ground and surface water, total water budget; (2) the different views in the countries should have mutual understanding on the basics of water use; (3) there are local, regional and national levels of water authorities in different countries. The management of transboundary issues should consider how to address them. Local management can be political on transboundary issues; (4) lack of data can prevent countries from agreeing on solutions; (5) boundaries can change and compound other issues associated with transboundary water; (6) "dueling experts" can hold back transboundary issues. For example, hydrologists can obtain different results depending on who is funding the study. When experts fail to come to consensus, politicians and other stakeholders can choose data and use them to their advantage; and (7) transboundary issues are expensive to achieve. Some issues can delay agreements by decades and cost millions of dollars.

14.7 Climate Adaptation Program in the HKH Region

To meet the challenges emerging in the HKH, a pioneering program called the Himalayan Climate Change Adaptation Program (HICAP) has been launched from 2011. One of the core objectives of this program is to promote the understanding of the needs and gaps across the HKH. The ultimate goal is to foster joint dialogue on mountain ecosystems in the context of climate change and adaptation, and to encourage the HKH countries to work together at the regional and global levels on these issues. The HICAP seeks to develop an overview of the existing adaptation policy measures in the HKH countries, and address the needs of mountain landscape and people, and to identify critical issues that must be addressed to meet current and future risks associated with climate change (ICIMOD 2017).

Institutionally, the International Centre for Integrated Mountain Development (ICIMOD) is the intergovernmental organization to promote knowledge and action aimed at shared management on issues and priorities that are common across the HKH. The organization generates and shares knowledge across countries and communities on the issues of common concern, such as poverty alleviation, adaptation, and resilience building. It addresses resources that cross the boundaries, such as water and ecosystems (Molden et al. 2017).

14.8 Regional Policies, Institutions and Agreements

Although there seems to be a lack of comprehensive climate change policy for the entire HKH region, there are various policies/programs in the HKH countries for dealing with climate change (ICIMOD 2017). There are ongoing efforts for transboundary cooperation among the HKH countries, and multiple national policy responses within the countries themselves. For example, Afghanistan and Bangladesh have the Climate Change Strategy and Action Plan (CCSAP) and the National Adaptation Program of Action (NAPA); Bhutan has constituted NAPA; China has the National Climate Change Program (NCCP) and Five-Year Plan; India has the National Action Plan on Climate Change (NAPCC); Myanmar has NAPA; Nepal has NCCP, LAPA, NAPA and NAP; and Pakistan has the National Climate Change Policy (NCCP). These policies, plans and programs address climate related issues in HKH regions, directly or indirectly. The country-wise climate related policies and provisions are discussed below.

14.8.1 Afghanistan

(a) Climate Change Strategy and Action Plan (CCSAP, 2015; NEPA/UNEP 2015): The CCSAP of Afghanistan aims to: (1) integrate and mainstream climate change into the national development framework; (2) support the creation of a national framework for action on climate change adaptation; (3) identify low emission development strategies; (4) improve coordination and partnerships between government institutions, civil society, the international donor community, and the private sector; (5) increase availability and access to additional financial resources for effectively addressing climate change; and (6) identify policy initiatives to address climate change adaptation in the vulnerable sectors and areas of agriculture, food security, water, biodiversity, natural disasters, health, and infrastructure.

(b) National Adaptation Program of Action for Climate Change (NAPA, 2009) of Afghanistan aims to: (1) identify priority activities that respond to urgent and immediate needs with regard to adaptation to climate change; (2) identify the country's most vulnerable areas to climate change: agriculture; biodiversity and ecosystems; energy; forests and rangelands; natural disasters; and water; (3) identify Afghanistan's key priority areas for climate change- improved water management and use efficiency; and community-based watershed management; and (4) identify Afghanistan's key challenges for addressing climate change, including a lack of expertise within relevant government institutions as a result of a low level of education, poor financing, and the fact that most government institutions are relatively nascent.

14.8.2 Bangladesh

Climate Change Strategy and Action Plan (CCSAP), 2009 of Bangladesh is a part of the over-all development strategy of the country. The climate change constraints and opportunities are being integrated into the over-all plan and programs involving all sectors and processes for economic and social development (MoEF 2009). The major provision made in the plan include: (1) food security, social protection and health; (2) comprehensive disaster risk management; (3) infrastructure development; (4) research and knowledge management; (5) mitigation and low carbon development; and (6) capacity building and institutional development.

National Adaptation Program of Action (NAPA) (MoEF 2005) includes provisions like: (1) reduction of climate change hazards through coastal afforestation with community participation; (2) providing drinking water to coastal communities to combat enhanced salinity due to sea level rise; (3) capacity building for integrating climate change in planning, designing of infrastructure, conflict management and land water zoning for water management institutions; (4) climate change and adaptation information dissemination to vulnerable community for emergency preparedness measures and awareness raising on climatic disasters; (5) construction of flood shelter, and information and assistance centres to cope with recurrent floods in the major floodplains; (6) mainstreaming adaptation to climate change into policies and programs in different sectors (focusing on disaster management, water, agriculture, health and industry); (7) inclusion of climate change issues in curriculum at secondary and tertiary educational institution; (8) enhancing resilience of urban infrastructure and industries to impacts of climate change; (9) development of eco-specific adaptive knowledge (including indigenous knowledge) on adaptation to climate variability to enhance adaptive capacity for future climate change; (10) promotion of research on drought, flood and saline tolerant varieties of crops to facilitate adaptation in future; (11) promoting adaptation to coastal crop agriculture to combat increased salinity; (12) adaptation to agriculture systems in areas prone to flash flooding in northeast and central regions; (13) adaptation to fisheries in areas prone to flooding in northeast and central regions through adaptive and diversified fish culture practices; (14) promoting adaptation to coastal fisheries through culture of salt tolerant fish species in coastal areas of Bangladesh; and (15) exploring options for insurance and other emergency measures to cope with climatic disasters.

14.8.3 Bhutan

National Adaptation Program of Action (NAPA), 2006 of Bhutan aims to: (1) eradicate poverty, and achieve food security by increasing access to remote areas; (2) safeguard hydropower generation as the backbone of the economy; (3) promote gender equality and empower women (reduce gender disparity in tertiary education); (4) reduce child mortality; (5) ensure environmental sustainability; and (6) minimize loss of life and livelihoods due to natural and climate related disasters.

14.8.4 China

The Government of China has adopted the principle of 'equal treatment to development and conservation' since the early 1980s, with immediate emphasis on energy conservation as a matter of strategic importance (NDRC 2007). Energy conservation was promoted through development of legal provisions; in 1997, the Energy Conservation Law was introduced (Amended in 2007 and 2008) (Nachmany et al. 2015) and in 2000, the Law on the Prevention and Control of Atmospheric Pollution was enforced.

In 2004, the 'China Medium- and Long-Term Energy Development Plan Outlines, 2004–2020' was approved by the State Council and the first China Medium-and Long-Term Energy Conservation Plan was launched by National Development and Reform Commission (NDRC). In 2005, the National People's Congress adopted Renewable Energy Law and the State Council issued notification on the immediate priorities for building a conservation-oriented economy and for accelerating the development of a circular economy. Moreover, the State Council issued the decision to publish and implement the interim provisions on promoting industrial restructuring and to strengthen environmental protection by applying a scientific approach for development.

Further, in 2006, the State Council issued a decision to strengthen energy conservation. All these documents serve as the policy and legal guarantee to further enhance China's capability in addressing climate change. In 2006, National Guideline for Medium- and Long-term Plan for Scientific and Technological Development (2006–2020) and Renewable Energy Act (Amendment 2009) were issued (Nachmany et al. 2015) and in 2007, the China's Scientific and Technological Actions on Climate Change was issued (China 2012). In the same year (2007), China's National Climate Change Program (CNCCP) was introduced which strives to build a resource conservative and environmentally friendly society, enhance national capacity to mitigate and adapt to climate change, and make further contributions to the protection of the global climate system (NDRC 2007).

In 2013, the National Strategy for Climate Change Adaptation was issued. Moreover, in 2014, the government of China passed the Energy Development Strategy Action Plan (2014–2020) and the National Plan for Tackling Climate

Change (2014–2020) (Nachmany et al. 2015). The 12th Fiscal Year Plan (2011–2015) and 13th Five Year Plan (2016–2020) include a number of environmental related goals pertaining to climate change. All these initiatives show that China is building institutional, technical and societal capacities to deal with the climate change. For this purpose, institutions like National Coordination Committee on Climate Change (NCCCC) and China Council for International Cooperation on the Environment and Development (CCICED) were established (King et al. 2012).

14.8.5 India

To address challenges of climate change, in 2007, the Indian government established the Prime Minister's Council on Climate Change (PMCCC). The Council, published the National Action Plan on Climate Change (NAPCC) in 2008 (GoI, PMCCC 2008) with the following guiding principles: (1) protecting the poor and vulnerable sections of society through an inclusive and sustainable development strategy, sensitive to climate change; (2) achieving national growth objectives through a qualitative change in direction that enhances ecological sustainability, leading to further mitigation of greenhouse gas emissions; (3) devising efficient and cost-effective strategies for end-use demand side management; (4) developing appropriate technologies for both adaptation and mitigation of greenhouse gases emissions extensively as well as at an accelerated pace; (5) engineering new and innovative forms of market, regulatory and voluntary mechanisms to promote sustainable development; (6) effective implementation of programs through unique linkages, including with civil society and local government institutions and through public-private-partnerships; and (7) welcoming international cooperation for research, development, sharing and transferring of technologies enabled by additional funding and a global IPR regime that facilitates technology transfer to developing countries under the UNFCCC.

In addition, the NAPCC has visualized eight national missions to achieve sustainable development along with advances in economic and environmental objectives (GoI, PMCCC 2008). Among others, the National Water Mission and National Mission for Sustaining the Himalayan Ecosystem are directly linked with the water resources across the country, as well as the HKH region. The National Water Mission aims to ensure sustainable water supply by conserving water, minimizing waste and ensuring equitable distribution of water resources throughout India; and the National Mission for Sustaining the Himalayan Ecosystem aims to enhance understanding of climate change impacts and adaptations in the Himalayas. The information obtained from this mission will feed into policy formulation for suitable management practices for the Himalayan ecosystem. In order to decentralize the NAPCC, the Government of India has issued an order for all states to submit their respective State Action Plans on Climate Change (SAPCC), which have been reportedly prepared for almost all states and Union Territories across India.

The National Water Mission, approved by the Cabinet in 2011, ensures integrated water resource management, conserves water, minimizes wastage and ensures equitable distribution of water within states. The mission is run by the Ministry of Water Resources, River Development and Ganga Rejuvenation (2011). The five identified goals of the mission (MoWR 2010) include: (a) creating a comprehensive water data base in the public domain and assessing the impact of climate change on water resources; (b) promoting citizen and state action for water conservation, augmentation and preservation; (c) focusing attention to overexploited areas; (d) increasing water-use efficiency by 20%; and (e) promoting basin-level integrated water resources management.

The mission document also aims to formulate river-linking projects. Since water is a state subject, the mission identifies the need for states to prepare their state-specific plans of action. It envisages that the respective State Specific Action Plans (SSAP) would conduct critical assessments of current water policies, formulate water budgets, and create comprehensive and integrated water plans for water security, safety and sustainability till 2050 (MoWR 2010).

The National Environment Policy (NEP), 2006, provides relevant measures for conservation of mountain ecosystems (GoI, PMCCC 2008). The measures include: (1) adopt appropriate land-use planning and watershed management practices for sustainable development of mountain ecosystems; (2) adopt "best practice" norms for infrastructure construction in mountain regions to avoid or minimize damage to sensitive ecosystems and despoiling of landscapes; (3) encourage cultivation of traditional varieties of crops and horticulture by promotion of organic farming, enabling farmers to realize a price premium; (4) promote sustainable tourism through access to ecological resources, and multi-stakeholder partnerships to enable local communities to gain better livelihoods, while leveraging financial, technical, and managerial capacities of investors; (5) take measures to regulate tourist inflows into mountain regions to ensure that these remain within the carrying capacity of the mountain ecology; and (6) consider particular unique mountain-scapes as entities with "Incomparable Values", in developing strategies for their protection.

14.8.6 Myanmar

National Adaptation Program of Action to Climate Change (NAPA), 2012, of Myanmar (MoECF 2012) includes eight priority areas: (1) agriculture; (2) early warning system; (3) forest; (4) public health; (5) water resources; (6) Coastal Zone; (7) energy and industry; and (8) biodiversity.

The National Sustainable Development Strategy (NSDS), 2009, comprehensively incorporates environmental considerations into social and economic development to achieve the goal of sustainable development. The document identifies three main goals for the country: (1) sustainable management of natural resources; (2) integrated economic development; and (3) sustainable social development (NCEA, MoF, and UNEP-RC-AP 2009).

14.8.7 Nepal

In Nepal, the National Five Year Plan provides strategic directions and a policy framework. Moreover, there are several legal/policy provisions to address climate change issues. For instance; the National Adaptation Program of Action (NAPAs) (GoN 2010), the National Framework for Local Adaptation Plans for Action (LAPA) (GoN 2011a), the National Climate Change Policy 2011 (GoN 2011b) and the National Adaptation Plans (NAPs) (GoN 2017) aim to address adaptation needs, and the National Strategies/Action Plans on Climate Change focus on monitoring, warning and response to climatic events. Various national climate change policies aim to steer the country towards climate resilient development (ICIMOD 2017).

The National Adaptation Program of Action (NAPA), 2010, has identified nine urgent and immediate climate change adaptation priority programs related to six thematic sectors: agriculture, forest biodiversity, water resources, health, infrastructure, and disaster. It is the first comprehensive government response to climate change which has also specified a coordination mechanism and implementation modality for climate change adaptation programs in Nepal. It has provided a basis for developing and implementing adaptation projects in Nepal. The NAPA has established a multi-stakeholder Climate Change Initiatives Coordination Committee (MCCICC).

National Framework for Local Adaptation Plans for Action (LAPA), 2011, was developed by the Government of Nepal as an operational instrument to implement NAPA prioritized adaptation actions. Its goal is to integrate climate adaptation and resilience into local and national planning, and to incorporate the four guiding principles of being bottom-up, inclusive, responsive and flexible. The aim of the LAPA is to integrate climate adaptation activities into local and national development planning processes, and to make development more climate-resilient.

The National Climate Change Policy (NCCP), 2011, was approved by the Government of Nepal in 2011 (GoN 2011b). The goal of the NCCP was to improve livelihoods by mitigating and adapting to the adverse impacts of climate change, adopting a low-carbon emissions socio-economic development path, and supporting and collaborating in the spirit of the country's commitments to national and international agreements related to climate change. It has time-bound targets to address climate risks and vulnerability in the country.

The National Adaptation Plan (NAP), 2017 (MoPE 2017) was constituted based on the experiences of NAPA and LAPA which provided opportunities for designing effective governance mechanisms that facilitate climate change adaptations and livelihood activities at the local level. The NAP has two key objectives: (1) reducing vulnerability to the impact of climate change by building adaptive capacity and resilience; and (2) facilitating the integration of climate change adaptation in a coherent manner into relevant new and existing policies, programs and activities, in particular development planning processes and strategies, within all relevant sectors and at different levels.

14.8.8 Pakistan

Pakistan is among the highly vulnerable countries with respect to the adverse impact of climate change, and requires policy formulation and enforcement. The National Climate Change Policy (NCCP) was instituted in 2012 which provides a framework for addressing the climate change issues faced by Pakistan at present and/or in the future (MoCC 2012; NCCP 2012). The over-all goal of the NCCP is to ensure that climate change is mainstreamed to the economically and socially vulnerable sectors of the economy and to steer Pakistan towards climate resilient development. The major objectives of the NCCP are to: (1) pursue sustained economic growth by appropriately addressing the challenges of climate change; (2) integrate climate change policy with other inter-related national policies; (3) focus on poverty and gender sensitive issues while promoting mitigation in a cost-effective manner; (4) ensure water, food and energy security in the face of challenges posed by climate change; (5) minimize risks arising from the expected increase in frequency and intensity of extreme weather events such as floods, droughts and tropical storms; (6) strengthen inter-ministerial decision-making and coordination mechanisms on climate change; (7) facilitate effective use of opportunities, particularly financial, available both nationally and internationally; (8) foster the development of appropriate economic incentives to encourage public and private sector investments in adaptation measures; (9) enhance the awareness, skill and institutional capacity of relevant stakeholders; and (10) promote conservation of natural resources and long term sustainability.

The Water Resources Related Policies and Provisions are inextricably linked with climate; thus, the projected climate change has serious implications for the HKH region's water resources. Freshwater resources in the HKH depends on snow/glacial melt (Table 14.1) and monsoon rains, both of which are sensitive to climate change. Country specific climate change projections strongly suggest a decrease in glacier volume and snow cover leading to alterations in the seasonal flow pattern of the Indus River System (IRS); increased annual flows for a few decades followed by a decline in flow in subsequent years; increase in the formation and outburst of glacial lakes; higher frequency and intensity of extreme climate events coupled with irregular monsoon rains causing frequent floods and droughts; and greater demand on water due to higher evapo-transpiration rates at elevated temperatures.

14.9 A Case of the Koshi River Basin *The Policy and Institutional Mechanism from Flood Risk Perspective*

The Koshi River basin is shared by China, India and Nepal and there seems to be no preventative plan for water induced disasters such as flood risk management from a river basin perspective. The management of rivers is largely driven by water policies, while disasters of all types are governed by disaster management policies

(Neupane et al. 2015). Despite being endowed with fertile land resources, water resources, potentiality of fisheries, hydro-power development and tourism, a higher incidence of poverty and low per-capita income is observed in the Koshi basin districts than in the corresponding national figures (Neupane et al. 2015). According to Neupane et al. (2015), to support the livelihoods of the Koshi basin, water related interventions are an entry point. For this purpose, the governments should focus on basin/sub-basin level policies which should include investments on water related infrastructure. Institutional capacity building from local to trans-national scales and establishment of good water governance (local to trans-national level) can address the major issues of the Koshi basin and improve the livelihoods.

Neupane et al. (2015) reviewed the water policies and institutional mechanisms from the perspective of flood risk in the Koshi river basin. For China, the history of water legislation is short and the laws governing river basin management are recent (Shen 2009). In 2002, the Chinese government amended the Water Law 1988, to establish a legal foundation for integrated water resource management and demand management.

In Nepal, traditional water resources management practices were focused on the supply side, where only technical solutions were considered to meet the growing demand for water. There were sectoral agencies focusing on isolated projects on irrigation, drinking water supply and sanitation, hydropower, flood control, and other uses. Mostly independent sector authorities controlled these projects on the basis of command and control (WECS 2005). The result was inter-sectoral and inter-regional conflicts over water use, which led to problems in efficiency, equity and environmental considerations. To overcome these problems, Nepal has realized that development and management of water resources have to move from the sectoral approach to an integrated and holistic approach with greater participation of the community as well as other relevant stakeholders. Accordingly, the Water Resources Act, 1992; the Water Resources Strategy (WRS), 2002; and the National Water Plan (NWP) 2005 were developed by the government of Nepal which are long-term planning of water resources in Nepal (Fig. 14.3). The WRS 2002 was formulated and based on identified policy principles with an IWRM approach.

Likewise, India, in the national water policy of 2012, laid down the principle of equity and social justice, which considers the utilization and allocation of water through informed decision-making and established good governance systems. This principle ensures the participation of women and disadvantaged groups of people in planning, implementation and decision-making processes of water resource management. Inter-basin water transfers are important not only for ensuring food security but also for meeting basic human needs and achieving equity and social justice.

The importance of having a unified perspective in planning, management and use of water resources has been recognized. Planning, development and management of water resources need to be governed by integrated perspectives considering local, regional, state and national levels, and keeping in view human, environmental, social and economic needs (MoWR 2012). The policy also recognized that river basins are to be considered as the basic unit of all-hydrological planning. There is a separate section on 'institutional arrangements' under the new water policy of India.

Fig. 14.3 The Karnali River in Nepal: About 210.2 million liters per day (MLD or km^3/year) water flow out of the country from all the rivers of Nepal. (Photograph by S. Thakuri)

The policy has strongly made two suggestions related to institutional aspects such as IWRM by taking river basin/sub-basin as a unit for planning, development and management of water resources. For this, the departments and organizations at the centre/state government levels should be restructured and made multi-disciplinary. The policy also suggests for appropriate institutional arrangements for each river basin to collect and collate all data on a regular basis with regard to rainfall, river flows, area irrigated for crops, utilization by both surface and ground water and to publish water budgeting and accounting based on the hydrologic balances for each river basin. In addition, an appropriate institutional arrangement for each river basin should also be developed for monitoring water quality.

Water Related Policy and Institutional Gap: All three countries- China, India and Nepal, of the Koshi River basin are facing increasing pressures of too much and too little water as a result of climatic factors, poor water governance and institutional deficit. The situation calls for harmonizing the water related laws, regulations and acts with the policies and programs for flood and drought risk reduction in the basin. Moreover, water resources development is a multi-sector and multi-faceted concern and it calls for coordinated planning and management of irrigation, hydropower generation, water supply, industrial use, disaster risk reduction and environmental protection.

The review carried out on the Koshi Basin Area showed that the coordinating role of existing institutions is very limited as most water management activities are being carried out by different water use sectors and sectoral agencies working at different administrative units (Neupane et al. 2015). It seems that most of the legal and institutional arrangements in China, India and Nepal are first targeted at the central level where several ministries, commissions, authorities and departments are involved for the development and regulation of water resources, although in most cases they are uncoordinated and fragmented.

Moreover water induced disaster risk reduction policy is not adequately connected with water management policies and institutions in the basin countries. This policy is greatly needed for the Koshi basin, as it is among the most disaster prone of the ten rivers that flow from the Hindu Kush Himalayan region.

14.9.1 Stand Alone Approaches of Flood Risk Management

The current disaster risk management approaches and practices by the three countries in the Koshi River basin have adopted independent paths focused merely on mitigating floods, while ignoring the holistic IWRM approach of water storage, water management and water-based livelihoods. In the downstream of the Koshi River basin, particularly, in Bihar, the flood risk can be reduced primarily by the construction of embankments. While, embankments helped reduce flood risk for some locations in the Koshi River basin in Bihar, they also proved to have adverse effects such as interference with drainage and the inability to handle erosion.

In the changing mountain context, communities in the Koshi basin have been responding to environmental uncertainties and hazards through diversifying livelihood options, changing land use patterns, seasonal migration and by changing food habits and sanitation practices (Eriksson et al. 2009). To strengthen this process, the national policies and institutions have an important role to play by making sure that local needs, priorities and concerns are reflected in the broader and integrated decision-making on a river basin scale. A study conducted by ICIMOD in China (Yunnan), India (Bihar) and Nepal (Koshi Basin Area) showed that effective and integrated use of existing policies, institutional frameworks capabilities and enabling conditions coupled with access to livelihood options and opportunities can enhance the capacity to respond successfully to environmental uncertainties, including water stress and hazards (Eriksson et al. 2009).

14.10 Conclusions and Perspectives

The HKH region is rich in water resources, but they are unevenly distributed among the countries. Water resources could be used for economic prosperity in the region by maximizing the benefits of this renewable resource. Poverty reduction and food security should be the major goals of better water management. We underline the needs of international cooperation, institutional framework, and regional thinking.

In all eight countries of the region, climate change related policies and plans are well covered in the country and regional governance. Some countries have more pronounced programs, such as China's National Climate Change Program and Five-Year Plan. India, Nepal and Pakistan seem well prepared to respond to climate change in the Himalayan environment. Bhutan has strategies by maintaining the forest coverage and biodiversity conservation. These policies, plans and programs help address climate related issues in the HKH region.

The HKH region is known as water resource rich region; however, its use is not properly optimized. People from this region depend on water resources for drinking, irrigation, food, hydropower, sanitation, and industry, as well as for the functioning of important ecosystem services. Prosperity of the region lies in maximizing the use of water resources and ensuring efficient use of the resources. Further, the region is prone to water induced disasters such as landslides, floods, glacial lake outburst floods, and droughts. Due to its physical setting, the HKH region is prone to various water-induced hazards (e.g., landslides, floods, glacial lake outburst floods, and droughts). Responses to address these potential disasters are needed to ensure the living quality of the people.

Considering the high rural poverty in Bangladesh, Pakistan, Nepal, Afghanistan, Myanmar and Pakistan, water management becomes meaningful to address the poverty issues of the region. There are examples of proper use of water in agriculture, food security and energy in parts of Bhutan, China and India.

Water, being linked with livelihoods and development should be properly researched; the water body types, Himalayan ecosystem services, require scientific and policy data.

The human population in the HKH depends on water resources for multiple purposes, which could be achieved only through regional cooperation. Thus, regional cooperation in transboundary waters and related issues should be met through water resources research, management, and efficient water use. Different bilateral and multilateral institutions are working on these issues and they need to be fostered further for the efficient use of water.

References

Ahmed, A.U., R.V. Hill, L.C. Smith, et al. 2007. *The world's most deprived: Characteristics and causes of extreme poverty and hunger*. Washington, DC: International Food Policy Research Institute.

Amarasinghe, U.A., M. Giordano, Y. Liao, et al. 2005. *Water supply, water demand and agricultural water scarcity in China: A basin approach*. CPSP Report, 11.

Biggs, E.M., E. Bruce, B. Boruff, et al. 2015. Sustainable development and the water–energy–food nexus: A perspective on livelihoods. *Environmental Science & Policy* 54: 389–397.

Chalise, S.R., and N.R. Khanal. 2001. An introduction to climate, hydrology and landslide hazards in the Hindu Kush-Himalayan region. In *Landslide hazard mitigation in the Hindu Kush-Himalayas*, ed. L. Tianchi, S.R. Chalise, and B.N. Upreti, 51–62. Kathmandu: ICIMOD.

Chen, Y.N., W.H. Li, C.C. Xu, et al. 2007. Effects of climate change on water resources in Tarim River Basin, Northwest China. *Journal of Environmental Sciences* 19: 488–493.

China. 2012. *Second national communication on climate change*. The People's Republic of China.

Choudhury, E. 2015. Nature of transboundary water conflicts: Issues of complexity and the enabling conditions for negotiated cooperation. *Journal of Contemporary Water Research & Education* 155: 43–52.

Dyurgerov, M.D., and M.F. Meier. 2005. *Glaciers and changing earth system: A 2004 snapshot*. Boulder: Institute of Arctic and Alpine Research, University of Colorado.

Endo, A., I. Tsurita, K. Burnett, et al. 2017. A review of the current state of research on the water, energy, and food nexus. *Journal of Hydrology: Regional Studies* 11: 20–30.

Eriksson, M., J. Xu, A.B. Shrestha, et al. 2009. *The changing Himalayas: Impact of climate change on water resources and livelihoods in the greater Himalayas*. Kathmandu: ICIMOD.

Gerlitz, J.Y., S. Banerjee, B. Hoermann, et al. 2014. *Poverty and vulnerability assessment: A survey instrument for the Hindu Kush Himalayas*. Kathmandu: International Centre for Integrated Mountain Development.

GoN, 2010. *National Adaptation Program of Action (NAPA)*. Kathmandu: Government of Nepal.

GoN, 2011a. *National Framework on Local Adaptation Plans for Action (LAPA)*. Kathmandu: Government of Nepal, Ministry of Environment.

GoN, 2011b. *National Climate Change Policy, 2011*. Kathmandu: Ministry of Environment, Government of Nepal.

Government of India, Prime Minister's Council on Climate Change. 2008. *National Action Plan on Climate Change, 2008*. http://www.moef.nic.in/sites/default/files/Pg01-52_2.pdf. Accessed 4 Aug 2018.

ICIMOD. 2017. In: *Proceedings of the regional policy workshop on adaptation outlook for the Hindu Kush Himalaya. ICIMOD Proceedings 2017/5*. Kathmandu: ICIMOD.

ICIMOD, 2018. *Hindu Kush Himalaya Region*. Kathmandu: ICIMOD. http://www.icimod.org/?q=1137. Accessed 14 July 2018.

Immerzeel, W.W., L.P. Van Beek, and M.F.P. Bierkens. 2010. Climate change will affect the Asian water towers. *Science* 328: 1382–1385.

Immerzeel, W., F. Pellicciotti, and M. Bierkens. 2013. Rising river flows throughout the twenty-first century in two Himalayan glacierized watersheds. *Nature Geoscience* 6: 742–745.

IUCN, IWMI, Ramsar Convention, WRI. 2003. *Water resources atlas*. http://multimedia.wri.org/watersheds_2003/index.html. Accessed 12 June 2007.

Jarvis, A., H.I. Reuter, A. Nelson, et al. 2008. *Hole-filled SRTM for the globe Version 4*.

Kayastha, R.L. 2001. Water resources development of Nepal: A regional perspective. In *Sustainable development of the Ganges-Brahmaputra-Meghna Basin*, ed. A.K. Biswas and J.I. Uitto, 122–144. Tokyo: United Nationals University Press.

Karki, M., R.S. Tolia, T.J. Mahat, et al. 2012. *Sustainable mountain development in the Hindu Kush–Himalaya from Rio 1990 to Rio 2012 and beyond*. Kathmandu: ICIMOD.

King, D., M. Cole, S. Tyldesley, et al., 2012. *The response of China, India and Brazil to climate change: A perspective for South Africa*. Oxford: Smith School of Enterprise and Environment, University of Oxford.

Kumar, V., P. Singh, and V. Singh. 2007. Snow and glacier melt contribution in the Beas River at Pandoh Dam, Himachal Pradesh, India. *Hydrological Sciences–Journal des Sciences Hydrologiques* 52 (2): 376–388.

Merz, J. 2004. *Water balances, floods and sediment transport in the Hindu Kush-Himalayas*. Berne: Institute of Geography, University of Berne.

Mi, D., and Z. Xie. 2002. *Glacier inventory of China*. Xi'an: Xi'an Cartographic Publishing House.

Ministry of Water Resources, River Development and Ganga Rejuvenation. 2011. *Mission document*. http://wrmin.nic.in/writereaddata/nwm28756944786.pdf. Accessed 30 Oct 2017.

MoCC. 2012. *National Climate Change Policy*. Islamabad: Ministry of Climate Change, Government of Pakistan.

MoECF. 2012. *Myanmar's National Adaptation Program of Action (NAPA) to climate change*. Myanmar: National Environmental Conservation Committee, Ministry of Environmental Conservation and Forestry.

MoEF. 2005. *National Adaptation Plan of Action (NAPA)*. Dhaka: Ministry of Environment and Forest, Government of the People's Republic of Bangladesh.
MoEF, 2009. *Bangladesh climate change strategy and action plan 2009*. Dhaka: Ministry of Environment and Forest, Government of the People's Republic of Bangladesh.
Molden, D., E. Sharma, A.B. Shrestha, et al. 2017. Advancing regional and trans-boundary cooperation in the conflict-prone Hindu Kush-Himalaya. *Mountain Research Development* 37 (4): 502–508.
MoPE, 2017. *Synthesis of the stocktaking report for the National Adaption Plan (NAP) formulation process in Nepal*. Kathmandu: Government of Nepal.
MoWR. 2010. *Mission document*. Ministry of Water Resources. http://documents.gov.in/central/15658.pdf. Accessed 3 Oct 2017.
MoWR. 2012. *National Water Policy (2012)*. Government of India, Ministry of Water Resources, India.
Mukherji, A., D. Molden, S. Nepal, et al. 2015. Himalayan waters at the crossroads: Issues and challenges. *International Journal of Water Resources Development* 31 (2): 151–160.
Nachmany, M., S. Fankhauser, J. Davidová, et al. 2015. *Climate change legislation in China: An excerpt from the 2015 Global Climate Legislation Study—A review of climate change legislation in 99 countries*. London: Grantham Research Institute on Climate Change and the Environment, Globe and Inter Parliamentary Union (IPU).
NCCP. 2012. National Climate Change Policy of Pakistan.
NECA, MoF, UNEP-RC-AP. 2009. National Sustainable Development Strategy of Myanmar. National Commission for Environmental Affairs, Ministry of Forestry, United Nations Environment Program, Regional Resource Center for Asia Pacific.
NEPA/UNEP. 2015. *Climate change and governance in Afghanistan*. Kabul: National Environmental Protection Agency (NEPA) and United Nations Environment Program (UNEP).
Neupane, N., H.K. Nibanupudi, and M.B. Gurung. 2015. Interlacing of regional water policies, institutions and agreements with livelihoods and disaster vulnerabilities in the HKH region: A case study of Kosi River Basin. In *Mountain hazards and disaster risk reduction. Disaster risk reduction (methods, approaches and practices)*, ed. H. Nibanupudi and R. Shaw. Tokyo: Springer.
Qin, D.H. 2002. *Glacier Inventory of China (Maps)*. Xi'an: Xi'an Cartographic Publishing House.
Rasul, G. 2014. Food, water, and energy security in South Asia: A nexus perspective from the Hindu Kush Himalayan region. *Environmental Science & Policy* 39: 35–48.
Rasul, G. 2016. Managing the food, water, and energy nexus for achieving the sustainable development goals in South Asia. *Environmental Development* 18: 14–25.
Sandhu, H., and S. Sandhu. 2015. Poverty, development, and Himalayan ecosystems. *Ambio* 44 (4): 297–307.
Schreier, H., and P.B. Shah. 1996. Water dynamics and population pressure in the Nepalese Himalayas. *GeoJournal* 40 (1–2): 45–51.
SEI, UNDP. 2006. *Linking poverty reduction and water management*. Poverty-Environment Partnership. 82.
Sharma, P., and T. Pratap. 1994. *Population, poverty, and development issues in the Hindu Kush Himalayas, development of poor mountain areas*. Kathmandu: ICIMOD.
Sharma, E., N. Chhettri, and K.P. Oli. 2010. Mountain biodiversity conservation and management: a paradigm shift in policies and practices in the Hindu Kush-Himalayas. *Ecological Research* 25: 909–923.
Shen, D. 2009. River basin water resources management in China: A legal and institutional assessment. *Water International* 34 (4): 484–496.
Shrestha, A.B., N.K. Agrawal, B. Alfthan, et al. 2015. *The Himalayan Climate and Water Atlas: Impact of climate change on water resources in five of Asia's major river basins*. Kathmandu: ICIMOD, GRID-Arendal and CICERO.
Sivakumar, M.V.K., and R. Stefanski. 2011. Climate change in South Asia. In *Climate change and food security in South Asia*, 13–30.

Tarar, R.N. 1982. Water resources investigation in Pakistan with the help of Landsat imagery snow surveys, 1975–1978. In *Hydrological aspects of alpine and high mountain areas. Proceedings of the Exeter Symposium, July 1992, IAHS Publication No 138*, 177–190. Wallingford: IAHS.

WECS. 2005. *National Water Plan 2005—Nepal*. Kathmandu, Nepal: Water and Energy Commission Secretariat (WECS).

NDRC. 2007. China's National Climate Change Programme. Prepared under the Auspices of National Development and Reform Commission (NDRC), People's Republic of China. 63.

Chapter 15
Indigenous Practice in Agro-Pastoralism and Carbon Management from a Gender Perspective: A Case from Nepal

Rashila Deshar and Madan Koirala

Abstract Indigenous knowledge is the means making the practice possible in livelihood activity of HKH region. Pasturelands management and agropastoral activities carried out by indigenous people produce enough carbon and sequester large quantities of aboveground and belowground carbon. Such activities by indigenous people in Nepal Himalaya may have widespread effects on regional climate and global carbon cycles. This chapter showed the evaluating of indigenous gender perspective in the carbon management in Gatlang VDC of Rasuwa District, Nepal. The findings revealed that most of the labor related to agropastoral activities carried out by women contributed to carbon input and output, but their role was hardly recognized and valued. In the major decision-making process, women had either no or little say. Women contributed more than men to carbon input and output activities and. Therefore, their role in carbon management should be given proper attention.

Keywords Agropastoral · Carbon sequestration · Decision making · Gender equality · Nepal Himalaya · Women contribution

15.1 Introduction

In the Himalayan region, the relationship of local people with its environment and natural resources has evolved through arduous experiences of different survival (Samal et al. 2000, 2003, 2004). These experiences helped evolve tools, technologies and practices for sustenance of the production systems in balance with economic conditions and ecological specificities (Samal and Dhyani 2005; Samal et al. 2000, 2004). These eco-culturally evolved ecosystem specific tools, technologies and practices constitute integral parts of appropriate innovative strategies, otherwise called the indigenous knowledge system that effectively conserves resources and

---

R. Deshar (✉) · M. Koirala
Central Department of Environmental Science, Tribhuvan University,
Kirtipur, Kathmandu, Nepal
e-mail: rdeshar@cdes.edu.np

also allows options for their optimal use (Singh 2006; Berkes et al. 2000; Gadgil et al. 1993; Chambers et al. 1989). Indigenous knowledge, therefore, is of crucial significance if one wishes to introduce a cost-effective, participatory and sustainable development process (Warren 1991). The indigenous knowledge in the Himalayan ecosystem, therefore, serves as a cultural and natural capital (Berkes and Folke 1992), assisting the native societies to 'live in harmony with nature' (Gadgil et al. 1993). This knowledge is passed on from generation to generation, in which the women in the region have a major role to play in view of their responsibility (Samal and Dhyani 2005).

In Nepal, 12% of the country is classified as rangelands (LRMP 1986; Rajbhandary and Pradhan 1990; Rai and Thapa 1993; Shrestha 2001), with most being located in the hilly and mountainous areas of northern Nepal (LRMP 1986). Although not large in terms of land area, these rangelands have numerous functions that provide significant ecological and livelihood values for Nepal's mountain societies. Nepal's rangelands and their biological and physical resources play a critical role in the region's overall economic development and in the people's well-being. Animal husbandry depends on rangeland grazing and makes up a small but important part of farming practices for ethnic groups living in northern Nepal (Rai and Thapa 1993). Apart from hosting herders, livestock and wildlife, Nepal's rangelands provide numerous ecosystem services, including watershed and biodiversity conservation and carbon sequestration, as well as sites for tourism (Miller 1997).

Transhumance pastoralism is the seasonal migration of livestock and humans between many agro-ecological zones. It is an age-old practice in many mountain regions (Byers 1996; Rota and Sperandini 2009). In the Himalaya, pastoralists have transformed the ecosystem into economically productive assets for their livelihoods for more than 1000 years, and even today the region provides a home for a large number of people dependent on livestock (Miller 1999; Byers 1996; McVeigh 2004; Kreutzmann 2012). In grassland-livestock production in many mountain areas and in other pastoral communities (Radel and Coppock 2013), the role of women and men differ substantially (Shang et al. 2016; Khadka and Verma 2012) and their contribution to carbon balance can differ greatly (Cecelski 2000; IFAD 2004; OECD 2008). In the Hindu Kush Himalayan region, women's contributions to the conservation and management of forests, ecologically sensitive areas, water springs, and biodiversity resources are immense (Karki and Gurung 2012; Gurung et al. 2011; Khadka and Verma 2012; Khadka et al. 2014). However, their participation in decision-making and benefit sharing is poor throughout the region (Parajuli et al. 2010; Bhasin 2011). The majority of women have unequal access to productive resources such as land, enterprise, education, skills, information, and decision-making power (Gurung et al. 2011). Many activities in this region in agriculture and household chores are largely done by women. Large scale migration of men from the Himalayan region has attenuated the women's socio-economic responsibility, enforcing them to be heavily involved in food production. Women in the Himalayan region generally work long hours, attend to cattle, collect fuel, fodder and water in addition to performing normal duties at home, apart from managing agriculture. This study was undertaken to assess gender differences in agro-pastoral activities and in contribution to carbon management and decision-making of the Tamang ethnic community in the Himalayan region of Nepal.

15.2 Methods and Methodology

15.2.1 Description of the Study Area

The study was carried out at the Gatlang Village Development Committee (VDC) of Rasuwa district of Central Nepal (Fig. 15.1). The area lies between 27° 55′ to 28° 25′ N latitude and 85° 00′ to 85° 50′ E longitudes with an altitude ranging from 617 to 7227 m within an area of 1512 km². Gatlang, one of the VDCs of Rasuwa district, is geographically located in the mid-hills (2200 m) in a temperate zone. According to the Centre Bureau of Statistics (CBS 2011), there are 400 households with 1805

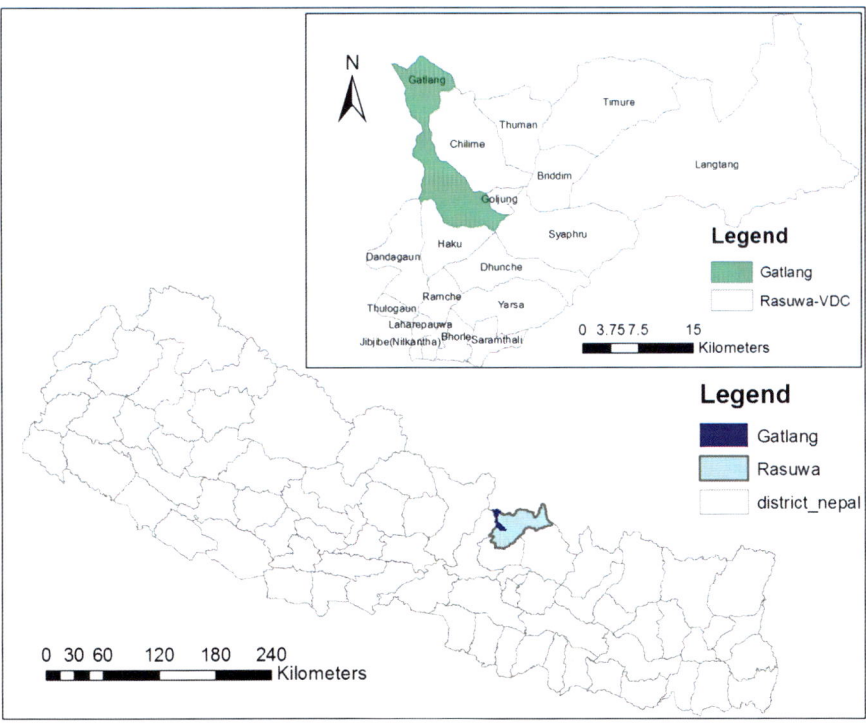

Fig. 15.1 Map of study area in Nepal

inhabitants of which 888 are men and 917 are women. The small population of the village is dominated by Tamang group, one of the indigenous tribes of Nepal and Tibetan speakers of specifically Tibetan descent living in Nepal (Goldstein 1975). The transhumance High Himalayan grasslands livelihood activities are found in the study area.

15.2.2 Research Approach

Thirty families of the Tamang ethnic group living in Gatlang VDC who are directly and indirectly linked with agro-pastoral livelihood were selected by using purposive sampling method for the survey. Data were collected one time from field study comprising semi-structured interviews in local language with herders and farmers of Gatlang by research team. All pastoral related livelihood activities were considered. Similarly, four focus group discussions were carried out with women, men and mixed groups. Key informants survey and observation of rangeland management were carried out. For carbon management, carbon input and output activities by men and women were studied.

15.3 Results and Discussion

15.3.1 Transhumance Pastoralism Practice

The Tamang ethnic groups of Gatlang VDC own some land and at least a few cattle for their subsistence. This mode of life is achieved through a division of labour within the family and the system of temporary accommodations called *goth* (animal shed). Two members of the family, usually an older parent (wife and husband) live either permanently or in rotation with other family members in the *goths* of *Kharkas* (higher pasture land). A minimal collection of household utensils is stocked, together with equipment for butter making. In the past a whole family used to live in the *goth* leaving the main village house for storage purposes. But recently, only old parents and daughter-in-law of each house live in a *goth* because the son of the family either works outside the village or owns a business in the village. Different composition of the herds such as only sheep (Fig. 15.2), only yak, only Chauri (cross between cow and yak) alone or a mixture of any two or three are maintained by the herders. Yaks do not thrive at low altitudes and, consequently, they are not kept in high numbers. The main purpose of keeping yak is for breeding. They prefer the Chauri (dzo-mo), yak-cow crossbreds, which provide more milk (Dong et al. 2009). Many people also keep sheep principally for their wool, which the women and men card, spin and weave to make jackets worn by men and *Syama* or black cloth worn by women.

Fig. 15.2 Sheep grazing in pasture land of Gatlang VDC, Rasuwa (Photography by Rashila Deshar 2018)

15.3.2 Sustainable Transhumance Pastoralism and Role of Women

Every month, the herds of Chauri, yaks and sheep are moved to fresh pastures, usually on a collectively agreed date, and the herder and cattle in the course of the year travel considerable distances over a defined course. Livestock reach the highest point during June to September. With the approach of winter they decend to Gatlang.

Herders of Gatlang move in a definite route and time pattern throughout the year to graze their livestock (Fig. 15.3). Their movement starts from the main permanent settlement of Gatlang village, at about 2000 m, where they stay for about 1 month, mid-April to mid-May. In summer for 4 months (June–September) in the rangelands above Gatlang village and about 7 months (mid-October to mid-May) in the grazing areas. Herders reach to the lowest point in mid-November where they spend about two and half months during the peak winter season. They move upward with the onset of spring season and reach to the high elevation rangelands at about 4000 m a.s.l. in mid-June and stay till mid-September. After mid-September (with the onset of winter, herders gradually move to low altitude areas resting for a few days in stops at different elevations before they reach to the lowest point in mid of November. Herders use the same route every year. According to Dong et al. (2009), in indigenous transhumance grazing system, different herds are grazed in different sites by the farmers according to their adaptability.

Normally, Chauri and sheep are moved gradually from alpine pastures at Peak Mountain as high as 4000–5000 m in summer to forest areas in the downstream valley as low as 1500–2000 m in winter. The Yaks never go down below 3000 m and

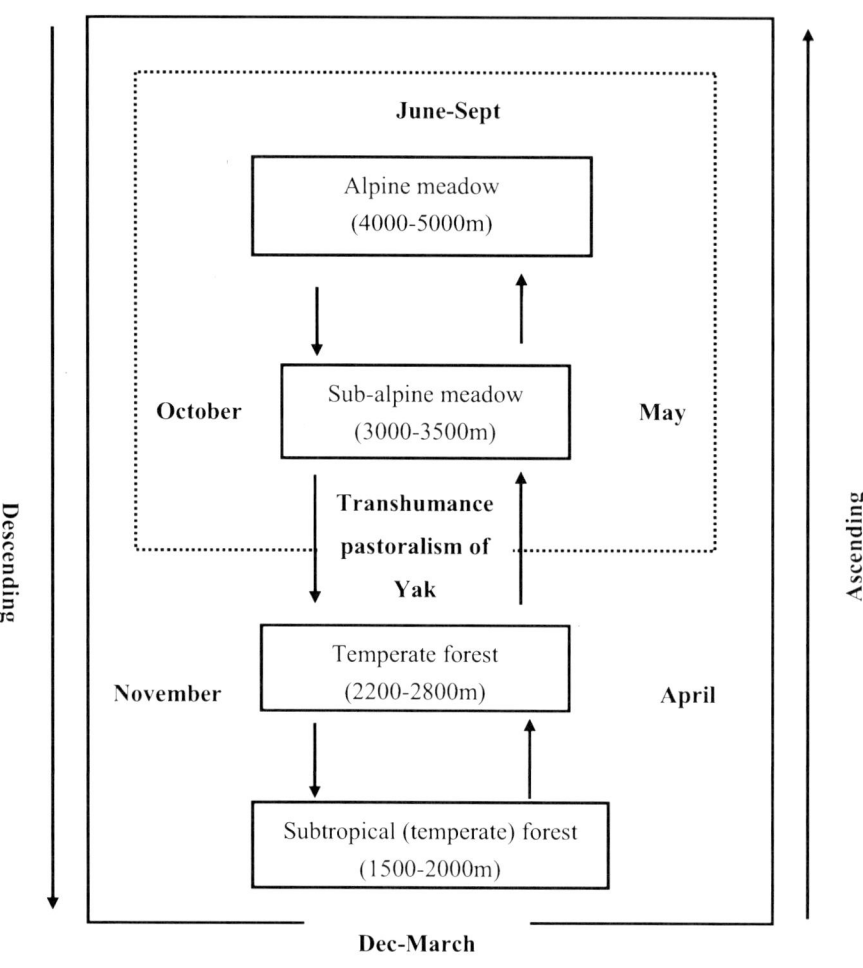

Fig. 15.3 Indigenous transhumance pastoralism of Chauri and sheep (–) and Yak (•••) in Gatlang VDC, Rasuwa (adapted from Dong et al. 2007)

spend the winter in sub-alpine pastures. The cattle, are herded in summer and graze on village scrubland or the stubble-field of cultivated zone in winter (Fig. 15.3). This seasonal shifting of livestock from one pasture land to another has maintained sustainability of pasture land of Gatlang village.

The rotational grazing of the pasturelands, according to feed availability, is an example of the deployment of indigenous knowledge adapted by local farmers of the village. With a high out-migration of men, women now play even a more crucial role in the rotational grazing management in the village. The Mother's Group, which is actively involved in such management, has developed regulations that even impose penalties for breaching the rules. A local herder said that such rotational grazing management by women has protected pasture land from being overgrazed and helped increase forage production. The ownership of traditionally practiced rotational rangeland management has now been taken by women. It has become a key strategy to make use of the scattered rangeland resources on a large spatial scale. Furthermore, such pastoralists' indigenous knowledge about ecology and social organization provided the information base for rangeland management strategies appropriate to deal with climate change (Homann et al. 2004). There is increasingly robust scientific evidence to show that pastoralism supports extensive livestock production in the rangelands which is one of the most sustainable food systems in the world. More than two decades of research has provided evidence that pastoralism is economically rational and viable, and is a vital tool for poverty alleviation, and large-scale conservation and ecosystem management (McGahey et al. 2015). In all these activities, women have immense roles and responsibility.

15.3.3 *Men and Women Labor in Agropastoral Activities*

The occupation of Gatlang people are mostly agro-pastoral based. Gatlang has a diverse geography revealing various land uses (agriculture, forest, built in area, water bodies). Cultivated fields are generally in sloped terraces in the vicinity of the main villages. Climatic conditions differ markedly over short distances owing to the wide altitude variation, permitting several growing seasons and variety of crops. The severe cold climate throughout the year are favorable for specific crops such as potatoes, maize, beans, barley, lentil and finger millet. The staple diet consists of maize flour and potatoes. Most families own land, however, the size of holding varies. Whatever crop is produced from the land is enough for about 6 months. Thus animal husbandry is of importance to supplement the food requirements. Each household has an average of two cattle, 15 sheep and goats and ten yaks and chauri. A total of 23 different agro-pastoral related activities were studied including milking, fuel firing, cooking, collecting firewood and litter, grazing livestock, collecting

Table 15.1 Men and women labor in agro-pastoral activities at Gatlang VDC

Agropastoral labors	Gender labor
Cooking lunch and dinner, cooking food of livestock, fetching water, washing cloths and dishes, dung collection, and labour exchange culture (*Parma*) between neighbors for wool weaving and preparing woolen cloths mostly done at night, planting and weeding	Women
Firewood, grass and leaf litter collection from forest, scouring and spinning wool, weaving, seeding, timber collection, chopping wood, milking, transporting milk to cheese factory, grazing livestock, raising and collecting livestock, sheep shearing, fertilizing, cultivation and harvesting	Men and women
Ploughing	Men

Table 15.2 Daily work calendar of family, Gatlang VDC

Time	Husband	Wife
4:00	Sleeping	Get up from bed
5:00	Get up from bed	Milking, cooking food for livestock, fetching water, cooking breakfast and lunch, washing cooking utensils and clothes
6:00	Drink tea and go to sell milk in cheese factory	
7:00		
8:00		
9:00	Returning home for lunch	Eating lunch only after serving food to all family member
10:00	**Take rest**	Washing cooking utensils, collecting firewood, grass and leaf litter from forest and looking after livestock and collet dung from grazing area. During the season, planting and harvesting. Labour exchange (*Parma*) for seeding, planting and harvesting with neighbours
11:00	Looking after livestock or planting and harvesting	
12:00		
13:00		
14:00		
15:00		
16:00	Returning home and resting	Returning from grazing land or agriculture land and scouring and spinning wool and weaving
17:00		
18:00		
19:00		Fetching water and cooking dinner
20:00	Eating dinner and going to bed	Serving dinner to all family members and eating dinner. Washing cooking utensils and cleaning kitchen
21:00-		Scouring, spinning and weaving of wool and preparing woolen cloths. *Parma* (labor exchange) between neighbors for wool weaving and preparing woolen cloths
22:00		
23:00		Going to bed

and drying manure, shearing, scouring, spinning of wool and weaving woolen cloths (Table 15.1). In all these activities, female are more responsible than male.

At the time of crop planting and harvesting, both men and women work together in field. Most work related to agriculture is done by family members. Ploughing is done by the man with personally-owned oxen and plough or borrowed oxen from the few families who own them. However, some particularly labour intensive tasks such as weeding, planting millet, and reaping are carried out in small groups usually of kin or friends. The need for labor force is fulfilled in three ways: family members, hired people and labor exchange known as "*parma*". Weeding is generally done by women, reaping by mixed groups and ploughing always by men. Women undertake almost all of the agriculture activities such as planting, weeding, manuring and harvesting. Men are busy with livestock and other work during day time. The typical day of Gatlang women consists of at least 18 hours of continuous work (Table 15.2). In comparison to women, although men performed hard work, working time is much less (two third).

Table 15.3 Carbon output activities done by men and women

Carbon output activities	Gender role
Cooking, burning, water boiling, milk warming, wool shearing, scouring, spinning and weaving of woolen cloths	Women
Collecting firewood, dead tree cutting, and cultivation and harvesting	Men and women
Tree felling, chopping of wood, making cheese, sale of milk and meat	Men

Table 15.4 Carbon input activities done by men and women

Carbon input activities	Gender role
Grassland management, planting of crops, wool shearing, scouring, spinning and weaving of woolen cloths	Women and men
Dung collection for manure from pasture land	Women

Fig. 15.4 The forest near rangeland is degraded due to over extraction of fuelwood (Photography by Rashila Deshar 2018)

15.3.4 Men and Women Labor Contribution to Carbon Management

The agro-pastoral activities of Gatlang VDC that are directly related to carbon management in terms of carbon input (Table 15.3) and output activities (Table 15.4) were studied. The carbon output activities such as cooking, burning, boiling, milk warming, shrubs and tree felling, litter collection, dung collection from rangeland and firewood collection are mainly carried out by women. Similarly, another source of carbon output in the carbon cycle was seen from the sale of livestock products such as milk, meat, leather and woolen cloths to areas inside and outside the

Fig. 15.5 A woman of Gatlang VDC cooking on a traditional open firewood stove (Photography by Rashila Deshar 2018)

grassland ecosystem. Different traditional cloths are made from wool of sheep, Chauri and yak. Manure storage methods and the amount of exposure to oxygen and moisture can produce greenhouse gases such as methane (CH_4). In the study area, the dung was collected from the grazing area by women and piled in one place which was later used as fertilizer in the field. This activity releases methane gas into the atmosphere. Capturing CH_4 from manure decomposition to produce biogas, a renewable energy, was not practiced in the village. Livestock dung was used only as fertilizer, and for cooking fuelwood was used which has caused deterioration of many forest areas (Fig. 15.4).

The households are still typified by the traditional open fire wood stove (a tripod type) without a chimney and good ventilation, symbolizing primitive cooking practices (Fig. 15.5). From focus group discussions, it was found that for cooking purposes, the required energy was obtained from fuelwood. According to the Central Bureau of Statistic (2011), among 400 total households in Gatlang, 398 households used firewood for cooking. Some houses have switched to liquefied petroleum gas (LP gas) for cooking but still they are dependent on firewood. In average 0.6 of a cylinder of LP gas was consumed by a household per month.

Furthermore, in average, 10 kg of fuelwood was used by a household per day. The cooking activities include: cooking food, warming water and food for livestock, warming milk and occassionally making alcohol. Dead trees were used as fuelwood, which was collected from the nearby forests by men and women when they visited forest with grazing livestock. In the pastoral regions of the Tibetan Plateau of China, collection of livestock dung for the use of fuel instead of fuelwood is very common. Yak dung is an important source of fuel in an area where firewood is not available (Miller 1999). Most Tibetan families use only yak dung for cooking and heating, as their limited income does not allow them to purchase fossil fuels. Families live in either tents or stone homes and, for economic reasons, use mainly simple stoves without chimneys (Xiao et al. 2015).

Carbon emission can lead to health deterioration and cause cardiovascular diseases and respiratory disorders. Women are most vulnerable to these health hazards as they do the cooking and spend the most time near the fire. The dependency of households on LP Gas is very minimal. One cylinder LP Gas is enough for 2–3 months in average.

15.3.5 Men and Women in Household Decision Making

Women make household decisions such as purchasing daily household goods. However, for major purchases such as land, cattle and other property, the decision of women are not considered. Literate male respondent (30%) allowed their wives to participate in major household decisions, while literate female respondents (3%) made major decisions either alone or jointly with their husbands. However, with illiterate male (30%) and illiterate female (37%) respondents wives were not allowed or didn't participate in major decision making processes. Furthermore, it was revealed that literacy plays a major role in decision making ($\chi^2 = 5.625$, df = 1, $p = 0.017$) in both males and females. Educated men allowed their wives to participate in household decision and educated females participated with their husbands in decision making. Therefore, education of women should be promoted so that women can play an equal role in household decision making.

Interestingly, despite the qualification of women, they are still found active in different household as well as social activities. Women were found to be more confident in managing agro-pastoral activities than men as was reported in the Tibetan region of China (in review of Shang et al. 2016). Due to traditions and religious beliefs in Tibet, women undertake most of the heavy labour associated with agriculture and livestock production (Dong et al. 2003) while men handle most of the decision-making. Nepal, like most societies in South Asia, was a rigidly patriarchal society in which most women received little or no formal education and had limited decision-making power in the household (Tamang 2000; WHO 2009; Paudel 2011). The condition of women was strictly controlled by patriarchal norms of the society (Acharya 1994) and women were generally subordinate to men.

15.4 Conclusion

In Gatlang VDC of Rasuwa District, most of the labor related to agro-pastoral activities was carried out by women. Women were found more social and engaged in collective works. Women were able to make minor household purchases, but not major household purchase. Women have a played major role in carbon output and input. The physical activities of women are integrally involved in carbon balance through their roles in livestock husbandry and fuelwood management, much more so than men. In essence, their daily activities are the key to the carbon cycle in high altitude grassland ecosystems. The study recommends that education of women be promoted so that they can participate equally in household decisions. Women should be given more opportunities for involvement in social development and ecosystem management. The important role of women in carbon management should be given greater prominence. Women should receive more credit for their roles in livestock production and management for the promotion of women's economic and social empowerment. Furthermore, governments should acknowledge the key role of indigenous high land people, especially women, in maintaining grasslands through their sustainable landscape management and livestock husbandry systems.

References

Acharya, M. 1994. *The statistical profile on Nepalese women: an update in the policy context.* Kathmandu, Nepal: Institute for Integrated Development Studies.

Byers, A.C. 1996. Historical and Contemporary Human Disturbance in the Upper Barun Valley, Makalu-Barun National Park and Conservation Area, East Nepal. Mountain Research and Development 16 (3):235

Berkes, F., and C. Folke. 1992. A systems perspective on the interrelations between natural, human-made and cultural capital. *Ecological Economics* 5: 1–8.

Berkes, F., J. Colding, and C. Folke. 2000. Rediscovery of traditional ecological knowledge as adaptive management. *Ecological Applications* 10 (5): 1251–1262.

Bhasin, V. 2011. Status of women in transhumant societies. *Journal of Sociology and Social Anthropology* 2 (1): 1–22.

Cecelski, E. 2000. *The role of women in sustainable energy development.* Golden, CO: National Renewable Energy Laboratory. https://doi.org/10.2172/758755.

Centre Bureau of Statistics (CBS). 2011. *National population and housing census.* Kathmandu: Central Bureau of Statistics.

Chambers, R., A. Pacey, and L.A. Thrupp. 1989. *Farmer first: Farmer innovation and agricultural research.* London: Intermediate Technology Publications.

Dong, S.K., R.J. Long, and M.Y. Kang. 2003. Milking and milk processing: Traditional technologies in the yak farming system of the Qinghai-Tibetan Plateau, China. *International Journal of Dairy Technology* 56 (2): 86–93.

Dong, S., J. Lassoie, K.K. Shrestha, et al. 2009. Institutional development for sustainable rangeland resource and ecosystem management in mountainous areas of northern Nepal. *Journal of Environmental Management* 90 (2): 994–1003.

Gadgil, M., F. Berkes, and C. Folke. 1993. Indigenous knowledge for biodiversity conservation. *Ambio* 22: 151–156.

Goldstein, M.C. 1975. Preliminary notes on marriage and kinship. *Contributions to Nepalese Studies* 2 (1): 57–69.

Gurung, J., K. Giri, A.B. Setyowati, et al. 2011. *Getting REDD+ right for women: an analysis of the barriers and opportunities for women's participation in the REDD+ sector in Asia,* 1–113. Washington, DC: United States Agency International Development.

Homann, S., G. Dalle, and B. Rischkowsky. 2004. *Potentials and constraints of indigenous knowledge for sustainable range and water development in pastoral land use systems of Africa: A case study in the Borana Lowlands of Southern Ethiopia (No. 333.74 H763)*. Eschborn: GTZ.

IFAD. 2004. *Enhancing the role of indigenous women in sustainable development. Third session of the permanent forum on indigenous issues*. Rome: International Fund for Agricultural Development.

Karki, S., and M. Gurung. 2012. Women's leadership in community forestry in the middle hills of Nepal. *Gender and biodiversity management in the Greater Himalayas: Towards equitable mountain development*, 25–27. Kathmandu, Nepal: ICIMOD.

Khadka, M., and R. Verma. 2012. *Gender and biodiversity management in the Greater Himalayas: Towards equitable mountain development*. Kathmandu: International Centre for Integrated Mountain Development (ICIMOD).

Khadka, M., S. Karki, B.S. Karky, et al. 2014. Gender equality challenges to the REDD+ initiative in Nepal. *Mountain Research and Development* 34 (3): 197–207.

Kreutzmann, H. 2012. Pastoral practices in transition: Animal husbandry in high Asian contexts. In *Pastoral Practices in High Asia*, 1–29. Dordrecht: Springer.

LRMP. 1986. Summary report of LRMP. Government of Nepal and Government of Canada. Land Resource Mapping Project: Kathmandu.

McGahey, D., J. Davies, N. Hagelberg, et al. 2015. *Pastoralism and the green economy—a natural nexus*. Nairobi: UNEP and IUCN. x + 58p.

McVeigh, C, 2004. Himalayan Herding is Alive and Well: The Economics of Pastoralism in the Langtang Valley. Nomadic Peoples 8 (2):107–124

Miller, D.J. 1997. Rangelands. ICIMOD Newsletter, 27, Kathmandu: ICIMOD.

Miller, D.J. 1999. *Rangeland management consultancy*, 24. Kathmandu, Nepal: Nepal-Australia Community Resource Management Project.

Organisation for Economic Co-operation and Development (OECD). 2008. *Gender and sustainable development: Maximising the economic, social and environmental role of women*. Paris: OECD Publishing.

Parajuli, R., R.K. Pokharel, and D. Lamichhane. 2010. Social discrimination in community forestry: Socio-economic and gender perspectives. *Banko Janakari* 20 (2): 26–33.

Paudel, S. 2011. Women's concerns within Nepal's patriarchal justice system. *Ethics in Action* 5 (6): 30–36.

Radel, C., and D.L. Coppock. 2013. The world's gender gap in agriculture and natural resources: evidence and explanations. *Rangelands* 35 (6): 7–14.

Rai, N., and M.B. Thapa. 1993. *Indigenous pasture management systems in high-altitude Nepal: A review. HMG Ministry of Agriculture/Winrock International Policy Analysis in Agriculture and Related Resource Management, Research Report Series, No. 22*. Kathmandu: Winrock International.

Rajbhandary, H.B., and S.L. Pradhan. 1990. Livestock development and pasture management. In *Background papers to the national conservation strategy for Nepal*, ed. National Planning Commission/The World Conservation Union, 1–10. Kathmandu: The World Conservation Union.

Rota, A., and S. Sperandini. 2009. *Livestock and pastoralists. Livestock thematic papers: Tools for project design*. Rome: International Fund for Agricultural Development.

Samal, P.K., and P.P. Dhyani. 2005. Indigenous soil-fertility maintenance and insecticides practices in traditional agriculture in Indian Central Himalaya: Empirical evidences and issues. *Outlook on Agriculture* 36 (1): 49–56.

Samal, P.K., R. Fernando, and D.S. Rawat. 2000. Influences of economy and culture in development among mountain tribes of Indian central Himalaya. *The International Journal of Sustainable Development and World Ecology* 7 (1): 41–49.

Samal, P.K., A. Shah, S. Tiwari, et al. 2003. Indigenous animal health care practices in Indian Central Himalaya: Empirical evidences. *Indian Journal of Traditional Knowledge* 2 (1): 40–50.

Samal, P.K., A. Shah, S.C. Tiwari, et al. 2004. Indigenous healthcare practices and their linkages with bioresource conservation and socio-economic development in Central Himalayan region of India. *Indian Journal of Traditional Knowledge* 3 (1): 12–26.

Shang, Z., A. White, A.A. Degen, et al. 2016. Role of Tibetan women in carbon balance in the alpine grasslands of the Tibetan Plateau: A Review. *Nomadic Peoples* 20 (1): 108–122.

Shrestha, T.B. 2001. *Status review national strategies for sustainable development forestry/rangeland/biodiversity*. Kathmandu: The World Conservation Union.

Singh, J.S. 2006. Sustainable development of the Indian Himalayan region: Linking ecological and economic concerns. *Current Science* 90 (6): 784–788.

Tamang, S. 2000. Legalizing state patriarchy in Nepal. *Studies in Nepali History and Society* 5 (1): 127–156.

Warren, K. 1991. Shifting cultivators: Local technical knowledge and natural resources management in the humid tropics. *FAO Community Forestry Note 8*.

World Health Organization (WHO). 2009. *Perspectives on sexual violence during early years of marriage in Nepal: Findings from a qualitative study: Social science research policy briefs*. Geneva: World Health Organization. http://www.who.int/iris/handle/10665/70231.

Xiao, Q., E. Saikawa, R.J. Yokelson, et al. 2015. Indoor air pollution from burning yak dung as a household fuel in Tibet. *Atmospheric Environment* 102: 406–412.

Chapter 16
Adaptation by Herders on the Qinghai-Tibetan Plateau in Response to Climate Change and Policy Reforms: The Implications for Carbon Sequestration and Livelihoods

Haiying Feng and Melissa Nursey-Bray

Abstract There are changes in livelihood strategies of five Tibetan herder communities in the face of climate change and government policies such as adjustments to use rights and restricted herder mobility. Data collection relied on a mixed-method approach, including household surveys and rural rapid appraisals (PRA). Results indicated that yak husbandry is the main source of livelihood and households have a restricted range of livelihood activities. Major coping strategies varied with production system and resource availability and options for mobility of herds. The perception of a majority of respondents was that land tenure reforms had led to creation of more bureaucracies, forced sedentarization, livelihood insecurity, collapse of pastoral adaptation, poverty, resource use conflicts and hindrance to long-term planning and permanent developments. There is need to amend strategies that threaten the environment and instead promote integration of community best practices initiatives in proven concepts of adaptation to climate change and livelihood vulnerability.

Keywords Yaks · Livelihoods · *Ophiocordyceps sinensis* · Climate change · Carbon sequestration · Local ecological knowledge · Climate variability · Sedentarization · Resettlement · Adaptation

---

H. Feng (✉)
Research Institute for Qinzhou Development, Beibu Gulf University, Qinzhou, China

M. Nursey-Bray
Department of Geography, Environment and Population, University of Adelaide, Adelaide, SA, Australia

16.1 Context and Setting

China's vast western regions include some of the world's highest mountains and driest deserts. The Qinghai-Tibetan Plateau (QTP) is an area of about 2.5 million km² that is the location of the headwaters of many large and important rivers. These include transboundary ones like the Brahmaputra, the Ganges, the Mekong, the Yellow and the Yangtze rivers (Fig. 16.1).

Inaccessibility and marginality make mountains one of the toughest environments for agriculture in terms of diversification of pastoral activities, resulting in a higher prevalence of vulnerability and warranting particular attention for sustainability of the overall mountain systems (Rashid et al. 2005). In spite of the complex challenges, pastoralists adopt suitable strategies to sustain their livelihoods, with their ability to transform the extensive marginal rangelands into economically productive areas (Mishra et al. 2010).

16.2 Scope and Purpose of this Chapter

Our goal was to understand how future climate change would affect the alpine rangeland community and the present (potential) adaptive mechanisms of the herder community to climate change. We were also interested in the likely impact of both climate change and the herder's adaptation strategies on carbon sequestration. Some adaptive strategies that have proven to be successful in the past may no longer be viable options. This has implications for the retention and transfer of local

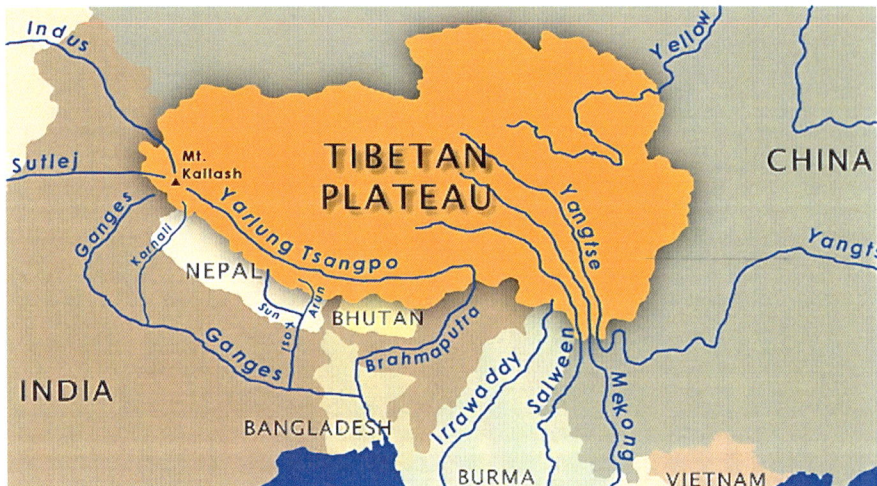

Fig. 16.1 The Qinghai-Tibetan Plateau nurtures nine major rivers in Asia, providing 1.3 billion people with freshwater

ecological knowledge (LEK) as discussed below. The shift from subsistence to a market-economy, is characterized by the necessity now for cash, the changing markets, and new consumer demands for eco-cultural tourism. All of these impinge on daily life of herders. Cash is required to pay for schooling of the children, for transport of goods to and from market and so on.

It is critical to gain a better understanding of pastoralists' strategies in light of increasing climatic variability, growing competition for land, rising population and decentralization. These facts have informed our research. This study assessed the changes in livelihoods as coping strategies to climate variability and restricted mobility in the cold and arid lands of the QTP with particular focus on traditional yak-based including: pastoral, systems to gain a better understanding of: (1) changes in livelihood activities; (2) coping strategies to avert vulnerability associated with changes; (3) challenges and vulnerabilities of livelihoods associated with transformation; (4) the likely consequences for carbon losses and gains.

16.3 Methodology

A cross section exploratory study was conducted to understand the changes in livelihood strategies of five Tibetan herder communities in the face of climate change and government policies such as adjustments to use rights and others, including restricted mobility. Data collection relied on a mixed-method approach, including household surveys and rural rapid appraisals (PRA). Questionnaire-based qualitative surveys were conducted in five herding communities to understand how the recent changes in the legal framework and changing climate have caused specific changes in yak husbandry practices. A sample of the major grouping of the issues canvassed in the survey is shown in Box 16.1.

The five herder communities were spread across the QTP from sites in western Qinghai to sites with in the Tibetan Autonomous Region (TAR). Altitudes ranged from 2800 m a.s.l. to 4300 m a.s.l. All the herder communities surveyed were ethnic Tibetans, all relied for their livelihoods on yak husbandry and seasonal work involved in collection and sale of *Ophiocordyceps sinensis* – a remarkable combination of a grass, fungus and caterpillar. It is known locally as 'God grass'. The fungus that invades the body of the caterpillar has medicinal properties and the product has high monetary value (Fig. 16.2).

Box 16.1 Principal Issues Explored in the Five Tibetan Herder Communities in this Study
Demography, kinship and household structure

Age distribution, gender balance, family structure, kinship

Income sources and expenditure

Contribution from herding, alternative income sources, migrant labor
Role of *Cordyceps sinensis*
Cultural and eco-tourism benefits and impacts
Expenditure on festivals, weddings, funerals, gifts to the temples

Perceptions about government policy

Impact of grazing bans
Consequences of restricted access to high altitude summer pastures
Attitudes toward conservation efforts, national parks and reserves
Experience with sedentarization and re-settlement

Perceptions about climate change

Impact of warmer winters, longer autumns
Rainfall and snow—changing patterns impacting herding practices
Glacial retreat—pros and cons

Adaptive strategies to cope with change

Herding practices
Impact of cash economy
Arrangements about schooling, health, welfare

Herder outlook and future prospects

Impact of ageing on herders
Attitude of younger generation toward traditional herding
Perceptions of grassland degradation and prospects for recovery

Fig. 16.2 Picking *Ophiocordyceps sinensis* from the grassland (Photography by Yongqiang)

16.3.1 Pastoralism on the Qinghai-Tibetan Plateau: Yak Husbandry As a way of life and means of livelihood

Pastoralists on the QTP have sustained their economies through efficient utilization of grassland resources by high altitude livestock species, such as yaks (*Poephagus grunniens*). However, pastoralism is evolving more rapidly due to changes in the socio-economic, institutional and policy environments (Shreshtha 1994). Several authors (Roder et al. 2002; Tulachan and Neupane 1999; Wangchuk et al. 2006, 2014) noted that socio-economic development was the main driver of change in the highland pastoral systems of the QTP. Yak husbandry is the main livelihood source for the high-altitude communities (Hua and Squires, 2015).

Pastoralism based on exploitation of dry lands and often some form of herd mobility has been historically practiced in many parts of the world, including the QTP. Pastoralists live in a context of environmental uncertainty albeit worsened by climate change. Consequently, pastoralists developed a diverse range of strategies, institutions and networks to exploit this unpredictability and risks to their advantage. Exploiting common natural resources such as grazing land or forest can provide income, food, medicine, tools, fuel, fodder, construction materials, and so on. The means by which poor households gain income and meet their basic needs are often met by multiple livelihood activities and survival strategies that may include employment as migrant workers but in the mountainous regions of QTP that option is not readily available. Livestock mobility, controlled breeding and grazing, were the three major critical strategies although these have not been fully taken advantage in government policy.

The global perceptions about pastoralism in these marginal lands are however rapidly changing. Thus it is critical now to gain a better understanding of the pastoralist strategies in light of increasing climatic variability, growing competition for land, rising population and decentralization to provide and sound policy advice. Livestock mobility was the principal means by which pastoralists coped with and took full advantage of natural resource variability in dry lands (Behnke et al. 1993; Scoones 1994). However, over the last three decades, scholars working with pastoral groups from both the northern and southern hemisphere have reported an almost universal decline in herd mobility, caused by commoditization of the pastoral economy and villagization policies (Fratkin 1997). Sedentarization has been particularly impeded by declines in communal grazing land, widespread land use change, and wildlife conservation.

Although mobile pastoralism is economically viable, environmentally friendly, securing livelihood for many people and causing less ecological impact (Scoones 1994), many countries have taken up structural adjustments including: land individualization, public asset privatization leading to loss of communal grazing resources and limiting pastoral mobility. Livelihood diversification has been one of the strategies to manage changing climate extreme events. According to Ellis (1998), livelihood diversification is the process by which rural families construct a diverse portfolio of activities and social support capabilities in order to survive and to improve their standards of living. This has forced pastoralists to seek alternative means of production, sources of livelihoods and in some cases where appropriate, destocking, regulated mobility, renting of grazing resources, seeking for cash-based livelihoods through employment in eco- and cultural tourism, harvesting *Cordyceps* have been alternatives. Unfortunately, these coping strategies have not been assessed to inform policy for sustainable human livelihoods and environmental protection for contribution to the QTP's socio-economic transformation and environmental resilience.

16.3.2 Climates are Changing on the Qinghai-Tibetan Plateau

The Qinghai-Tibetan Plateau, the largest and highest plateau in the world, is known as the "third pole" on the Earth and is highly sensitive to climate change (Qiu 2008). Climate warming has become an undoubted fact (Vitousek 1994; Ma et al. 2014). Over the past six decades, the rate of climate warming on the Tibetan Plateau has been more than twice the global average (Chen et al. 2013). Along with the change in the precipitation pattern (Shen et al. 2015), how this dramatic climate change affects the sensitive alpine ecosystem of the Tibetan Plateau has long been taken as an important question in ecological research. The Tibetan Plateau is a typical alpine region where alpine meadow vegetation serves as a major natural resource for local herdsmen's livelihoods and sustains livestock production (Haynes et al. 2014).

The QTP is a critical area of plant diversity that is sensitive to global climate change, and plants in this alpine ecosystem are particularly vulnerable to climate change (Ma et al. 2014). Climate warming has changed the species composition and community structure in alpine ecosystems and exerted both positive and negative effects on alpine plant diversity (Zhao et al. 2015; Shang et al. 2017). Climate warming may lead to the loss and fragmentation of suitable habitats, and whether plant diversity will change as predicted depends on whether there is sufficient time for the migration of alpine plants and their substitution between species. It has been shown that the rate of warming at high altitudes is much higher than that at low altitudes (Chen et al. 2013).

16.3.3 Temperature Change in China's Grassland Regions

The IPCC has predicted that the temperature would increase by an average of 1.1–6.4 °C on the Earth's surface by the end of this century. Analysis of meteorological data from the study area for 65 years (1951–2016) also showed an average temperature increase of 0.229 °C per decade in this area.

From 1951 to 2009, the ground surface temperature increased on an average by 1.38 °C in China, the warming rates is 0.23 °C/10a, which is similar to the level of global warming. The temperature in the north, north-west and south-west China has obviously increased over the 50 years ending 2012. The warming rate is 0.22 °C/10a, 0.37 °C/10a and 0.24 °C/10a in north, north-west and south-west regions, respectively (Liang et al. 2014). Temperatures rose in the mid-and-late 1980s, in particular since 1978. The warming in north China and on the QTP was greater than in other regions, especially in winter and spring. Most regions of QTP, including Xizang (Tibet) and Qinghai province, have a greater warming rate compared to other regions. In the three-river headwaters region of Qinghai (Yellow, the Yangtze and the Lancang), from 1961 to 2010, the annual and seasonal average temperature underwent several cold and warm fluctuations, but the average temperature had a significant rising trend at statistical significance level, especially since 2001.

16.3.4 Precipitation Change in Qinghai-Tibetan Plateau

On the Qinghai-Tibetan Plateau, the precipitation falls mainly during the summer and autumn (from June to September), with winters and springs relatively drier. In contrast with the temperature trends, the precipitation trend since 1960 has shown less seasonal and spatial fluctuation but an overall slight increase and high interannual variation at the whole-plateau scale (Xu et al. 2010; Kang et al. 2010; Li et al. 2010). The precipitation trends show a significant increase during the winter and spring, but non significant decreases during the summer and autumn (Li et al. 2010). Future projections of precipitation using the IPCC models indicate that the wetting trend will continue on the plateau.

16.4 Results

Even though yak farming faces challenges due to climate changes, Yak husbandry is still a pillar and foundation of the livelihood of most of Tibetan households.

Survey results revealed that women were increasingly involved in yak husbandry and household work as men pursued other revenue raising endeavours some of which, like collection and sale of *Cordyceps* (caterpillar fungus—used in traditional medicine) called "soft gold" are quite lucrative (Fig. 16.3).

The *Cordyceps* business overtook yak farming as the main income-earning activity for many households. In contrast to yak farming, *Cordyceps* harvest time is only 12 up to 26 days from the beginning of May to the latter half of June, depending on the current weather, the climate of the same period last year, the location and altitude of grasslands etc. Therefore, when harvest time is coming, all groups—the men, women, the elderly, the children, 'outsiders' are involved in *Cordyceps*-picking. Because of the limited time and geographic location of Cordyceps-growing grassland, the pickers camp on site and have to relocate often and carry all foodstuffs and fieldwork tools with them.

Fig. 16.3 The *Cordyceps* in (**a**) the soil, (**b**) out of the ground and (**c**) on the street market (Photography by Haiying Feng)

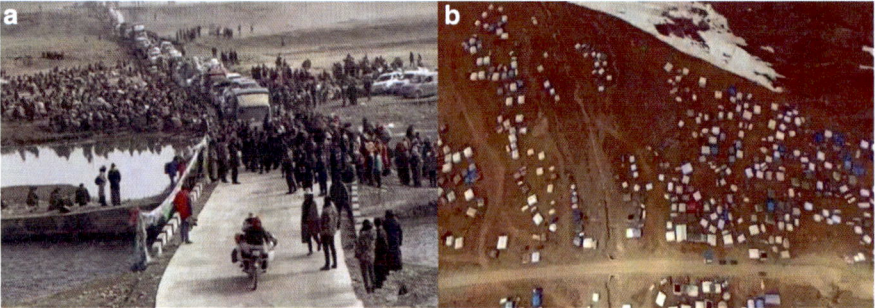

Fig. 16.4 (**a**) The groups on the way to harvest Cordyceps; (**b**) the tents of pickers scattered in the camp base (Photos from http://www.sohu.com/a/232058606_374114)

There are so many ways to sell the *Cordyceps* like selling to dealers or companies, individuals etc. But for most of the freelance pickers, generally they take them to the street market to sell (see the Fig. 16.3c). In some places of QTP, where *Cordyceps* grow, there is a new industry, called the *Cordyceps-cultural tourism*, being fostered by local governments and the people. The *Cordyceps* picking is natural and so profitable without direct cost. The *Cordyceps* business involving picking, selling, processing, and eco- and cultural tourism is a large proportion of the revenue stream for many local households (directly or indirectly). It is a big boost for local people and helps them cope with the vulnerabilities of livelihood in this remote and harsh environment (Fig. 16.4).

Neglect of yak herding and the intensity of *Cordyceps* harvesting has seen a decline in the overall grassland condition and most herders migrated a month earlier to the summer grazing land because of lack of forage on rangelands at the lower levels (Fig. 16.5). yak mortality increased and fodder scarcity became more acute, which is a major constraint to yak raising. Despite the good income from the *Cordyceps* business, our survey revealed that yak husbandry was the preferred earning activity (over *Cordyceps)* due to herders' confidence in yak husbandry as a

Fig. 16.5 Winter pastures at lower elevations are often severely overgrazed (Photography by Haiying Feng)

reliable source of livelihood. The every-day needs of the people can be met from yak husbandry.

The significance of 'own use' of milk and other yak products is often underestimated. On the upside, extra cash income from sale of *Cordyceps* also led to increase in the number of households buying commercial feeds for over-wintering yaks (Commonly between November and May yaks, without supplementary feed, lose 30–40% of their body weight.). Of several measures proposed by yak herders to improve yak farming, increasing grassland productivity and provision of subsidies for feed purchases were the most important measures. Yak husbandry practices have undergone a few positive, but many more undesirable, changes in recent years, some of them are related to climate change or perceptions of climate change.

16.4.1 Changes in Gender Roles

While women are considered the most vulnerable group in most societies worldwide, women pastoralists are doubly vulnerable because they are members of the largely marginalized communities. Gender also matters in adaptation decision making—for example, studies in Ghana show that women are more affected by the impacts of climate change, and thus need to be incorporated more in adaptation policy (Owusu et al. 2019). Gender imbalances and inequalities prevent the society as a whole from realizing the full potential of women in social, economic, legal and political spheres. The changes in the roles played by men and women observed in current study could be attributed to capacity building and empowerment and global sensitization on equity and equality that have been promoted by different organizations including governments which helps women pastoralists to transform their impoverished communities.

Both men and women participate in different livelihood-related activities. This portrays the abilities of women despite their limited access to and control of resources and exposure to more risks. Workload of women increased under new land fragmentation processes in rangelands due to increased responsibilities in yak herding and income generation (from sales of surplus milk, cheese, yoghurt). Women have gained higher influence in household decisions making than before and some are engaged in small-scale business (including handicrafts and sales of souvenirs to eco-cultural tourists) hence getting financial resources under their own control.

16.4.2 Sedentarization and 'Ecological Settlement'

It is inevitable that droughts as a form of climate change increase the proportion of land unsuited even for pastoralism hence eventual competition for land. In many instances, this has led to sedentarization and, what is termed in China as 'ecological settlement'. In reality the people may be thought of as ecological refugees. Sedentarization in the dry lands has been a result of sharp economic, political, demographic, and environmental changes. Sedentarization leads to an increase in social interaction, demand for and access to social services from government. Indeed, the settled communities have access to social services such as, infrastructure, health services and education services.

Compared with the mobile pastoral situation there should be a reduction in vulnerability. The reallocation of pastoral land to the outsiders remains an essential driver for livelihood insecurity, grievance and conflict, marginalization and spread of poverty among rural communities. Ecological resettlement is widely practiced throughout the QTP to alleviate poverty and reduce pressure on sensitive areas has been in force since 2005. Ecological settlement (see below) is rendered necessary to protect vulnerable areas such the headwaters of the Three Rivers (Mekong, Yellow and Yangzte) (Fig. 16.1) but is not without controversy (Tashi and Foggin 2012; Du 2012).

16.5 Discussion and Conclusions

Glaciated mountain areas such as the QTP are particularly susceptible to global warming and the inhabitants of these regions must change and adapt. Different livelihood strategies may suit different households; some may intensify, others diversify, and they may be some who are better off by migrating.

16.5.1 Adaptation and Mitigation, Coping Strategies vs. Adaptation, Typology of Measures, Time Frames

Adaptation is an integral part of the response to climate change. Of bigger interest to policy makers, however, is the ability of societies to implement adaptations. The IPCC reiterates that adaptation is more than just finding a technical 'fix', but should incorporate a combination of strategic and technical options (IPCC 2012) and build adaptive capacity. To date adaptation programs are implemented using social vulnerability approaches, resilience programs and climate risk management initiatives (Biagini et al. 2014).

Within these frameworks, a wide array of adaptation typologies does exist. These typologies occur in five broad areas: (1) timing (anticipatory, concurrent, reactive); (2) spatial scope (is it local, regional, national etc.); (3) its intent (is it autonomous, reactive or planned); (4) form (technological, behavioral, financial, institutional); (5) the degree to which it necessitates change (is it incremental, transformational) (Biagini et al. 2014). Lesnikowski et al. (2013) assert there are three kinds of adaptation (1) recognition, (2) groundwork, (3) adaptation action, while Dupuis and Biesbroek (2013) have created a four-pillar typology of adaptation, which includes symbolic adaptation policy, contiguous policy, contributive policy and concrete policy. In the agricultural context, adaptation typologies have been developed in Canada, and are clustered around technological developments, government programs and insurance, farm productions practices and in farm financial management (Smit and Skinner 2002). Adaptation can also be constructed as actions which create entry points for adaptation to occur (Eisenack and Stecker 2012). Adaptation may be reactive or proactive, coordinated by government, industry or the community or be a collaborative endeavor amongst multiple stakeholders.

Yet many barriers and factors exist that affect robust decision making about what adaptation pathway to follow (Bhave et al. 2016). Dealing with the uncertainty of exactly how climate change will manifest and therefore the extent, diversity, regularity, distribution and magnitude of its impacts (Ha-Duong et al. 2007; Petit 2005) is an additional problem. Policy makers need to consider how much climate change uncertainty they wish to adapt to, whether robust adaptation options are socially, environmentally and economically acceptable and the extent to which climate change uncertainties compare with other uncertainties (e.g., changes in demand) (Dessai and Hulme 2007)? The solution to dealing with uncertainty lies in ensuring that adaptation policy is robust, and anticipates future impacts based on a wide array of understandings about climate change.

As such, managing risk plays in important role in engaging with uncertainty and achieving adaptation. Other factors affecting farmer adaptation decision making included, levels of education, farm size, exposure to extensions services, household income, and perception of impacts (Mi et al. 2017). Behavioral change is also influenced by the perception of risk associated with hazards (Arbuckle et al. 2015).

Adaptive management is a related concept that embeds greater fluidity and flexibility within conventional environmental management systems and takes place in two phases: (1) the institutionalization of a framework in which intentional and varied policies may be implemented; (2) learning over time by monitoring the responses of the system on which the varied adaptations have been enacted (Arvai et al. 2006). In this context, adaptive management can be documented in various regions.

The role of local knowledge is another important factor in the adoption of successful climate adaptation, specifically, adaptations that align with traditional practice and are based on local knowledge (Li et al. 2013). For example, in the QTP, the documentation of local ecological knowledge when combined with scientific research can, by its strengths in being able to reflect knowledge at local scales (that is integrated over a range of variables and time scales), assist in facilitating adaptation efforts (Klein et al. 2014). Yet, local knowledge alone is not enough and subsidies or support from the central government can potentially play a part in limiting vulnerability: while it is necessary for local farmers to build a system of adaption, those that combine reliance on their own knowledge with scientific research and government support are successful (Li et al. 2013). Chen et al. (2018) in a study of farmer's livelihood adaptation in the arid regions of Minqin Oasis, China, shows that their success is conditioned by government policy as well as the limits of their own resources.

Finally, the success of adaptation is affected by scale and time and how and where they align with modes of livelihood production and maintenance (Burnham and Ma 2018): smallholder farmers in the Loess Plateau region of China highlight how historical multi scalar and social ecological processes shape decision making. The success of adaptations depends on the extent to which they enable landholders to manage risk and uncertainty over time and place. Where adaptation options create obstacles to effective livelihood production they will be dropped (Burnham and Ma 2018). Ultimately, successful adaptation is determined by a multitude of factors, including cross scale interaction and the other stressors on households which play a role in "facilitating or delimiting a households' ability to adapt to change" (Burnham and Ma 2018). This is important given that adaptation in this sense, is not a simple reaction to the physical stimuli of climate change *per se* but a result of how the multiple and interacting contradictions, stresses and tensions within the socio-ecological domain can be resolved: "the development of effective adaptation governance requires in depth, situated understanding of how adaptation is embedded in particular environmental, social, political, economic, and institutional contexts" Burnham and Ma (2018).

16.5.2 *Implications of Herder Perceptions, Behavior and Adaptive Strategies for Carbon Gains and Losses*

As herding practices change in response to changing vegetation and soils, with permafrost melting, retreating glaciers retreat and biogeochemical cycles are altered, the carbon flux has changed. Overgrazing has reduced the above-ground biomass (AGB) in many alpine grasslands (Yang et al. 2009) and trampling has contributed

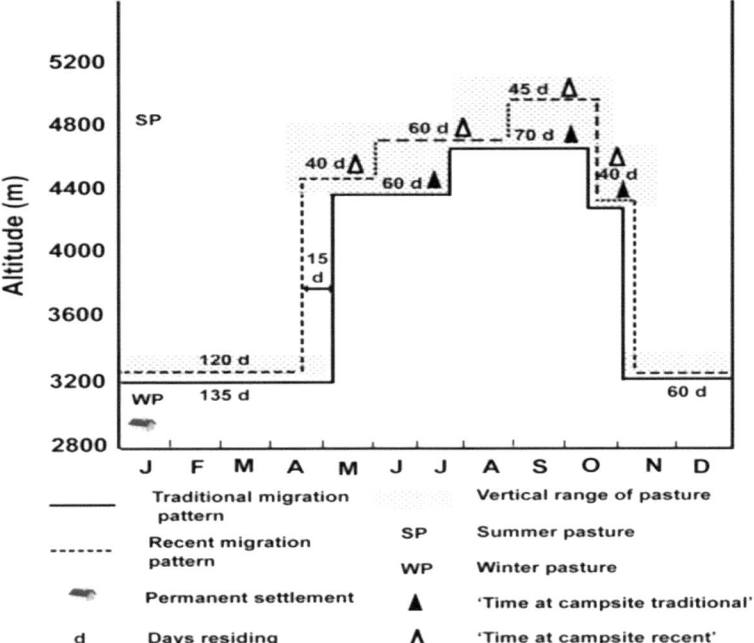

Fig. 16.6 Temporal–spatial diagram of annual pastoral migration patterns in the case-study area showing recent changes from the traditional pattern. Duration days are the average over the five villages interviewed, all of which follow similar migration patterns; (1) about 15 days earlier departure from winter pasture (WP) in early summer; (2) more frequent moving of camp sites in the summer pastures (SP)

to oxidation of soil organic carbon. In the past huge amounts of soil carbon were stored and, because of the low temperatures, oxidation and decay was at a slow rate. Gains of soil carbon exceeded losses. This trend was reversed from the 1990s when livestock numbers on the QTP rose and some restrictions on mobility of people and their livestock were imposed. Later, fencing and the imposition of grazing bans and the excision of some areas of prime grassland to form national parks and nature reserves led to greater concentration of livestock on the remaining grassland. Herders moved their livestock away from the over-utilized lowlands a month earlier in spring than was customary and, because of the warming trend, remained in the uplands for a longer time (Fig. 16.6).

There are other adaptive strategies that herders may use to overcome some of the constraints imposed by policy and regulations and by changing climatic conditions (Squires and Hua 2015). These include *multi-household grazing management* patterns as an adaptation. The government, through the Animal Husbandry Bureau is pushing the advantages of *modernization and intensification* (in terms of productivity) over traditional herding practices. To modernize, means to run fewer livestock and look after them better—'more from less' (Michalk et al. 2010). Housing yaks and sheep in warm pens reduces energy expenditure in winter, supplementary feedstuffs (like hay and corn stalks etc.) means heavier birth weights and greater survival of neonates, the culling of older and/or barren females to reduce herd/flock size. Shang et al. 2016 advocate adoption of *alternative mobility schemes* where seasonal movements of people and their herds is more rational. The scheme involves the establishment of semi-permanent winter homes (with warm pens for the livestock, watering facilities and other improvements related to veterinary needs etc. This suggested pattern of mobility maximizes the utilization of grasslands but puts in place some safeguards.

The government favors *resettlement and re-location* as a way to adapt to changing climate and the new vision for an "ecological civilization" and so far over $US 1 billion has been spent. There are problems associated with the people who are uprooted and re-settled in sites away from the grassland where lack of language and job skills (despite the provision of some training) reduce the prospects of getting gainful employment (Tashi and Foggin 2012; Wang et al. 2006). Social problems like alcoholism are also rife. The government felt compelled to act though because the burgeoning livestock inventories and the degradation of grasslands is serious ecological threat, especially in critical areas like the Three Rivers region where the headwaters of the Mekong, Yangtze and Yellow rivers occur (Tashi and Foggin 2012). Ecological resettlement has been initiated by the Chinese government on a large scale and aims to help degraded landscapes to recover and to improve the living standards of local people in western China.

Since 2003, the government has invested RMB 7.5 billion (Chinese yuan, over U.S.$1 billion) in Qinghai Province to establish the world's second-largest nature reserve around the headwaters of the Yangtze, the Yellow and the Mekong rivers (Sanjiangyuan). The resettlement of Tibetan herders from the Sanjiangyuan grasslands to urban areas is one of the project activities (Wang et al. 2006; Du 2012). *Land tenure and use rights reform* is under way. Land fragmentation is being countered by provision for land amalgamation where grasslands can be rented or use rights re-assigned to allow more 'commercial' rangeland-based operations. Use rights have been extended from 30 to 50 years and the legal right to transfer is being enabled. Banks and rural credit providers are now more inclined to allow herders to borrow money using the use right certificate as collateral.

There is much debate about the balance between *conservation vs. utilization role of recently proclaimed national parks*. The world's largest national park has been established by the Chinese government and as in other countries (e.g. France) the model adopted allows the continuation of herding, hunting and gathering of fuel wood and medicinal plants etc. within the park. The concern though is with the balance between traditional herding and the objectives of conservation of the unique resources within the park. Rare species like Tibetan antelope, wild sheep, and wild donkeys and camels and predators like snow leopards and raptors like eagles, owls and vultures that need protection.

From the viewpoint of the herder the proclamation of national parks is a further blow to the herders' freedom of mobility and a loss of a potentially valuable source of forage. The government (both at State level and in the Provinces) is under pressure to conserve the dwindling inventories of wildlife, including endangered plants, reptiles, birds, amphibians and mammals. What everybody seeks is to foster *land stewardship—to balance improved livelihoods with biodiversity conservation and land protection* but to achieve this is no easy feat (Squires 2012).

The emerging issues of yak husbandry systems on the QTP include interventions to strengthen yak husbandry and help herders make informed choices in the high-altitude rangelands including the introduction of schemes to make yak raising attractive to the mountain youth. Essentially, yak husbandry is at a crossroads where a firm decision is needed to either encourage and strengthen the husbandry practices or witness the gradual extinction of the age-old tradition. De Haan (2016) acknowledged that pastoralism everywhere is at the verge of disappearance as pursuit of a purely pastoralist life has become increasingly difficult. Pastoral livelihood has been considered by many as archaic and undesirable and there is poor understanding of the benefits of pastoralism amongst politicians, policy makers and technocrats often leading to unfriendly decisions.

Acknowledgement HYF is a Research Fellow in the School of Social Sciences, University of Adelaide, Australia and support during her term in the University of Adelaide is much appreciated. One of us (HYF) is a recipient of The National Social Science Fund of China Project: "Grassland degradation and herder sustainable livelihood in the Qinghai-Tibetan Plateau"(Grant number:17BMZ106). The cooperation of herders in five communities is greatly appreciated as is help from local Bureau officials in Qinghai and in the Tibet Autonomous region.

References

Arbuckle, J.G., L. Wright Morton, and J. Hobbs. 2015. Understanding farmer perspectives on climate change adaptation and mitigation: The roles of trust in sources of climate information, climate change beliefs and perceived risk. *Environment and Behaviour* 47 (2): 205–234.

Arvai, J., G. Bridge, N. Dolsak, et al. 2006. Adaptive management of the global climate problem: Bridging the gap between climate research and climate policy. *Climatic Change* 78 (1): 215–225.

Behnke, R.H., I. Scoones, and C. Kerven. 1993. *Range ecology at disequilibrium: New models of natural variability and pastoral adaptation in African Savannas*. London: Overseas Development Institute.

Bhave, A.J., D. Conway, S. Dessai, et al. 2016. Barriers and opportunities for robust decision making approaches to support climate change adaptation in the developing world. *Climate Risk Management* 14: 1–10.

Biagini, B., R. Bierbaum, M. Stults, et al. 2014. A typology of adaptation actions: A global look at climate adaptation actions financed through the Global Environmental Facility. *Global Environmental Change* 25: 97–108.

Burnham, M., and Z. Ma. 2018. Multi-scalar pathways to smallholder adaptation. *World Development* 108: 249–262.

Chen, H., Q. Zhu, C. Peng, et al. 2013. The impacts of climate change and human activities on biogeochemical cycles on the Qinghai-Tibetan Plateau. *Global Change Biology* 19: 2940–2955.

Chen, J., S. Yin, H. Gebhardt, et al. 2018. Farmers' livelihood adaptation to environmental change in an arid region: A case study of the Minqin Oasis, northwestern China. *Ecological Indicators* 93: 411–423.

De Haan, C., ed. 2016. *Prospects for livestock-based livelihoods in Africa's dry lands. World Bank Studies*. Washington, DC: World Bank.

Dessai, S., and M. Hulme. 2007. Assessing the robustness of adaptation decisions to climate change uncertainties: A case-study on water resources management in the East of England. *Global Environmental Change* 17 (1): 59–72.

Du, F. 2012. Ecological resettlement of Tibetan herders in the Sanjiangyuan: A case study in Madoi county of Qinghai. *Nomadic Peoples* 16 (12): 116–133.

Dupuis, J., and G.R. Biesbroek. 2013. Comparing apples and oranges: The dependent variable problem in comparing and evaluating climate change adaptation policies. *Global Environmental Change* 23 (6): 1476–1487.

Eisenack, K., and R. Stecker. 2012. A framework for analyzing climate change adaptation as actions. *Mitigation and Adaptation Strategies for Global Change* 17: 243–260.

Ellis, F. 1998. Household strategies and rural livelihood diversification. *The Journal of Development Studies* 35 (1): 1–38.

Fratkin, E. 1997. Pastoralism: Governance and development issues. *Annual Review of Anthropology* 26: 235–261.

Ha-Duong, M., R. Swart, L. Bernstein, et al. 2007. Uncertainty management in the IPCC: Agreeing to disagree. *Global Environmental Change* 17 (1): 8–11.

Haynes, M.A., K.J.S. Kung, J.S. Brandt, et al. 2014. Accelerated climate change and its potential impact on Yak herding livelihoods in the eastern Tibetan plateau. *Climatic Change* 123 (2): 147–160.

Hua, L.M., and V. Squires. 2015. Managing China's pastoral lands: Current problems and future prospects. *Land Use Policy* 43: 129–135.

IPCC. 2012. In *Managing the risks of extreme events and disasters to advance climate change adaptation. A Special Report of Working Groups I and II of the Intergovernmental Panel on Climate Change*, ed. C.B. Field, V. Barros, T.F. Stocker, et al., 582. Cambridge: Cambridge University Press.

Kang, S., Y. Xu, Q.You, et al. 2010. Review of climate and cryospheric change in the Tibetan Plateau. *Env Res Lett* 5(1).

Klein, J.A., K. Hopping, E. Yeh, et al. 2014. Unexpected climate impacts on the Tibetan Plateau: Local and scientific knowledge in findings of delayed summer. *Global Environmental Change* 28: 141–162.

Lesnikowski, A., J. Ford, L. Berrang-Ford, et al. 2013. Adapting to health impacts of climate change: A study of UNFCCC Annex I parties. *Environmental Research Letters* 6 (4): 044009.

Li, C., Y. Tang, H. Luo, et al. 2013. Local farmers' perception of climate change and local adaptive strategies: A case study from the middle Yarlung Zangbo River Valley, Tibet, China. *Environmental Management* 52: 894–906.

Li, L., S. Yang, Z. Wang, X. Zhu, and H. Tang, 2010. Evidence of warming and wetting climate over the Qinghai-Tibet Plateau. *Arct Antarct Alp Res* 42, 449–457. https://doi.org/10.1657/1938-4246-42.4.449.

Liang, Y., Ganjurjav, W.N. Zhang, et al. 2014. A review on effect of climate change on grassland ecosystem in China. *Journal of Agricultural Science & Technology* 16 (2): 1–8.

Ma, Y., Y. Wang, and X. Wang. 2014. Classification of the snow disasters and circulation features of the blizzard in Altay region, China. *Journal of Arid Land Resources and Environment* 28 (8): 120–124.

Mi, Z., Y. Wei, C. He, et al. 2017. Regional efforts to mitigate climate change in China: A multi criteria assessment approach. *Mitigation and Adaptation Strategies for Global Change* 22: 45–66.

Michalk, D., L. Hua, D. Kemp, et al. 2010. Redesigning livestock systems to improve household income and reduce stocking rates in China's western grasslands. In *Towards sustainable use of rangelands in North-west China*, ed. V.R. Squires, L.M. Hua, D.G. Zhang, et al., 301–324. Dordrecht: Springer.

Mishra, C., S. Bagchi, T. Namgail, et al. 2010. Multiple use of Trans-Himalayan rangelands: Reconciling human livelihoods with wildlife conservation. In *Wild rangelands: Conserving wildlife while maintaining livestock in semi-arid ecosystems*, vol. 6, 291–311. Chichester: Wiley-Blackwell.

Owusu, M., M. Nursey-Bray, and D. Rudd. 2019. Gendered perception and vulnerability to climate change in urban slum communities in Accra, Ghana. *Regional Environmental Change* 19 (1): 13–25.

Petit, M. 2005. Scientific uncertainties and climate risks. *Comptes Rendus Geoscience* 337: 393–398.

Qiu, J. 2008. China: The third pole. *Nature* 454 (7203): 393–396.

Rashid, H., S. Robert, and A. Neville. 2005. Mountain systems. In *Ecosystems and human well-being: Current state and trends. Millennium ecosystem assessment*, 681–716. Washington, DC: Island Press.

Roder, W., G. Gratzer, and K. Wangdi. 2002. Cattle grazing in the conifer forests of Bhutan. *Mountain Research and Development* 22: 1–7.

Scoones, I. 1994. *Living with uncertainty: New directions in pastoral development in Africa*. London: Intermediate Technology Publications.

Shang, Z.H., Q. Dong, A. Degen, et al. 2016. Ecological restoration in the Qinghai-Tibetan Plateau: Problems strategies and prospects. In *Ecological restoration: Global challenges, social aspects and environmental benefits*, ed. Victor Squires, 151–176. New York: NOVA Science Publishers.

Shang, Z.H., R. Zhang, A. Degen, et al. 2017. Rangelands and grasslands in the Tibetan Plateau of China: Ecological structure and function at the top of the world. In *Rangelands along the Silk Road: Transformative adaptations under climate and global change*, ed. Victor R. Squires, Zhanhuan Shang, and Ali Ariapour, 65–102. New York: NOVA Science Publishers.

Shen, M., S. Piao, T. Dorji, et al. 2015. Plant phenological responses to climate change on the Tibetan Plateau: Research status and challenges. *National Science Review* 2 (4): 454–467.

Shreshtha, S. 1994. *Evolution of mountain farming systems*. In: *Proceedings of the FAO/ICIMOD Seminar at Lumle, Pokhara, Nepal*. International Centre for Integrated Mountain Development (ICIMOD).

Smit, B., and M.W. Skinner. 2002. Adaptation options in agriculture to climate change: a typology. *Mitigation and Adaptation Strategies for Global Change* 7: 85–114.

Squires, V.R. 2012. Rangeland Stewardship in Central Asia: Balancing livelihoods, biodiversity conservation and land protection. Springer: Dordrecht

Squires, V.R., and L.M. Hua. 2015. On the failure to control overgrazing and land degradation in China's pastoral lands: Implications for policy and for the research agenda. In *Rangeland ecology, management and conservation benefits*, ed. Victor R. Squires, 19–42. New York: NOVA Science Publishers.

Tashi, G., and M. Foggin. 2012. Resettlement as development and progress? Eight years on. Review of emerging social and development impacts of an ecological resettlement project in Tibet Autonomous region. *Nomadic Peoples* 16 (1): 134–151.

Tulachan, P.M., and A. Neupane. 1999. *Livestock in mixed farming systems of the Hindu Kush-Himalayas: Trends and sustainability*. Kathmandu, Nepal: International Centre for Integrated Mountain Development (ICIMOD).

Vitousek, P.M. 1994. Beyond global warming: ecology and global change. *Ecology* 75 (7): 1862–1876.

Wang, X.M., F.G. Liu, Q. Zhou, et al. 2006. On the formation reasons of eco-refugee from a geography perspective in the source region of Yellow river, Lancang river and Yangtze river. *Ecological Economy* 9: 131–134.

Wangchuk, K., D. Karma, and L. Ugyen. 2006. *Crushing the bone: Minimizing grazing conflicts in community Tsamdro—A case study from Dhur Village*. Bumthang, Bhutan: Choekhor Geog.

Wangchuk, K., M. Wurzinger, A. Darabant, et al. 2014. The changing face of cattle raising and forest grazing in the Bhutan Himalaya. *Mountain Research and Development* 34 (2): 131–138.

Xu, K., J. D. Milliman, and H. Xu, 2010. Temporal trend of precipitation and runoff in major Chinese Rivers since 1951. *Global Planet Change* 73: 219–232. https://doi.org/10.1016/j.gloplacha.2010.07.002.

Yang, Y.H., J.Y. Fang, Y.D. Pan, et al. 2009. Above ground biomass in Tibetan grasslands. *Journal of Arid Environment* 73 (1): 91–95.

Zhao, H.D., S.L. Liu, S.K. Dong, et al. 2015. Analysis of vegetation change associated with human disturbance using MODIS data on the rangelands of the Qinghai-Tibetan Plateau. *The Rangeland Journal* 37: 77–87.

Chapter 17
Developing Linkages for Carbon Sequestration, Livelihoods and Ecosystem Service Provision in Mountain Landscapes: Challenges and Opportunities in the Himalaya Hindu Kush (HKH) Region

Victor R. Squires

Abstract A number of critical issues have emerged in the Hindu Kush Himalayan (HKH) region as reflected in the downward spiral of resource degradation, increasing rural poverty, and food and livelihood insecurity in mountainous regions. New and comprehensive approaches to mountain development are needed to identify sustainable resource development practices such as Sustainable Land Management (SLM), to strengthen local institutions and knowledge systems, and increase the resilience of both mountain environments and their inhabitants. The key objectives of this chapter are to reflect upon the links between carbon retention and capture, the provision of other ecosystem services and livelihood impacts, consider the challenges in developing payment systems for carbon storage and outline the key forward-looking interdisciplinary and multi-stakeholder opportunities to advance progress towards pro-poor, climate-smart development in the mountainous regions of the HKH. These are among the measures that can be seen as opportunities.

Keywords Understandings · Carbon · Land degradation · Climate · Poverty · Common pool resources · Glaciers · Irrigation

V. R. Squires (✉)
Institute of Desertification Studies, Beijing, China

University of Adelaide, Adelaide, SA, Australia

17.1 Introduction

Mountain regions encompass nearly 24% of the total land surface of the earth and are home to approximately 12% of the world's population (Huber et al. 2005). Their ecosystems play a critical role in sustaining human life both in the cold arid highlands and the arid lowlands and deserts through which rivers flow to bring life and provide livelihoods for about ten million people. During recent years, resource use in high mountain areas has changed mainly in response to the globalization of the economy and increased world population (Kreutzmann 2012). As a result, mountain regions are undergoing rapid environmental change, exploitation, and depletion of natural resources leading to ecological imbalances and economic unsustainability (Kreutzmann 2012). Moreover, the changing climatic conditions have stressed mountain ecosystems through higher mean annual air temperatures and the melting of glaciers and snow (Hagg 2018). Altered precipitation patterns have also had an impact (Hua et al. 2018).

A number of critical issues have emerged as reflected in the downward spiral of resource degradation, increasing rural poverty, and food and livelihood insecurity in mountain regions. New and comprehensive approaches to mountain development are needed to identify sustainable resource development practices like Sustainable Land Management (SLM), to strengthen local institutions and knowledge systems, and to increase the resilience of both mountain environments and their inhabitants. Their important role in capture and storage (sequestration) of carbon, especially soil organic carbon (SOC) needs to be protected. The mountains in Hindu Kush Himalayan (HKH) region are the water towers such as that from the Tibetan Plateau (Fig. 17.1).

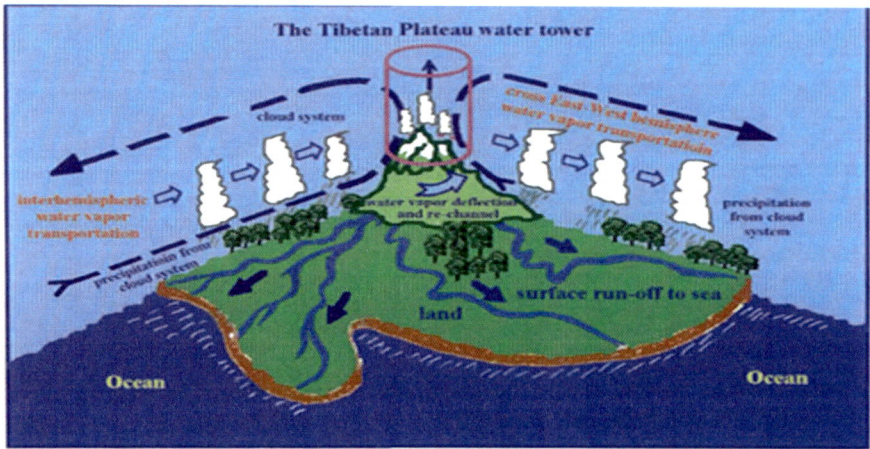

Fig. 17.1 Topographic and orographic features combine to create a situation where the HKH uplands act as water towers such as in this example from the Tibetan plateau

Due to orographic effects, the mountains receive an annual precipitation of 1000–2000 mm. The HKH region consists of several independent nations which have different needs and requirements for water resources. While downstream riparian nations irrigate in summer, the mountainous upstream countries have an interest in hydropower generation during the cold season. This inevitably leads to water use conflicts along transboundary rivers. It is obvious that water is the most precious and conflict-prone natural resource in the region. In Kyrgyzstan, Tajikistan and western China, especially the Tibetan Plateau, large quantities of water are stored in mountain glaciers, which again are strongly influenced by climate change. The lowlands of the Hindu Kush Himalayan (HKH) region are dry: steppe, semi-deserts and deserts. The main reasons for the aridity are predominant westerly winds and lee-side effects, which cause very low precipitation sums (less than 100 mm per year) in the desert regions e.g. Gobi and Taklamakan.

17.2 Glaciers: A Water Storage System Under Threat

The HKH contains the largest glacier area outside the polar regions, and is, therefore, referred to as the "Third Pole". According to the Randolph Glacier Inventory, 30,200 glaciers with a total area of 64,497 km^2 exist in the HKH region. A large area of glaciers and ice fields exists in the Tibetan Plateau, the eastern Tian Shan and the Qilian mountains of China (Shi and Liu 2000). Glacier retreat caused by climate warming is a worldwide effect and also affects the HKH glaciers. As in many other mountain systems of the mid-latitudes, the majority of Tian Shan glaciers, for example, were more or less in equilibrium from the late 1950s into the 1970s (Makarevich and Liu 1995).

All recent catchment-scale studies revealed a loss of glacier area over several decades and most of them showed accelerating loss. From the Qilian Mountains in West China, very high retreat rates of −0.73% per year have been reported (Huai et al. 2014). The other regional studies from China (Lu et al. 2002 reported rates between −0.51% per year from the source of the Yellow River between 1967 and 2000 (Liu et al. 2002) to −0.05% per year at the source of the Yangtze River between 1969 and 2000 (ibid). Farinotti et al. (2015) have assessed the area retreat for the whole Tian Shan as −18% from 1961 to 2012, equaling a rate of −0.45% per year. A spatially closer look shows larger area changes in the outer ranges (−0.38 to −0.76% per year) compared with the inner (−0.15 to −0.40% per year) and eastern ranges (−0.05 to −0.31% per year) since the middle of the twentieth century (Sorg et al. 2012).

In general, small and low-lying glaciers in the more humid margins of the mountains suffered more than the large and high-lying ice masses in the interior. Based on four different global climate models, Aizen et al. (2007) made projections for the whole Tian Shan and projected volume losses of up to 43% as a worst case scenario until 2070–2099. In the best case scenario, meaning a most glacier-friendly climate, there will be almost no volume losses. Based on the output of 14 global climate models, Radic and Hock (2014) simulated glacier volume changes for the entire HKH and found a mean loss of 49% until 2100. From 1961 to 2012, the Tian Shan glaciers have lost 5.4 Gt mass per year (Farinotti et al. 2015), additionally contributing to runoff.

Changes in glacier melt water amount have a huge hydrological impact in the HKH, because, here, the mountain ranges are surrounded by arid lowlands where water is a precious resource. This preserves the signal of snow and ice melt far downstream because it is not superimposed by precipitation, as in more humid climates. In the Tarim basin, for example, glacier melt contributed 41.5% to total runoff (Gao et al. 2010). In the Tarim river, up to 20% of total runoff can be attributed to an increase in glacial melt in recent decades (Pieczonka and Bolch 2014; Duethmann et al. 2016). By storing solid precipitation and releasing it in summer, glaciers control the seasonal water availability.

In arid regions, glacier meltwater is sometimes almost the only source of water during the main period of plant growth. As they build up their ice masses in cold, wet seasons and reduce them again during hot, dry seasons, glaciers have a compensating effect on streamflow: rivers with glaciers in their catchment have a guaranteed minimum flow, either from rainfall or from melt, whereas non-glaciated catchments rely completely on rainfall during summer. This effect of glaciations is to reduces the year-to-year variability of runoff in moderately glacierized river basins, which is highly beneficial for drinking water security, agriculture and water resources management. Already today, the lowland countries within the HKH region face serious water stress, caused by increasing aridity of the climate and—so far much more important—the growing demand for water. By simulating the hydrological cycle of the catchment with and without water withdrawals, Aus der Beek et al. (2011) quantified the water use impact as high as 86%, thus only 14% is caused by climate change!

Until the end of the twenty-first century, a further increase in temperature is very likely (Lioubimtseva and Henebry 2009). Prolonged glacier retreat will transform melt-dominated runoff regimes into rain-dominated regimes. This will reduce water availability during summer months dramatically. Furthermore, evapo-transpiration losses will increase in the warmer atmosphere, especially in low-lying downstream regions, further tightening water availability during the growing season (Hou et al. 2007).

17.3 Land-Use Change: A Major Driver of CO_2 Emission and Climate Change

In the last decades, due to climate changes, soil deterioration and land use/land cover changes (LULCCs), land degradation risk has become one of the most important ecological issues in the HKH region. Land degradation involves two interlocking systems—the natural ecosystem and the socio-economic system (Fig. 17.2).

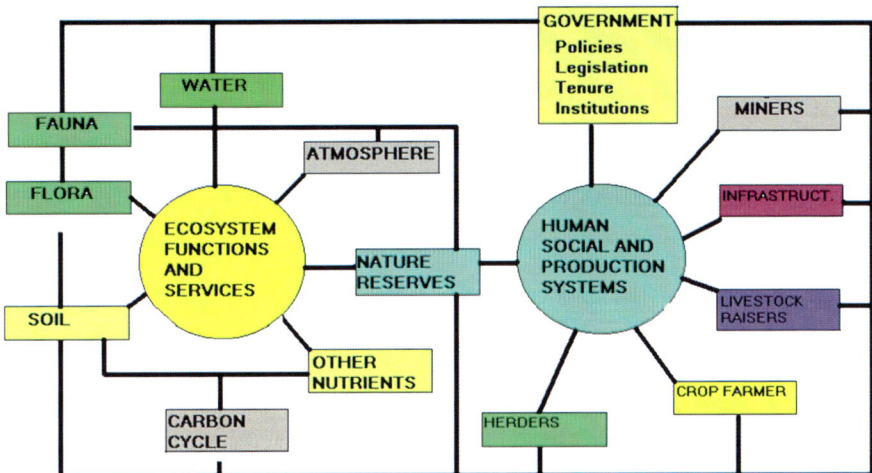

Fig. 17.2 There are two interlocking systems—the natural ecosystem and the socio-economic system

The complexity of land degradation processes should be addressed using a multidisciplinary approach. The principal aim of ecosystem management is to help develop and promote strategies to protect functional integrity and balance social/economic and biodiversity values (Squires 2016).

Sustainable land management (SLM) is the desired outcome. Sustainability is an elusive goal. The notorious vagueness of the term and its scope for varied and seemingly legitimate interpretation by different parties appear to make it all but useless as an operational guide (O'Riordan 1988). One has only to consider simple questions (Sustain what? How? For whom? Over what time period? Measured by what criteria?) to appreciate that sustainability can never be defined precisely. Regardless of the ambiguity of the term, however, there appears to be a general consensus that achieving sustainability will place new demands on individuals, society and science. The challenge facing science is how best to structure and undertake research to meet the diverse, and often apparently conflicting, needs of society, local communities and individual natural resource users (Squires 2015). This may be very difficult in the HKH region.

Land-use changes and degradation of forests, meadows, steppes and the soils in the HKH region have important implications for CO_2 emissions and global climate change (Lioubimtseva et al. 2005). Estimates of land-use categories have been documented for various regions within the HKH. In specific areas of Nepal, Bhutan and western China, these changes were quite large. Some evidence exists to show that the land-use changes and forest/soil degradation affect C pools significantly. For example, the net emissions of C due to land-use changes were reported to be as low as 6.9×10^6 mg but up to 42.1×10^6 mg year^{-1} depending on plant community type. Emissions from fuelwood consumption alone were estimated to be 1.47×10^6 mg year^{-1} (Upadhyay et al. 2005).

Control of a piece of land shapes the land-use and the willingness of land users to incur costs in implementing land management practices. In much of the HKH, the poor own very small plots while communal tenure arrangements may limit access, use and benefit-sharing. Diverse land tenure systems in the HKH, characterized by customary and statutory land rights, legal pluralism, land claims through tree planting and the misconception of "abandoned" land means addressing these challenges is difficult. Roncoli et al. (2007) highlighted that in systems with a mixture of open access and common property regimes, the multifunctional, fragmented and dynamic characteristics of land-use by pastoralists and farmers requires a holistic approach that, besides carbon capture, also integrates crop and livestock production (Poyatas et al. 2003).

Bennett et al. (2010) highlighted a general inability to define and enforce rights to particular grazing resources and inadequate local institutions responsible for management in open access community rangelands. Carbon sequestration projects can thus pose a collective action problem (cf. Ostrom 1990). Strong but flexible local institutions embedded in multi-level governance structures are key to addressing legal ambiguity, social tensions, social inequalities and overlapping resource-use rights (Bennett et al. 2010). The prior existence of active local organizations and ensured participation may serve as criteria for establishing community-oriented carbon sequestration projects. However, Schnegg (2018) points out that the fundamentals on which community based management (CBM) is predicated may be flawed. One of the dominant frameworks—CBM, was informed by common-pool theory, and emphasized principles such as fixed boundaries, proportional cost sharing, and formal sanctioning (Saunders 2014).

In common-pool theory, those principles have been recognized as salient factors for successful, sustainable, and just natural resource management. Common-pool theory itself largely applies a rational actor model and typically assumes that the actors interact with a single issue in mind. Schnegg (2018) argues that a better understanding of the concept of institutional multiplexity would help promote more durable and beneficial outcomes in CBM. Institutional multiplexity describes the number of transactions between two households in a social network. Institutional multiplexity implies that people cannot separate the sharing of a resource like access to stock and domestic water or grazing land from sharing in other domains. Institutional multiplexity hinders the implementation of design principles such as fixing boundaries, sharing costs proportional to use, and formal sanctioning (all pillars of CBM). However, it also opens other means for governing nature through social control (Schnegg 2018).

Changes in land use and management practices to store and sequester carbon are integral to global efforts that address both climate change and alleviate poverty. Knowledge and evidence gaps nevertheless abound. Many of the knowledge gaps in understanding carbon storage in mountain ecosystems suffer from a lack of empirical data and scientific evidence, which limits the utility of scientific knowledge for research users such as policy makers and NGOs. Measurement challenges restrict the number of studies focusing on processes and trade-offs, impeding development of accurate carbon accounting methodologies. Incomplete knowledge of carbon cycles makes it difficult to up-scale plot or field-level studies to inform regional or global model development, hindering accurate prediction of how land, non-carbon ecosystem services and livelihoods may be affected by climatic, environmental and other changes.

17.4 C Sequestration, Payments for Ecosystem Services and the Poor

Climate change mitigation efforts linked to land-use and land management generally seek to increase the amount of carbon stored in soils and biomass. However, a trade-off exists in that to realize many livelihood and ecosystem service benefits from SOC requires its depletion (e.g. through crop production) and thus a net release of CO_2 (Janzen 2006). Understanding how different land use and management systems can both maintain and enhance carbon storage and other ecosystem services (the "hoard it or use it" conundrum identified by Janzen (2006), as well as identifying where the trade-offs between these goals are situated, are key research challenges, especially in relation to how SOC can be increased without suppressing decomposition rates so that nutrient cycling is not adversely affected (Powlson et al. 2011).

For the poor to benefit from carbon storage, through both climate finance streams and the collateral ecosystem service benefits delivered by carbon-friendly land management, a number of serious adjustments need to be made (Upadhyay et al. 2005; Schnegg 2018).There is the need to draw together understanding from different disciplinary bases to develop applied research, grounded in sound science, to deliver policy-relevant outcomes of practical value. Working with multi-stakeholders across scales from the local to the regional is necessary to ensure that scientific advances can advance policy and practice to deliver carbon, ecosystem service and poverty alleviation benefits.

Community-based land management projects within the voluntary carbon sector increasingly apply standards and protocols designed to reduce trade-offs and deliver multiple benefits across carbon storage, poverty alleviation, community empowerment, and biodiversity conservation dimensions. However, accurate accounting methodologies underpinning the carbon components of such assessments are lacking due to an absence of scientific data, models, appropriate local monitoring methods and regional measurement protocols, particularly in drylands, where methods need to address inherent spatial and temporal dynamism (Schmidt et al. 2011; Reynolds et al. 2011). These challenges mean carbon sequestration gains (or prevented losses) are difficult to quantify and need to be integrated with assessments of livelihood costs, benefits and trade-offs. The lack of coherent and credible science accessible to practitioners remains a significant obstacle to the development of integrated practice.

17.5 Knowledge Gaps and Methodological Challenges in Understanding Carbon Storage

Efforts to increase the size of the terrestrial carbon store are perhaps most commonly associated with climate change mitigation. However, the presence of carbon in both soils and biomass is hugely beneficial for a range of ecosystem functions and services, assisting in the provision of adaptation options and the maintenance of natural resource-based livelihoods. Simply by being present, SOC improves soil structural stability and water holding capacity (Holm et al. 2003). The decomposition of organic carbon generates further direct benefits through the recycling of nutrients and maintenance of soil fertility (Stursova and Sinsabaugh 2008; Scholes et al. 2009). This, in turn, contributes to other supporting, provisioning and regulating services, particularly food and timber production.

17.5.1 Below Ground Carbon: SOC Stores and Fluxes

The need to include SOC storage in payment schemes is long recognized (Lal 2004), but only simple models are used at present, based on changes in soil organic matter (SOM) measurements through time (e.g. Wildlife Works Carbon 2010, https://theredddesk.org/countries/actors/wildlife-works-carbon). A greater range and depth of field data are essential to enable monitoring of changes in SOC storage (Powlson et al. 2011) and the development of a new generation of soil carbon models (Schmidt et al. 2011). These need to be linked to the development of methodologies that local communities can use to monitor SOC.

Changes in land management practices (e.g. reduced tilling, reduced grazing and prevention of deforestation) can reduce heterotrophic respiration losses, preserving the SOC store. However, scientific data gaps limit our ability to include SOC stores and fluxes in the evaluation of benefits accruing from land management practices and reduce the accuracy of future predictions of SOC store changes under different land management and climatic scenarios.

Climatic warming is increasing the global flux of CO_2 from soils to the atmosphere. Most meta-analyses do not distinguish between CO_2 from microbial decomposition of SOC and that from plant roots. It is thus impossible to determine if any increase in soil CO_2 efflux is due to accelerated SOC decomposition (and therefore represents a decline in SOC stores) or greater primary productivity (with no associated decline in SOC). It is challenging to separate the two sources in the field. Consequently, there are few *in situ* data, particularly in the HKH. Reliable assessment of processes affecting CO_2 efflux rates requires *in situ* chamber monitoring systems to collect gases from remote field locations.

Processes affecting SOC stores in mountainous areas have some fundamental differences to those in mesic ecosystems. First, because of low precipitation, low mean annual temperatures, presence of permafrost etc., microbial activity is low and rate of decomposition of above ground litter is much lower (Shang et al. 2017). Information on the amount, distribution and species composition of microbes in soils is critical to respiration and the fate of SOC, yet empirical information is lacking on how enzymes are affected by disturbance and climatic changes. Fungi and bacteria largely control SOC respiration processes, influencing the residence time of SOC storage.

17.5.2 Above Ground Biomass (AGB) Stores and Fluxes

AGB stores are determined by the balance between carbon accumulation from primary production and carbon losses related to mortality, human use and land use change. AGB influences settlement patterns across a landscape, and plays a vital role in rural livelihood activities such as livestock grazing, timber harvesting, and fuelwood production. It is therefore crucial to understand drivers of AGB, projected future trends, and their implications, in order to develop policies that protect resources and promote livelihood options. Significant knowledge gaps nevertheless remain (Abson and Termansen 2010) relating to: (limited observational data on the spatial distribution and temporal variability of AGB; and (Aizen et al. 2007) poor understanding of the natural and human drivers of AGB and the links to changes in other ecosystem services.

Addressing these gaps requires integration of forestry and grassland expertise with ecological and remote sensing techniques alongside livelihoods and resource use assessments grounded in the social sciences.

17.6 Lack of Observational Data for Present Day AGB Storage

AGB varies considerably across the HKH at a range of scales, complicating mapping and monitoring. AGB can be estimated from plot studies, where tree biomass is related to standard forestry observations such as tree diameter. In this way, change in AGB storage can be monitored through re-sampling of permanent vegetation plots. Such efforts are necessary but labor intensive, restricting the achievable spatial and temporal coverage. Recently, remote-sensing studies have been used to estimate biomass and these offer potential for developing regional AGB estimates. However, it is difficult to determine the accuracy of these estimates due to limited observational data (ground truthing).

Earth observation (EO) studies offer considerable scope for extending monitoring and understanding of AGB on national and regional scales (Mitchard et al. 2009) and first need to be calibrated against *in situ* observations. Baccini et al. (2008) used optical data from the moderate resolution imaging spectroradiometer (MODIS) on the Terra and Aqua satellites, trained and tested against plot-based biomass data to predict above ground biomass at 10 × 10 m resolution. Uncertainties remain substantial and relate to deforestation and degradation rates, biomass storage and the cursory treatment of the impacts of logging, livestock grazing, fires and shifting cultivation which are difficult to identify and quantify by remote sensing.

17.6.1 Lack of Quantitative Assessments of Natural and Human Drivers of AGB Storage

Global change affects AGB storage largely through shifts in precipitation—a major uncertainty in climate projections—and through poorly understood responses to rising atmospheric CO_2 concentrations.

Additional direct human impacts are important determinants of AGB stores; increased demand for fuelwood leads to forest degradation, smallholder agriculture contributes to deforestation, whereas farm abandonment allows AGB accumulation (Poyatas et al. 2003). Clearance of woodlands for agriculture reduces both AGB and SOC, resulting in a release of up to 30 t C ha^{-1} and after cessation of agriculture, AGB recovers at 0.7 t C ha^{-1} year^{-1} reaching pre-disturbance levels after 20–30 years, whereas SOC shows no significant change over these timescales. Further assessments across agro-ecological settings are essential to widen the significance of these plot-based case studies, enabling development of national and regional-scale analyses. AGB is vital in meeting domestic requirements for energy, with fuelwood collection. The role of fuelwood collection in determining regional forest quality and AGB storage is uncertain, although unsustainable extraction in peri-urban locations has been documented (FAO 2009).

17.7 Linking Scientific Evidence Gaps and Ecosystem Service Evaluation Challenges

Carbon store and flux dynamics are physical changes to an ecosystem's structures and processes, resulting in changes in the bundle of services flowing from an ecosystem and the benefits that humans derive from interactions with that ecosystem (Daily 1997; Tongway and Ludwig 2011; Squires 2016). Ecosystem services associated with carbon are numerous.

17.7.1 Lack of Quantitative Assessments of Natural and Human Drivers of AGB Storage

Separating human and climatic drivers requires sub-sampling of regions of similar climatic influence but different human impacts, for example across protected areas or national/regional boundaries, and also requires links to specific model classes. The exact nature of relations is poorly quantified in the HKH and needs further testing. Such knowledge is vital if payments for carbon sequestration are to capture all potential impacts that changing land management practices can have on the bundle of ecosystem services drawn on by the rural poor in pursuit of their livelihoods.

Ecosystem services are often interdependent, so optimization of a single service may have unforeseen impacts on other ecosystem services (Abson and Termansen 2010). For example, optimization of climate regulation through payments for carbon sequestration may affect food provision and water regulation, at worst, limiting successful adaptation to climate change. Broader ecosystem service impacts of carbon sequestration schemes, therefore, require careful consideration.

The key challenges are: (1) need to better understand the relationships between carbon, ecosystem service provision, and drivers of future change; (2) lack of nuanced understanding of the links between poverty and land tenure and the implications this has for the design and implementation of carbon payment schemes; and (3) shortage of appropriate decision support tools for land management decisions and adaptation strategies, alongside the thresholds at which land users will shift towards carbon mitigation scenarios, particularly in rangelands and alpine meadows.

Cost-benefit analyses that consider trade-offs and synergies across carbon and ecosystem service dimensions as well as across different cultural logics represent vital assessment tools in further advancing the understanding of trade-offs. While scientific and process-based evidence for carbon-ecosystem service relationships is lagging, changes to land management practices to deliver carbon sequestration and other ecosystem services benefits are already being implemented by some stakeholders.

17.8 Poverty, Institutions and Land Tenure: Implications for Carbon Payment Schemes

Delivering carbon payment and ecosystem service benefits to the poorest groups in society first requires identification of who is poor and where they are located. Large-scale datasets permit comparisons across different areas and can target climate finance as a poverty alleviation mechanism using analyses of current and future climate risk and vulnerability mapping. However, for the poor to benefit requires a context-specific understanding of what poverty is and how it is managed. Existing datasets use multiple indicators to determine what poverty is and who is poor (e.g. Thornton et al. 2002), reflecting the multi-dimensional nature of poverty, taking into account lack of choice or capability, as well as material living standards and an inability to meet basic needs. However, those living in poverty have their own ideas about what it means to be poor, based on what is socially and culturally important to them.

17.9 Conclusion: Key Steps Towards Climate-Smart Pro-Poor Investments in Carbon Sequestration

Improved data and knowledge on the spatial distribution of carbon storage and release, whilst important in its own right, will not directly create poverty alleviation, carbon storage, adaptation and ecosystem service benefits without new forms of collaborative working across academic disciplines and with partners at the community level in the private sector and in national government (Fig. 17.3).

Reflections on the experiences of such multi-stakeholder, multi-level partnerships will be essential to the wider uptake of carbon-friendly land management projects with support from international bodies and the private sector in ensuring the full valuation of benefits and their trading in the emerging climate finance sector.

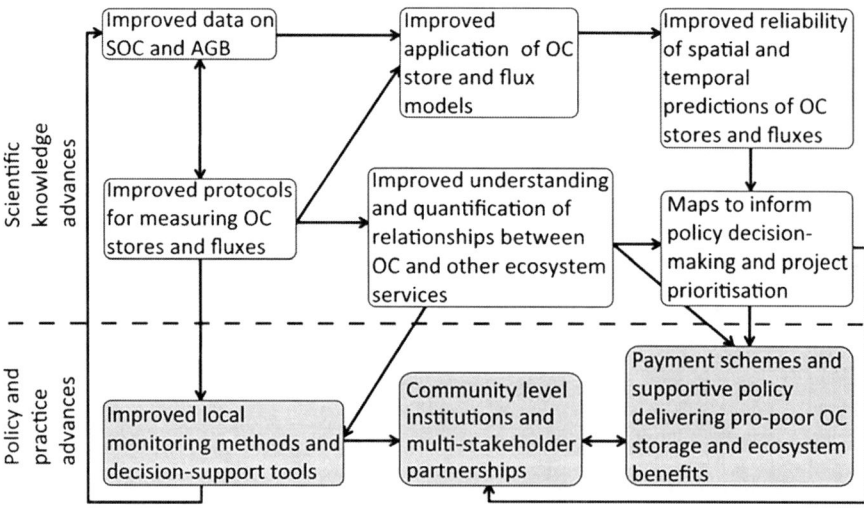

Fig. 17.3 Possible route to delivering pro-poor carbon storage and ecosystem service benefits based on an improved scientific evidence base (Stringer et al. 2012)

References

Abson, D.J., and M. Termansen. 2010. Valuing ecosystem services in terms of ecological risks and returns. *Conservation Biology* 25 (2): 250–258.

Aizen, V.B., E.M. Aizen, and V.A. Kuzmichonok. 2007. Glaciers and hydrological changes in the Tien Shan: Simulation and prediction. *Environmental Research Letters* 2 (2007): 045019.

Aus der Beek, T., F. Vol, and M. Flörke. 2011. Modelling the impact of global change on the hydrological system of the Aral Sea basin. *Physics and Chemistry of the Earth, Parts A/B/C* 36 (13): 684–695.

Baccini, A., N. Laporte, S.J. Goetz, et al. 2008. A first map of tropical Africa's above-ground biomass derived from satellite imagery. *Environmental Research Letters* 3 (2008): 045011.

Bennett, J., A. Ainslie, and J. Davis. 2010. Fenced in: Common property struggles in the management of communal rangelands in central Eastern Cape Province. South Africa. *Land Use Policy* 27: 340–350.

Daily, G.C. 1997. *Nature's service: Societal dependence on natural ecosystems*. Washington D.C: Island Press.

Duethmann, D., C. Menz, T. Jiang, et al. 2016. Projections for headwater catchments of the Tarim River reveal glacier retreat and decreasing surface water availability but uncertainties are large. *Environmental Research Letters* 11 (5): 054024.

FAO. 2009. *State of world forests*. Rome: Food and Agriculture Organization.

Farinotti, D., L. Longuevergne, G. Moholdt, et al. 2015. Substantial glacier mass loss in the Tien Shan over the past 50 years. *Nature Geoscience* 8: 716–722.

Gao, X., B.S. Ye, S.Q. Zhang, et al. 2010. Glacier runoff variation and its influence on river runoff during 1961–2006 in the Tarim River Basin, China. *Science China Earth Sciences* 53: 880–891.

Hagg, W. 2018. Water from the mountains of Greater Central Asia: A resource under threat. In *Sustainable land management in Greater Central Asia*, ed. V.R. Squires and Qi Lu, 237–248. London: Routledge, New York.

Holm, A.M., I.W. Watson, W.A. Loneragan, et al. 2003. Loss of patch-scale heterogeneity on primary productivity and rainfall-use efficiency in Western Australia. *Basic and Applied Ecology* 4 (6): 569–578.

Hou, P., R.J.S. Beeton, R.W. Carter, et al. 2007. Response to environmental flows in the lower Tarim River, Xinjiang, China: An ecological interpretation of water table dynamics. *Journal of Environmental Management* 83 (4): 383–391.

Hua, L.M., Y. Niu, and V.R. Squires. 2018. Climatic change on grassland regions and its impact on grassland-based livelihoods in China. In *Grasslands of the world: Diversity, management and conservation*, ed. V.R. Squires, J. Dengler, H. Feng, and L. Hua, 355–368. Boca Raton: CRC Press.

Huai, B., Z. Li, S. Wang, et al. 2014. RS analysis of glaciers change in the Heihe River Basin, Northwest China, during recent decades. *Journal of Geographical Sciences* 24 (6): 993–1008.

Huber, U.I., H.K.M. Bugmann, and M.A. Reasoner, eds. 2005. *Global change and mountain regions: An overview of current knowledge. Advances in global change research*. Vol. 23. Dordrecht: Springer.

Janzen, H.H. 2006. The soil carbon dilemma: Shall we hoard it or use it? *Soil Biology and Biochemistry* 38: 419–424.

Kreutzmann, H., ed. 2012. *Pastoral practices in high Asia: Agency of 'development' effected by modernisation, resettlement and transformation*. Dordrecht: Springer.

Lal, R. 2004. Soil carbon sequestration impacts on global climate change and food security. *Science* 304 (5677): 1623–1627.

Lioubimtseva, E., and G.M. Henebry. 2009. Climate and environmental change in arid Central Asia: Impacts, vulnerability, and adaptations. *Journal of Arid Environments* 73 (11): 963–977.

Lioubimtseva, E., R. Cole, J.M. Adams, et al. 2005. Impacts of climate and landcover changes in arid lands of Central Asia. *Journal of Arid Environments* 62 (2): 285–308.

Liu, S., A. Lu, and Y. Ding. 2002. Glacier fluctuations and the inferred climate changes in the A'nyemaqen Mountains in the source area of the Yellow River. *Journal of Glaciology and Geocryology* 24: 701–706.

Lu, A., T. Yao, and S. Liu. 2002. Glacier change in the Geladanong area of the Tibetan Plateau monitored by remote sensing. *Journal of Glaciology and Geocryology* 45: 559–562.

Makarevich, K.G., and C. Liu. 1995. In *Glacierization of the Tian Shan*, ed. M. Dyurgerov, C. Liu, and X. Zichu, 189–213. Moscow: VINITI.

Mitchard, E.T.A., S.S. Saatchi, I.H. Woodhouse, et al. 2009. Using satellite radar backscatter to predict above-ground woody biomass: A consistent relationship across four different African landscapes. *Geophysical Research Letters* 36: L23401.

O'Riordan, T. 1988. The politics of sustainability. In *Sustainable environmental management: Principles and practice*, ed. R.K. Turner. Boulder, Colorado: Westview Press.

Ostrom, E. 1990. *Governing the commons: The evolution of institutions for collective action*. Oxford: Oxford Press.

Pieczonka, T., and T. Bolch. 2014. Region-wide glacier mass budgets and area changes for the Central Tian Shan between 1975 and 1999 using Hexagon KH-9 imagery. *Global and Planetary Change* 128: 1–13.

Powlson, D.S., A.P. Whitmore, and K.W.T. Goulding. 2011. Soil carbon sequestration to mitigate climate change: a critical re-examination to identify the true and the false. *European Journal of Soil Science* 62: 42–55.

Poyatas, R., J. Latron, and L. Pilar. 2003. Land use and land cover change after agricultural abandonment. *Mountain Research and Development* 23 (4): 362–336.

Radic, V., and R. Hock. 2014. Glaciers in the earth's hydrological cycle: Assessments of glacier mass and runoff changes on global and regional scales. *Surveys in Geophysics* 35: 813–837.

Reynolds, J.F., G. Bastin, L. Garcia-Barrios, et al. 2011. Scientific concepts for an integrated analysis of desertification. *Land Degradation and Development* 22: 166–183.

Roncoli, C., C. Jost, C. Perez, et al. 2007. Carbon sequestration from common property resources: Lessons from community-based sustainable pasture management in north-central Mali. *Agricultural Systems* 94 (1): 97–109.

Saunders, F.P. 2014. The promise of common pool resource theory and the reality of commons projects. *International Journal of the Commons* 8 (2): 636–656.

Schmidt, M.W.I., M.S. Torn, S. Abiven, et al. 2011. Persistence of soil organic matter as an ecosystem property. *Nature* 478: 49–56.

Schnegg, M. 2018. Institutional multiplexity: Social networks and community-based natural resource management. *Sustainability Science* 13: 1017–1030.

Scholes, R.J., P.M.S. Monteiro, C.L. Sabine, et al. 2009. Systematic long-term observations of the global carbon cycle. *Trends in Ecology and Evolution* 24 (8): 427–430.

Shang, Z.H., R. Zhang, A. Degen, et al. 2017. Rangelands and grasslands in the Tibetan Plateau of China: Ecological structure and function at the top of the world. In *Rangelands along the Silk Road: Transformative adaptations under climate and global change*, ed. Victor Squires, Zhanhuan Shang, and Ali Ariapour, 65–101. New York: Nova Science Publishers.

Shi, Y., and S. Liu. 2000. Estimation of the response of the glaciers in China to the global warming of the 21st century. *Chinese Science Bulletin* 45: 668–672.

Sorg, A., T. Bolch, M. Stoffel, et al. 2012. Climate change impacts on glaciers and runoff in Tien Shan (Central Asia). *Nature Climate Change* 2: 725–731.

Squires, V.R. 2015. Sustainable rangeland management: An ecological and economic imperative. In *Rangeland ecology, management and conservation benefits*, ed. Victor R. Squires, 3–18. New York: Nova Science Publishers.

———. 2016. *Ecological restoration: Global challenges, social aspects and environmental benefits*, 309. New York: Nova Science Publishers.

Stringer, L.C., A.J. Dougill, A.D. Thomas, et al. 2012. Challenges and opportunities in linking carbon sequestration, livelihoods and ecosystem service provision in drylands. *Environmental Science and Policy* 19–20: 121–135.

Stursova, M., and R.L. Sinsabaugh. 2008. Stabilization of oxidative enzymes in desert soil may limit organic matter accumulation. *Soil Biology and Biochemistry* 40 (2): 550–553.

Thornton, P.K., R.L. Kruska, N. Henninger, et al. 2002. *Mapping poverty and livestock in the developing world*. Vol. 124. Nairobi, Kenya: ILRI.

Tongway, D.J., and J.A. Ludwig. 2011. *Restoring disturbed landscapes: Putting principles into practice*. Washington: Island Press, Covelo, London.

Upadhyay, TP., P.L. Sankhayan, and B. Solberg, A. 2005. A review of carbon sequestration dynamics in the Himalayan region as a function of land-use change and forest/soil degradation with special reference to Nepal Agriculture. *Ecosystems and Environment* 105:449–465.

Chapter 18
Experience for Future Good Practice and Policy of Combined Carbon Management and Livelihood in HKH Region

Zhanhuan Shang, A. Allan Degen, Devendra Gauchan, and Victor R. Squires

Abstract How can we improve carbon management at the field and household level? This is the key question to carry out implementation of climate control. Although traditional knowledge is important in guiding this practice, the modeling technique could be used to improve the traditional approach. In this chapter, we summarize experience of low carbon emission practices and cases of compensation in HKH areas to improve livelihood. The results, in some cases, could be recommended throughout the HKH by training and demonstration.

Keywords Good practice · Carbon management · Livelihood development · HKH region · Ecological compensation · Training and demonstration

Z. Shang (✉)
School of Life Sciences, State Key Laboratory of Grassland Agro-Ecosystems, Lanzhou University, Lanzhou, China
e-mail: shangzhh@lzu.edu.cn

A. A. Degen
Desert Animal Adaptations and Husbandry, Wyler Department of Dryland Agriculture, Blaustein Institutes for Desert Research, Ben-Gurion University of Negev, Beer Sheva, Israel
e-mail: degen@bgu.ac.il

D. Gauchan
Biodiversity International, Lalitpur, Nepal

V. R. Squires
Institute of Desertification Studies, Beijing, China

University of Adelaide, Adelaide, SA, Australia

18.1 Introduction

Undoubtedly, the quantification of carbon metrology and benefits is a big challenge for the carbon management campaign. It is usually based on models generated by scientific measurements and is important for strategy and policy decisions (Ashton et al. 2012). However, we should always stress that in HKH region, any environmental solution scheme should take into account the local resident's livelihood benefits, which is the basis for sustainable development (Banskota et al. 2007; Stringer et al. 2012; Kemp et al. 2013). Solutions to livelihood problems are very difficult at the current status of the HKH, and outside assistance is required (Sharma 2016). For example, the purpose of the REDD+ program is to improve the ecosystem's carbon sinking function through the implementation of techniques and policies that are converted into livelihood compensation (Atela et al. 2015; Newton et al. 2016; Marquardt et al. 2016; Castro-Nunez et al. 2016). Here we should reiterate the important suggestion in Chap. 1 that 'if we want to build up an integrated solution of benefit in HKH region through the carbon evaluation pathway, that should rely on a linking approach between carbon and livelihood in the HKH.

Livelihood improvement can be realized through carbon sinking benefits; but how to link this benefit to livelihoods and promote residents to participate in carbon management is a key step (Bass et al. 2000; Lebel 2007; Chhatre and Agrawal 2009; Benessaiah 2012). The most direct approach is to replace carbon compensation with livelihood assistance, which is the core content of the research program in carbon management and livelihood benefits (Food and Drink Federation 2008; Ojea et al. 2016). The Chinese government also attempted the ecological compensation strategy, but it is difficult to put into effective practice (Chang and Ghoshal 2014; Dong et al. 2016). One of the problems in the HKH region is that most carbon-livelihood strategies depend on some degree of transportation, which is not readily available (Pandey et al. 2016).

Until now, most projects on carbon evaluation in the HKH examined a specific region and not the whole area, and thus implementing policies on carbon management for the HKH is difficult (Marquardt et al. 2016). The typical program endorsed is REDD+, which focuses on cases and problem solutions. The important and useful practices links stakeholders to survey project benefits in the carbon management campaign in the HKH (McAfee 2016; Castro-Nunez et al. 2016; Ingalls and Dwyer 2016). More pathways of carbon management is required for carbon sinking benefits or compensation for local stakeholders (Ojea et al. 2016). For example, vegetation reclaiming in degraded land is based mainly on carbon evaluation, (Negi et al. 2015). The social-carbon-metrology system and its link with livelihood should be more fully developed. The implementation of the carbon management program should be tightly linked to the regional environment, politics, institutions, culture and education of the local residents (Stringer et al. 2012; Kemp et al. 2013; Lusiana et al. 2012, 2014).

The main problems faced by global climate change is poverty, gender equality, education and medical treatment. Carbon management could help fight poverty and gender equality (Stringer et al. 2012; Sharma 2012; Khadka et al. 2014) and its role should be promoted (Stewart et al. 2011, 2016). In gender equality, carbon management can strengthen the women's capacity in livelihood improvement (Shang et al. 2016a, b). For example, in the alpine meadow area in the Tibetan plateau, women contribute more to carbon management than men, and, therefore, should receive more carbon compensation to be used for their health and education (Shang et al. 2016b). As the HKH is a typical poor region in the world, environmental improvement and climate governance should emphasize anti-poverty because livelihood improvement of the local residents can contribute towards the mitigation of climate change (Xu et al. 2009; Sharma 2012; Negi et al. 2015; Pandey et al. 2016). In the implementation of these programs, it is important that the households, groups and villages, and the local residents accept the policies and benefits of the program (Bremer et al. 2016). Anti-poverty strategies should enhance special industries based on traditional culture, and for more benefits from the market. Then carbon management programs should examine the market requirements, and to promote marketing of the products (Hilson 2011; Benessaiah 2012).

18.2 The Pathway from Policy to Action

18.2.1 Assisting the Local Government to Make the Carbon Management Policy Effective

Global carbon management strategies should keep close links with local governments. Then, the strategies can be included in government policy making, and provide guidelines for reliable implementation (Benessaiah 2012; Stringer et al. 2012; Berry et al. 2014; Dong et al. 2016). Carbon management should be combined with livelihood improvement (Stewart et al. 2016) in the program and should be included in the daily activities of the government (Smith and Scherr 2003). In addition, some local governments could learn and implement appropriate methodologies from adjacent regions. Successful strategies could be distributed quickly among the regions of the HKH (Chhatre and Agrawal 2009; Dougill et al. 2012).

18.2.2 How to Incorporate Livelihood Development into the Carbon Trade System for Obtaining Carbon Compensation

Carbon storage is the basis for carbon management and compensation that can promote anti-poverty and gender equality. HKH is a typical non-industrial region, and it has great potential for carbon fixing. The policies should strive for the local resident participation in the scheme, which would increase the chances of success in promoting ecosystem management for carbon sinking. Direct benefit by local residents from carbon management is the first priority in the action. Two scenarios are envisioned. Firstly, carbon can be sold directly for cash or other goods, or assistance, and could also be accumulated. This kind of benefit should include assistance in education, infrastructure improvement and medical care. The second is in carbon emission. A key issue is determining the turning point of carbon emission to carbon sinking and what is the driving force. Then we can decide what assistance can promote this kind of conversion. In the policy, a clear detail of assistance and compensation should be prepared for the local government.

18.2.3 Promoting Priority Compensations for Anti-Poverty Programs

Because of global diversity, there are different ability and capacity levels for tackling climate change in different regions (Golub et al. 2013; Dong et al. 2016). In the fifth report by IPCC, a transformation of livelihoods was proposed, but, in many poor regions, this transformation is insufficient for global climate governance (Kates et al. 2012). In the HKH, the primary demand for some areas is assistance in fighting disasters and poverty. Here, perhaps food compensation is the first requirement in the carbon management campaign in HKH that needs a sustainable solution. Another example is the need for clean water facilities in some regions, and then compensation from carbon trade can cover infrastructure building. The carbon management program should join with other programs to promote common action for anti-poverty, livelihood development, and carbon neutralization and the carbon management benefits should be shared with many stakeholders, and a stable network of global climate governance should be established.

18.2.4 Playing the Traditional and Alternative Knowledge and Technique System

The transformation of livelihood initiative in developing countries faces many problems, because in these regions, livelihood depends heavily on the direct consumption of natural resources from land (path dependent). These areas need more assistance, investment, and talent for supporting transformations (Kates et al. 2012). An alternative method, which is using traditional knowledge would be necessary to replace the transformation action (Havlik et al. 2014). It is important to integrate the traditional customs in the regional climate governance and livelihood development in HKH (Pagányi et al. 2013). For example, in forest burning, we do not know the effect of burning on forest biodiversity, ecosystem function and forest cycling, and the traditional livelihood's positive impact on forest carbon sinking. This information is needed at all scales to providing proper community forest management (Diemont and Martin 2009).

18.3 Experience Summaries of Carbon Management and Livelihood Benefits

18.3.1 Livestock-Chicken Grazing System on Qilian Mountain

Grassland degradation in the Qilian mountain in the northern margin of HKH region, has led to carbon emission and livelihood decline (Long et al. 2010b). From 2007, the research team from Lanzhou University started a study of a mixed grazing system of livestock-chicken in the Qilian mountains, and developed and extended the system to other areas. In this grazing system, chicken grazed in the winter pasture in June-September, when the livestock are moved to the summer pasture. The system not only decreased grazing pressure, but also increased livelihood benefits. In addition, grasshopper numbers were controlled by the grazing chickens. The ecological theory of this system is that grazing chickens during the grasshopper egg production time allows chicken production and reduces grasshopper damage to the pasture. This system has proven to be much more profitable than just the grazing by sheep (Sun et al. 2016, Figs. 18.1 and 18.2).

In the livestock-chicken grazing mixing system, the chicken group will occupy winter pasture at June-September, at that time the livestock moved to the summer pasture. The result of 5 years monitoring experiment showed that the grasshopper density in grazing chicken pasture was less than control pasture. However, after sixth years of chicken grazing, the grasshopper density in grazing chicken pasture was not different to the control pasture, that showed the grazing chicken as effective to control the grasshopper population. Furthermore, the meat quality of the grazing chicken was better than traditionally raised chicken in a coop (Sun et al. 2013). At present, the livestock-chicken grazing system is very popular in rangeland areas and is replacing the traditional methods (Fig. 18.3). Carbon balance has not

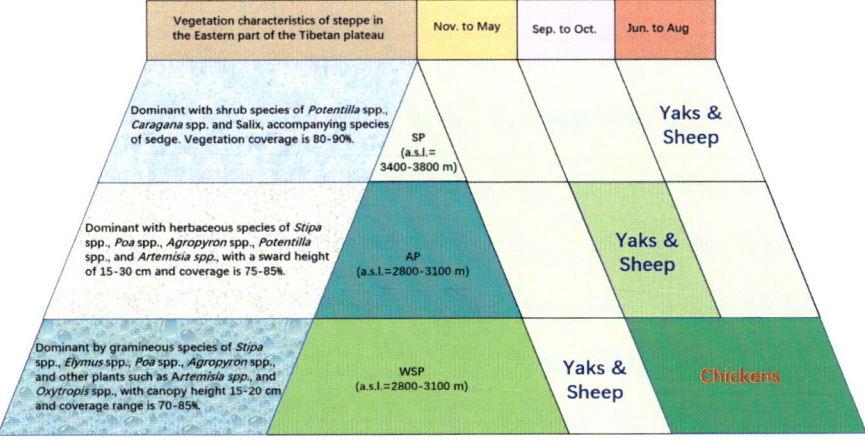

Fig. 18.1 The model of a mixed grazing system of livestock-chicken based on the ecological niche gap in Qilian mountain pasture (Sun et al. 2016)

Fig. 18.2 Grazing chicken experiment in Qilian mountain by Lanzhou University study team (Photography by Zhanhuan Shang 2008)

Fig. 18.3 Grazing chicken industry development in other areas of the Tibetan plateau (Guinan County of Qinghai province) in HKH region (Photography by Zhanhuan Shang 2018)

been determined, but moderate chicken grazing has not degraded the grassland, but has curbed the outbreaks of grasshoppers in dry years. The livestock-chicken grazing system has conformed to the transformation model proposed by IPCC for tackling climate change and improving livelihood.

18.3.2 The Case of Extremely Degraded Grassland Reclaiming

Serious land degradation on the Tibetan plateau resulted in the disappearance of vegetation and substantial carbon emission which led to livelihood security problems of the local residents (Shang and Long 2007). In the typical land degradation in the alpine grasslands, the top soil-root layer disappears and leaves a bare path, that results in a reduction of soil water content and of pasture biomass. This has occurred on 467×10^4 ha in the headwater area of Yangtze river, Yellow river and Lantsang river in Qinghai province of Tibetan plateau (named 'Sanjiangyuan') (Fig. 18.4). When 1 ha of alpine meadow becomes bare land, 10 cm of top soil is removed, and about 20–46 mg of soil organic carbon is lost (Shang 2006). For protecting the soil water content and reclaiming the bare land for ecological function and livestock production on Tibetan plateau, the Chinese government has made large investments to control and reclaim the bare-land in Sanjiangyuan. The local researchers used the building 'artificial-grassland' method. The 'artificial-grassland' method with local plant seeds was used to restore the grasslands. With proper management this method can establish good vegetation coverage quickly and provide much forage for livestock; however, there is a great risk that the 'artificial-grassland' will be degraded within 3–5 years and become 'bare-land' again (Shang et al. 2018). The key to success is good and sustainable management of the artificial-grassland to develop a stable grassland soil and vegetation regime.

A current study of more than 15 years demonstrates that the 'artificial-grassland' with good management can provide sustainable forage production and reclaim the soil nutrition and carbon gradually (Shang et al. 2018). Feng et al. (2010) reported

Fig. 18.4 Formation process of bare land degraded grassland on Tibetan plateau (Shang et al. 2018)

that artificial grassland established on bare land in Qinghai province had 33% higher soil carbon content when compared to bare land and also had stable vegetation. A study case at an altitude of 3700–4000 m showed that in 5 years of proper management, artificial grassland recovered the soil carbon level in a moderate degraded meadow. The soil carbon storage in the artificial grassland (107 t per hm^2, was not significantly different than the non-degraded meadow (137 t per hm^2), but was 38% higher than the bare land (78 t per hm^2). However, the artificial grassland degraded again after 4 years, and its soil carbon content (78 t per hm^2) declined to the level of bare land (Li et al. 2014). Zhang (2015) compared degraded artificial grassland, bare-land, and un-degraded meadow in three counties, that some artificial grassland degraded again and decreased vegetation biomass but the higher soil carbon level, while soil carbon decreased to a level below bare land (Zhang 2015). These responses depended on the area and background, and utilization history. These findings are useful for the artificial grassland establishment and management in bare land areas on the Tibetan plateau.

Under good management, artificial grassland can improve the ecological and productive functions (Fig. 18.5). A 6 year study on the economic benefits of artificial grassland showed that good management with investment that earn more income than without investment (Dong et al. 2011a, b; Table 18.1). The first investment required for artificial grassland establishment requires funds for seeds, fertilizer, machinery and labor and fencing. In addition, communication and participation by local residents should be encouraged (Shang et al. 2017). Whether from the improvement and the effective management resulting in improved carbon benefits and livelihoods is the core for ecological restoration engineering in the HKH (Shang et al. 2016b, 2017).

Fig. 18.5 Bare land degraded grassland (left) and artificial grassland established on bare land after 18 years (right) (Photography by Shang et al. 2009, 2017)

Table 18.1 Economic statistics of establishing sown grassland on bare soil land (study site a.s.l = 3700 m) (Dong et al. 2011a)

Treatment	Years	Hay forage production (kg/hm^2)	Stocking rate (sheep/hm^2)	Income (sheep/hm^2)	Invest cost (sheep/hm^2)	Net income (sheep/hm^2)	Net income/invest
Sown grassland without management and under traditional grazing system	1th	3959.8	7.59	2277	1537.5	739.5	0.48:1
	2th	4950.4	9.49	2847	22.5	2824.5	125:1
	3rd	3554.9	6.82	2046	22.5	2023.5	89:1
	4th	2641.3	5.07	1521	22.5	1498.5	66:1
	5th	1545.8	2.96	888	22.5	865.5	38:1
	6th	792	1.52	456	22.5	433.5	19:1
Sown grassland with management and planned grazing system	1th	3959.8	7.59	2277	1537.5	739.5	0.48:1
	2th	7496.1	14.38	4314	322.5	3991.5	12:1
	3rd	6010.3	11.53	3459	322.5	3136.5	10:1
	4th	5741.9	11.01	3303	322.5	2980.5	9:1
	5th	5183.2	9.94	2982	397.5	2584.5	7:1
	6th	5151.5	9.88	2964	397.5	2566.5	6:1

18.3.3 Multiple System Coupling on the Tibetan Plateau

Combining agriculture, forest, and pasture is common for developing sustainable livelihoods. For example, agro-grassland and agro-forest can increase productivity of the land area (Herrero et al. 2010; Dong et al. 2011a, b; Dong et al. 2016). In the HKH, in particular in the China-Himalaya region, the grain-agriculture, livestock production, forestry and specific industries all provide income in the mountain and river valley region (Dong et al. 2010). There is much water, mountain and valley land and forests in the region, which allows all these enterprises to be profitable. The marketing and commerce mechanism should be examined in these multiple systems, and the government should encourage multiple system coupling by policies and investments (Dong and Sherman 2015). The development of ecological protection and productivity is another important way for promoting multiple systems (Khan et al. 2014).

In a regional case for solving overgrazing and starvation by livestock in alpine grasslands, forage was produced for fodder and livestock was fattened at lower altitudes. The livestock was brought from higher altitudes, fattened and then marketed. This system reduced the stocking rate and reduced the pressure on the natural grassland and. In another case, four enterprises formed the multiple system (Zhao et al. 2018; Fig. 18.6). In the first part the herders-controlled stocking rate at the pasture growing stage to retain half the amount of above-ground biomass at each grazing period. In the second part, artificial forage supplement was provided when needed. In the third part, supplements were provided to prevent deficiencies in nutrition and elements. In the fourth part, local herders received ecological compensation based

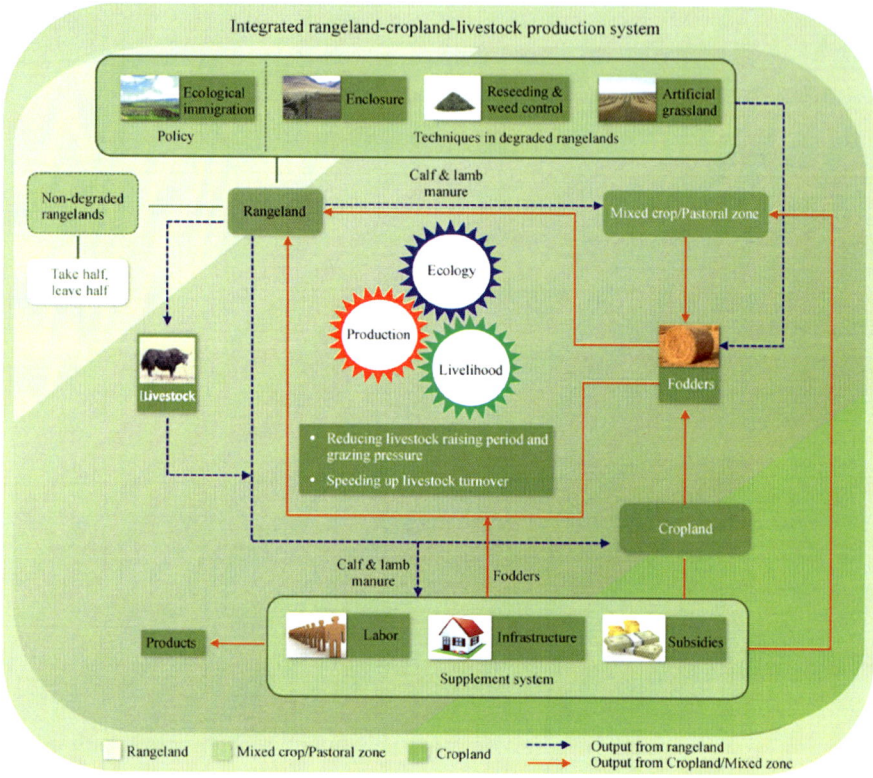

Fig. 18.6 The functions of rangeland, mixed crop/rangeland, cropland and its integrated production system on the Qinghai–Tibetan Plateau (Zhao et al. 2018)

on carbon benefits. Zhao et al. (2018) proposed a new model of "three zones coupled system" approach of pastoral livestock on the QTP (Fig. 18.6). In addition, an integrated model for increasing grassland carbon and livelihood benefit in pastoral areas on the Tibetan plateau was proposed by Shang et al. (2014, 2017). It was concluded that more attention should be given to the multi-system approach and that the model should be adapted to local conditions. It was also suggested that the participation of local residents should be encouraged in carbon management and livelihood improvement.

Another typical case of system coupling is in the Lhasa valley and northern Tibet (Fig. 18.7; Long et al. 2010a, b). In this case, forage or grain feed is transported from the Lhasa valley (<3500 m) to the northern Tibet alpine mountain pasture area (>4500 m) for supplementing livestock during forage shortage in the cold season. This system has not shown a significant ecological benefit as yet, but the forage producers in the Lhasa valley earned considerable income and the soil carbon content was increased (Shang et al. 2009). In northern Tibet, the forage supplement during the cold season (winter and spring) greatly increased the vegetation coverage and the carbon storage. (Shang et al. 2016a, b, c). An experiment of cropping-forage

Fig. 18.7 Valley cropping-mountain rangeland—livestock framing model (Long et al. 2010a, b)

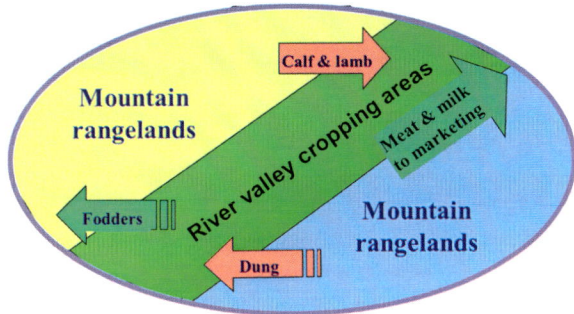

Table 18.2 The soil carbon storage and net-economic benefit analysis of cropping-forage rotation in the Lhasa valley region (Bianba 2006)

Crop-forage rotation system in Lhasa valley area	Soil organic carbon storage (0–20 cm) kg/ha	Net income (3 years) (RMB/ha)
Common vetch-barley-wheat	51.0*105 kg	28,079
Alfalfa-barley-wheat	53.5*105 kg	30,132
Wheat-barley-wheat	36.2*105 kg	15,468
Wheat-wheat-wheat	28.4*105 kg	14,559
Fallow-barley-wheat	41.0*105 kg	8683

rotation showed that soil carbon storage and livelihood benefit can be enhanced with legume forage (Table 18.2; Bianba 2006).

18.3.4 Community Forestry

Community forestry also known as social forestry, was proposed by Jack Westoby in 1968 at the British Commonwealth's Ninth Forest Conference. The concept of community forestry has been in use (Rana and Chhatre 2017). The forest is an important resource for tackling climate change and anti-poverty, particularly in the HKH region. India was the first country to develop community forestry, which enhanced forestry coverage, tree density, wood biomass and biodiversity, by the participating community (Banerjee 1999; Rossi 2007). Nepal was one of the earliest countries to enforce policies and laws of the community forestry. Community forestry projects have greatly enhanced forest carbon storage and tree density. In particular, the REDD+ project provided livelihood assistance and contributed to mitigate climate change effects in Nepal (Pandey et al. 2016; Anup et al., 2018). However, due to differences in topography, and management in carbon storage, the community forests require an overall evaluation to improve the ecological benefits (Luintel et al. 2018).

In the evaluation of community forestry project in Nepal, there were positive outcomes and good knowledge about forest coverage, carbon and livelihood of local resident in community forestry area. For example, in the two community forestry located in Pokhara and Hathmandu, the survey research did the comparison of vegetation carbon, soil carbon and tree from between 2011 and 2014, the result found out that, the project increased the carbon storage and tree density (Rossi 2007). However, in recent an evaluation study had been surveyed 620 community forestry in Nepal (Bluffstone et al. 2017; Luintel et al. 2018). More information is needed at the country level in for the projects to enhance the capacity building of local resident for biodiversity protection and carbon storage in the community (Bluffstone et al. 2017; Luintel et al. 2018).

Pandey et al. (2017) evaluated 105 community forests in the REDD+ program by using the carbon benefit method to account for the cost of community forest management and carbon fixing benefits. It emerged that there was very little benefit for carbon fixing after management costs were included in the balance. Therefore, the REDD+ project, should consider other ecological service functions in the policies for developing community forest projects (Pandey et al. 2017; Newton et al. 2016). Karki et al. (2018) concluded that there are four pathways to use forests for food security in Nepal: (1) income and employment; (2) inputs to increase food production; (3) directly for food; and (4) renewable energy for cooking. However, food security is a serious problem in community forests which has not be solved by the management of community forests (Karki et al. 2018). Firstly, there is a need for in-depth research to generate alternative methods and transform the conventional 'forests for soil conservation' to 'forests for food security' without compromising the environmental services provided by forests. Secondly, there is a need to engage critical researchers and policy makers to apprehend and recognize the pathways linking community forests and food security. Thirdly, in order to complement this critical policy research there is a need to better understand the quantitative dimensions and interactions among the four food security pathways. Finally, the integration of sectoral planning and actions, mainly between the agriculture and forestry agencies, needs to be reflected in order to mainstream priorities that would ensure community forest-food security linkages. This should be considered at the national level policy debates as well as policy formation at various government levels (Karki et al. 2018). Although the community forest policies were developed broadly in Nepal, there is still a large challenge to arrive at a solution for the local resident's livelihood demands and sustainable management of the forests (Khatri et al. 2018; Figs. 18.8 and 18.9).

18.3.5 Community Pasture

Grassland-livestock production is important for small households but has resulted in grassland degradation and livelihood problems. Community cooperation programs have developed community pasture and forage production under government financial support. In addition, factories were established for animal products and other local products, and for the sale of the products (Zhang and Li 2014; Fig. 18.10). To

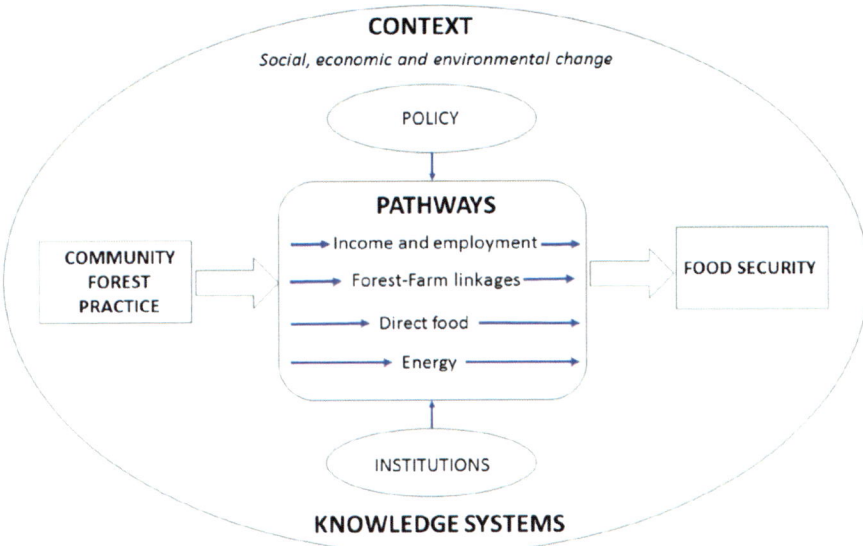

Fig. 18.8 Pathways linking community forests with food security (Karki et al. 2018)

Fig. 18.9 The husbandry industry in community forestry of Nepal (Photography by Yu Li 2017)

promote cooperative systems on the Tibetan plateau, the Chinese government launched a number of projects of community pasture to support combined agriculture and livestock, including research projects. This cooperation model of community pasture has been developing all over the Tibetan plateau. It has affected herders and companies, and it has improved industry, ecological compensations mechanism, and marketing standard. From 2012–2016, the Ministry of Agriculture of China launched community pasture demonstrations in Sichuan, Gansu, Qinghai,

Fig. 18.10 A Yak dairy factory in the cooperation unit (left) and grassland cooperation management area (right) (Photography by Shang et al. 2017)

Tibet, Yunnan, that demonstrated new improved techniques for resource management and production exploitation (Wei et al. 2015; Zeibai 2018).

Essentially, the community cooperation is a natural resource management system applied broadly to pastures, agriculture, livestock, forests and fisheries. It is generally equitable and egalitarian, efficient and comprehensively sustainable (Bullinger et al. 2010; Crewett 2012; Dörre 2015). In this regard, the strategy should also include the provision of specified financial, technical, and educational support for the transition period to the new regime when requested by the respective management bodies. Ignoring empirical evidence and local circumstances can lead to a repetition of the often-made mistake of implementing theoretically sophisticated approaches that, finally, fail on the ground (Dörre 2015). Now in the China-HKH region, family pasture and community pasture have been established and used for grassland-livestock systems which are important in carbon management and livelihood development.

Follow-up assessments are very important for the management of community pastures. In the assessment, ecological, livelihood, and social benefits (ELSBs) are the basic evaluation parameters. It is important for the assessments that the local residents' impressions of the benefits of these co-operative systems. In a number of surveys, it emerged that these impressions can vary widely, (Wei et al. 2015; Zhang 2014) which suggests that the cooperation system of grassland husbandry is still in the developing stage and needs improvement.

The link among the environment, livelihood, and society is important to assess the sustainable development of community pasture. Cheng et al. (2018) assessing the 'Panchen Project' of 2012–2016 for 8 community pastures, reported that the degree of linkage increased gradually significant difference among the parameters (Table 18.3). He considered that the investments and talents, with support from policies, culture and industries were the key factors in linking the variables. Cheng et al. (2018) concluded that community pasture management should satisfy local demands and take into account natural resource limitation, fragility, industry planning, and environment services. It should also encourage participation by local residents and enhance the ecological services and social-economic development.

Table 18.3 Coupling degree of coordination (ecological sector, livelihood sector and social sector) of community pasture development in the Qinghai-Tibet Plateau from 2013 to 2016 (Cheng et al. 2018)

Name of community pasture	2013	2014	2015	2016
Mozhugognka	*	**	**	***
Hongyuan	**	***	***	****
Henan	**	***	****	****
Shangri-la	**	**	**	****
Yambajan	*	*	*	**
Ganzi	*	*	**	***
Yushu	**	*	**	**
Xiahe	**	**	***	***

* Extreme incoordination, ** Incoordination, *** Quasi coordination, **** Coordination

18.3.6 Alliance for Transboundary Protection Areas in HKH

Trans boundaries are a very important area among the HKH's countries for natural protection reserves. In the transboundary area of HKH, there are unique landscapes, corridor channels, high culture diversity, and high biodiversity (Sandwith et al. 2001). In general, the transboundary area contains rare animal species' and acts as a migratory route for species of animals. The 'transboundary-protecting-alliance' has a very important role for species and ecosystems and landscape protection in the world. The transboundary protection needs coordinating efforts among the stakeholder countries for maintaining protection needs, policy making, and assistance. For example, in the Hindu Kush Karakoram Pamir region, there are six major reserves in transboundary areas among China, Pakistan, Tajikistan, and Afghanistan for protecting snow leopard, Marco Polo sheep and other rare species by (Fig. 18.11; Table 18.4). Another important aspect of transboundary protection area is the local resident's livelihood development and culture conservation.

The protection reserves development has considered the local residents' livelihood and development as well as their customs and culture. In the transboundary protection areas, communication among reserves is important for development of ecological and livelihood benefits. Periodic joint scientific surveys, ecological protection benefit evaluations, the effect of key species protection analysis, cost analysis and ecological compensations are parameters to be examined in transboundary protection areas (Long et al. 2018). It is important to share knowledge, traditional knowledge pertinent to the area, among transboundary areas to promote joint protection activities (Lewis 1996).

The HKH region needs more action in transboundary protection campaign in the world beyond the current global protection and development paradigm. For example, in September 13–17, 2018, Lanzhou University, ICIMOD and other organizations organized an international workshop titled "Harmonizing conservation and development along the Silk Road'. With growing challenges of climate change, globalization and increasing infrastructure projects, the maintenance of its

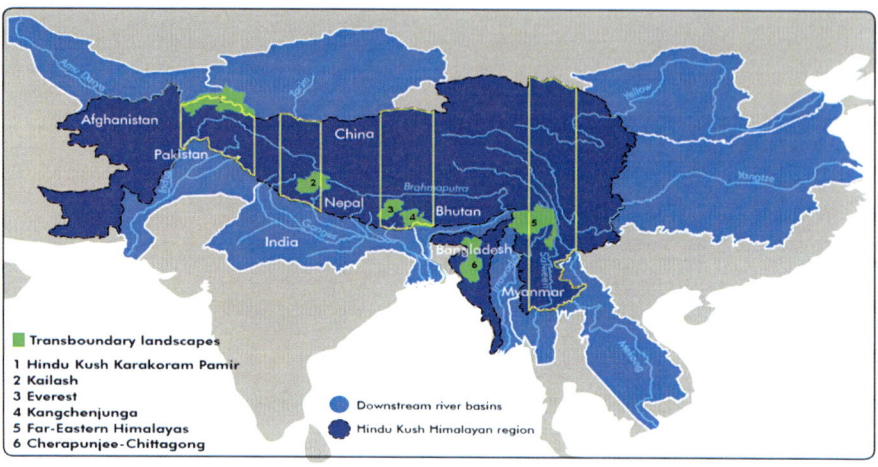

Fig. 18.11 Major transboundary areas in HKH region (source from ICIMOD)

fragile alpine ecosystem of HKH region would require truly transboundary management (Van Oosten 2018). Currently, the landscape of the transboundary area is managed as a series of "conservation islands," separated by a hardening of national borders, divergent national policies and increasingly fragmented infrastructural works. In the workshop, organizations and agencies expressed their concerns, and assessed the feasibility of creating a regional network of protected areas to harmonize conservation and development along the Silk Route (Van Oosten 2018). The network believes that sustainable and inclusive regional integration should be and can be achieved by sharing knowledge and information and by developing joint management plans to face the challenge of rapid infrastructural development (Van Oosten 2018). As a network, partners will be better positioned to contribute to a more ecologically sustainable and socially inclusive planning process. In this way, the proposed network will contribute to balancing conservation and development and set an example for other transboundary landscapes in the HKH region (Van Oosten 2018).

18.4 Conclusion

Until now, there is no clear and definite role of carbon management in the HKH, and its role in global carbon management. The action and plan of carbon management has been formulated under the Paris Agreement's formwork. As the biggest geographical unit in Eurasia, special attention should be given the HKH region for climate governance and carbon management. The lack of detailed and reliable data on carbon balance and livelihood in HKH has made it difficult in coordinating a carbon management campaign. Another important problem is the low economy and

Table 18.4 Six protection reserves in Hindu Kush Karakoram Pamir and their general status and condition (Long et al. 2018)

Protected area	Country	Area coverage (km^2)	Flagship species	Population	Geographical features	Nationality
Taxkorghan nature reserve (TNR)	China	15,000	Marco-polo sheep, Snow leopard, Ibex, Golden eagle, Tibetan vulture	30,000	China—Pakistan Economic Corridor	Wakhi, Tajik, Sarakoli, Kirghiz, Khow
Wakhan national park (WNP)	Afghanistan	10,878	Marco-polo sheep, Snow leopard, Ibex, Golden eagle, Tibetan vulture	14,000	Pakistan—Afghanistan trade corridor	Wakhi, Tajik, Sarakoli, Kirghiz, Khow
Broghil national park (BNP)	Pakistan	1348	Marco-polo sheep, Snow leopard, Ibex, Golden eagle, Tibetan vulture	1600	Pakistan—Afghanistan trade corridor	Wakhi, Tajik, Sarakoli, Kirghiz, Khow
Qurumbar national park (QNP)	Pakistan	142	Marco-polo sheep, Snow leopard, Ibex, Golden eagle, Tibetan vulture	500	Pakistan—Afghanistan trade corridor	Wakhi, Tajik, Sarakoli, Kirghiz, Khow
Khunjerab national park (KNP)	Pakistan	4500	Marco-polo sheep, Snow leopard, Ibex, Golden eagle, Tibetan vulture	7000	China—Pakistan Economic Corridor	Wakhi, Tajik, Sarakoli, Kirghiz, Khow
Zarkul national park (ZNR)	Tajikistan	1610	Marco-polo sheep, Snow leopard, Ibex, Golden eagle, Tibetan vulture	1000	Tajikistan—Afghanistan trade corridor	Wakhi, Tajik, Sarakoli, Kirghiz, Khow

education in the HKH region; consequently, implementation of an effective carbon management program is dependent on government assistance.

The final aim of carbon management and benefit improvement is the storage of carbon into the soil. However, in the HKH region, local residents' depend heavily on vegetation and natural resources, and it is difficult to ban activities related to livelihoods. Human activities are becoming more intensive in the HKH region and these activities have different impacts on carbon balance. Consequently, in considering livelihood transformations to alleviate poverty, consideration must be given to how the carbon benefits will be affected. Three aims can be adapted by the carbon management campaign in HKH. The first is an end hunger by achieving food security and improving nutrition and promoting sustainable agriculture. The second is to take urgent action to combat climate change and its impacts. The third is to protect, restore, and promote sustainable use of terrestrial ecosystems, manage sustainable forests, combat desertification, and halt and reverse land degradation and halt biodiversity loss (Timalsina et al. 2018). This three-priority in the carbon management and global climate campaign is appropriate in the development of the HKH region.

References

Anup, K.C., R. Maandhar, R. Paudel, et al. 2018. Increase of forest carbon biomass due to community forestry management in Nepal. *Journal of Forest Research* 29: 429–438. https://doi.org/10.1007/s11676-017-0438-z.

Ashton, M.S., M.L. Tyrrell, D. Spalding, et al. 2012. *Managing forest carbon in a changing climate*. Berlin: Springer.

Atela, J.O., P.A. Minang, C.H. Quinn, et al. 2015. Implementing REDD+ at the local level: Assessing the key enables for credible mitigation and sustainable livelihood outcomes. Journal of Environmental Management 157: 238–249.

Banerjee, A.K. 1999. Community forestry development in India. In *World forests society and environment*, ed. M. Palo and J. Unsivuori. Dordrecht: Springer.

Banskota, K., B.S. Karky, and M. Skutsch. 2007. Reducing carbon emission through community-managed forests in the Himalaya. Kathmandu: International Centre for Integrated Mountain Development.

Bass, S., O. Dubois, P. M. Costa, et al. 2000. Rural livelihoods and carbon management, London, IIED natural resource issues paper no. 1. London: International Institute for Environment and Development.

Benessaiah, K. 2012. Carbon and livelihoods in Post-Kyoto: Assessing voluntary carbon markets. *Ecological Economics* 77: 1–6.

Berry, N.I., R. Harley, and C.M. Ryan. 2014. Enabling communities to benefit from REDD+: Pragmatic assessment of carbon benefits. *Carbon Management* 4 (6): 571–573.

Bianba, Z. 2006. *A study on ecological and economic efficiency in forage-crop rotation system in Tibet farming area*. Thesis for Master's Degree Northwest A & F University.

Bluffstone, R.A., E. Somanathan, P. Jha, et al. 2017. Does collective action sequester carbon? Evidence from the Nepal community forestry program. *World Development* 101: 133–141.

Bremer, L.L., K.A. Farley, O.A. Chadwick, et al. 2016. Changes in carbon storage with land management promoted by payment for ecosystem services. *Environmental Conservation* 43 (4): 397–406.

Bullinger, A.C., A.-K. Neyer, M. Rass, et al. 2010. Community-based innovation contests: Where competition meets cooperation. *Creativity and Innovation Management* 19 (3): 209–303.

Castro-Nunez, A., O. Mertz, and M. Quintero. 2016. Propensity of farmers to conserve forest within REDD+ projects in areas affected by armed-conflict. *Forest Policy and Economics* 66: 22–30.

Chang, W., and S. Ghoshal. 2014. Respiratory quotients as a useful indicator of the enhancement of petroleum hydrocarbon biodegradation in field-aged contaminated soils in cold climates. *Cold Regions Science and Technology* 106–107: 110–119.

Cheng, C., A. Ren, Y. Wang, et al. 2018. An approach C degree model on ecological, social and economic coupling development in Qinghai-Tibet plateau research community. *Pratacultural Science* 35 (3): 677–685.

Chhatre, A., and A. Agrawal. 2009. Trade-offs and synergies between carbon storage and livelihood benefits from forest commons. *Proceedings of the National Academy of Sciences* 106 (42): 17667–17670.

Crewett, W. 2012. Improving the sustainability of pasture use in Kyrgyzstan. *Mountain Research and Development* 32 (3): 267–274.

Diemont, S.A.W., and J.F. Martin. 2009. Lacandon Maya ecosystem management: sustainable design for subsistence and environmental restoration. *Ecological Applications* 19 (1): 254–266.

Dong, S., and R. Sherman. 2015. Enhancing the resilience of coupled human and natural systems of alpine rangelands on the Qinghai-Tibetan plateau. *The Rangeland Journal* 37 (1): 1–3.

Dong, S., L. Wen, L. Zhu, et al. 2010. Implication of coupled natural and human systems in sustainable rangeland ecosystem management in HKH region. *Frontiers of Earth Science in China* 4 (1): 42–50.

Dong, Q.M., J.J. Shi, Y.S. Ma, et al. 2011a. Analysis on economic and ecological benefit of black-soil-beach sown grassland. *Acta Agrestia Sinica* 19 (2): 195–201.

Dong, S., L. Wen, S. Liu, et al. 2011b. Vulnerability of worldwide pastoralism to global changes and interdisciplinary strategies for sustainable pastoralism. *Ecology and Society* 16 (2): 10. http://www.ecologyandsociety.org/vol16/iss2/art10/.

Dong, S., K.-A.S. Kassam, J.F. Tourrand, et al. 2016. *Building resilience of human-natural systems of pastoralism in the developing world*. Berlin: Springer.

Dörre, A. 2015. Promises and realities of community-based pasture management approaches: Observations from Kyrgyzstan. *Pastoralism: Research, Policy and Practice* 5: 15. https://doi.org/10.1186/s13570-015-0035-8.

Dougill, A.J., L.C. Stringer, J. Leventon, et al. 2012. Lessons from community-based payment for ecosystem service schemes: From forests to rangelands. *Philosophical Transactions of the Royal Society B: Biological Sciences* 367 (1606): 3178–3190.

Feng, R.Z., R.J. Long, Z.H. Shang, et al. 2010. Establishment of Elymus natans improves soil quality of a heavily degraded alpine meadow in Qinghai-Tibetan Plateau, China. *Plant and Soil* 327: 403–411.

Food and Drink Federation (FDF). 2008. *Working for the economy: Our contribution to the economy*. London: Food and Drink Federation/University of Reading.

Golub, A.A., B.B. Henderson, T.W. Hertel, et al. 2013. Global climate policy impacts on livestock, land use, livelihoods, and food security. *Proceedings of the National Academy of Sciences* 110 (52): 20894–20899.

Havlik, P., H. Valin, M. Herrero, et al. 2014. Climate change mitigation through livestock system transitions. *PNAS* 111 (10): 3709–3714.

Herrero, M., P.K. Thornton, A.M. Notenbaert, et al. 2010. Smart investments in sustainable food production: revisiting mixed crop-livestock systems. *Science* 327: 822–825.

Hilson, M. 2011. Locking-in carbon, locking-out livelihood? Artisanal mining and REDD in sub-Saharan Africa. *Journal of International Development* 23 (8): 1140–1150.

Ingalls, M.L., and M.B. Dwyer. 2016. Missing the forest for the trees? Navigating the trade-offs between mitigation and adaptation under REDD. *Climatic Change* 136 (2): 353–366.

Karki, R., K.K. Shrestha, H. Ojha, et al. 2018. From forests to food security: Pathways in Nepal's community forestry. *Small-Scale Forestry* 17: 89–104.

Kates, R.W., W.R. Travis, and T.J. Wilbanks. 2012. Transformational adaptation when incremental adaptations to climate change are insufficient. *PNAS* 109 (19): 7156–7165.

Kemp, D., G. Han, X. Hou, et al. 2013. Innovative grassland management systems for environmental and livelihood benefits. Proceedings of the National Academy of Sciences 110 (21):8369–8374.

Khadka, D., M.S. Babel, S. Shrestha, et al. 2014. Climate change impact on glacier and snow melt and runoff in Tamakoshi basin in the Hindu Kush Himalayan (HKH) region. *Journal of Hydrology* 511: 49–60.

Khan, S.M., S. Page, H. Ahmad, et al. 2014. Ethno-ecological importance of plant biodiversity in mountain ecosystems with special emphasis on indicator species of a Himalayan Valley in the northern Pakistan. *Ecological Indicators* 37 (A): 175–185.

Khatri, D.B., K. Marquardt, A. Pain, et al. 2018. Shifting regimes of management and uses of forests: What might REDD+ implementation mean for community forestry? Evidence from Nepal. *Forest Policy and Economics* 92: 1–10.

Lebel, L. 2007. Adapting to climate change. *Global Asia* 2: 15–21.

Lewis, C. 1996. *Managing conflicts in protected areas*. IUCN, Gland, Switzerland, and Cambridge, UK. xii+100pp.

Li, Y.-Y., S.-K. Dong, L. Wen, et al. 2014. Soil carbon and nitrogen pools and their relationship to plant and soil dynamics of degraded and artificially restored grasslands of the Qinghai-Tibet plateau. *Geoderma* 213: 178–184.

Long, R.J., Z.H. Shang, L.M. Ding, et al. 2010a. *Regional workshop of 'pastroralism and rangeland management on the Tibetan plateau in the context of climate and global change'. (Workshop presentation)*. InWEnt, ICIMOD, TAAAS, NAC. Lhasa.

Long, R.J., Z.H. Shang, X.G. Li, et al. 2010b. Carbon sequestration and the implications for rangeland management. In *Towards sustainable use of rangelands in North-West China*, ed. Victor Squires, Hua Limin, Zhang Degang, and Guolin Li, 127–146. Berlin: Springer.

Long, R., Z. Shang, L. Zhang, et al. 2018. *Investigation report for natural and social-cultural science of natural reserve in Hindu Kush Karakoram Pamir*. Lanzhou: Lanzhou University.

Luintel, H., R.A. Bluffstone, and R.M. Scheller. 2018. The effects of the Nepal community forestry program on biodiversity conservation and carbon storage. *PLoS One* 13 (6): e0199526.

Lusiana, B., M.V. Noordwijk, and G. Cadisch. 2012. Land sparing or sharing? Exploring livestock fodder options in combination with land use zoning and consequences for livelihoods and net carbon stocks using the FALLOW model. *Agriculture Ecosystems & Environment* 159: 145–160.

Lusiana, B., M. van Noordwijk, F. Johana, et al. 2014. Implications of uncertainty and scale in carbon emission estimates on locally appropriate designs to reduce emissions from deforestation and degradation (REDD+). *Mitigation and Adaptation Strategies for Global Change* 19: 757–772.

Marquardt, K., D. Khatri, and A. Pain. 2016. REDD+, forest transition, agrarian change and ecosystem services in the hills of Nepal. *Human Ecology* 44 (2): 229–244.

McAfee, K. 2016. Green economy and carbon markets for conservation and development: A critical view. *International Environmental Agreements: Politics, Law and Economics* 16 (3): 333–353.

Negi, V.S., I.D. Bhatt, P.C. Phondani, et al. 2015. Rehabilitation of degraded community land in Western Himalaya: Linking environmental conservation with livelihood. *Current Science* 109 (3): 520–528.

Newton, P., J.A. Oldekop, G. Broding, et al. 2016. Carbon, biodiversity, and livelihoods in forest commons: synergies, trade-off, and implications for REDD+. *Environmental Research Letters* 11. https://doi.org/10.1088/1748-9326/11/4/044017.

Ojea, E., M.L. Loureiro, M. Alló, et al. 2016. Ecosystem services and REDD: Estimating the benefits of non-carbon services in worldwide forests. *World Development* 78: 246–261.

Pagányi, É.E., I. Van Putten, T. Hutton, et al. 2013. Integrating indigenous livelihood and lifestyle objectives in managing a natural resource. *PNAS* 110 (9): 3639–3644.

Pandey, S.S., G. Cockfield, and T.N. Maraseni. 2016. Assessing the roles of community forestry in climate change mitigation and adaptation: a case study from Nepal. *Forest Ecology and Management* 360: 400–407.

Pandey, S.S., T.N. Maraseni, K. Reardon-Smith, et al. 2017. Analyzing foregone costs of communities and carbon benefits in small scale community based forestry practice in Nepal. *Land Use Policy* 69: 160–166.

Rana, P., and A. Chhatre. 2017. Beyond committees: Hybrid forest governance for equity and sustainability. *Forest Policy and Economics* 78: 40–50.

Rossi, F.J. 2007. *Socio-economic impacts of community forest management in rural India*. Thesis of Doctor Degree in University of Florida.

Sandwith, T., C. Shine, L. Hamilton, et al. 2001. *Transboundary protected areas for peace and co-operation*. IUCN, Gland, Switzerland and Cambridge, UK.xi+111pp.

Shang, Z.H. 2006. *Studies on soil seed bank and regeneration of degraded alpine grassland in the Headwaters of Yangtze and Yellow rivers on Tibetan plateau*. Thesis of Doctor Degree in Gansu Agricultural University.

Shang, Z.H., and R.J. Long. 2007. Formation causes and recovery of the "Black Soil Type" degraded alpine grassland in Qinghai-Tibetan Plateau. *Frontiers of Agriculture in China* 1 (2): 197–202.

Shang, Z.H., Q.M. Ji, D.Z. Duoji, et al. 2009. Discussion on the development of crop-grass system in the 'three rivers' region of Tibet. *Pratacultural Science* 26 (8): 141–146.

Shang, Z.H., M.J. Gibb, F. Leiber, et al. 2014. The sustainable development of grassland-livestock systems on the Tibetan plateau: problems, strategies and prospects. *The Rangeland Journal* 36 (3): 267–296.

Shang, Z.H., Q.M. Dong, A. Degen, et al. 2016a. Chapter 8. Ecological restoration on Qinghai-Tibetan plateau: problems, strategies and prospects. In *Ecological restoration: Global challenges, social aspects and environmental benefits*, ed. Victor R. Squires, 161–176. New York: Nova Press.

Shang, Z.H., A. White, A.A. Degen, et al. 2016b. Role of Tibetan women in carbon balance in the alpine grasslands of the Tibetan plateau—a review. *Nomadic Peoples* 20: 108–122.

Shang, Z.H., B.J. Yang, C.H. Han, et al. 2016c. Chapter 4: Adaptive technology and demonstration in alpine grassland ecosystem. In *Climate change adaptive technology system for typical vulnerable terrestrial ecosystems*, ed. X.G. Lv et al., 130–138. Beijing: Science Press.

Shang, Z.H., R. Zhang, A. Degen, et al. 2017. Chapter 5. Rangelands and grasslands in the Tibetan Plateau of China: Ecological structure and function at the top of the world. In *Rangelands along the silk road: Transformative adaptation under climate and global change*, ed. Victor R. Squires, 65–101. New York: Nova Press.

Shang, Z.H., Q.M. Dong, J.J. Shi, et al. 2018. Research progress in recent ten years of ecological restoration for 'black soil land' degraded grassland on Tibetan plateau-concurrently discuss of ecological restoration in Sanjiangyuan region. *Acta Agrestia Sinica* 26 (1): 1–21.

Sharma, R. 2012. Impacts on human health of climate and land use change in the Hindu Kush–Himalayan region: Overview of available information and research agenda. *Mountain Research & Development* 32 (4): 480–486.

Sharma, J. 2016. Producing Himalayan Darjeeling: Mobile people and mountain encounters. *Himalaya, the Journal of the Association for Nepal and Himalayan Studies* 35 (2): 12.

Smith, J., and S.J. Scherr. 2003. Capturing the value of forest carbon for local livelihoods. *World Development* 31 (12): 2143–2160.

Stewart, J., M. Anda, and R.J. Harper. 2016. Carbon profiles of remote Australian indigenous communities: A base for opportunities. *Energy Policy* 94: 77–88.

Stewart, J., R.J. Harper, and M. Anda. 2011. Developing a model of carbon sources and sinks for indigenous communities in Australia. In *19th International congress on modelling and simulation*, ed. F. Chan, D. Marinova, and R.S. Anderssen, 3085–3091. Perth: Modelling and Simulation Society of Australia and New Zealand.

Stringer, L.C., A.J. Dougill, A.D. Thomas, et al. 2012. Challenges and opportunities in linking carbon sequestration, livelihoods and ecosystem service provision in drylands. *Environmental Science & Policy* 19–20 (5): 121–135.

Sun, T., R.J. Long, and Z.Y. Liu. 2013. The effect of a diet containing grasshoppers and access to free-range on carcase and meat physicochemical and sensory characteristics in broilers. *British Poultry Science* 54 (1): 130–137.

Sun, T., X. Liu, G. Sun, et al. 2016. Grasshopper plague control in the alpine rangelands of the Qilian Mountains, China: A socio-economic and biological approach. *Land Degradation & Development* 27: 1763–1770.

Timalsina, N., N. Agarwal, N. Bhattarai, et al. 2018. *Proceeding of joint BMUB-ICIMOD expert consultation workshop on Hindu Kush Himalayan Mountain soils*. Kathmandu, Nepal: ICIMOD.

Van Oosten, C. 2018. *Connecting landscapes along the Silk Route (Viewpoint)*. https://news.globallandscapesforum.org/viewpoint/connecting-landscapes-along-the-silk-route/.

Wei, X., X. Mao, and Z. Zhuoma. 2015. Preception analysis of 'Panchen Project' in proportioning the husbandry redevelopment in Tibetan area. *Journal of Qinghai Normal University (Philosophy and Social Sciences)* 37 (6): 14–17.

Xu, J.C., R.E. Grumbine, A. Shrestha, et al. 2009. The melting Himalayas: Cascading effects of climate change on water, biodiversity, and livelihoods. *Conservation Biology* 23 (3): 520–530.

Zeibai. 2018. *Report for the community pasture project in Tibetan plateau*. Beijing: The Ethnic Publishing House.

Zhang, B. 2014. *The research on performance evaluation of Qinghai ecological livestock husbandry cooperative*. The Master Dissertation of Qinghai University.

Zhang, R. 2015. *Comparison of nitrogen utilization by alpine plants and the influence of grassland change on soil/root carbon/carbonhydrate in alpine meadow*. Thesis of Master Degree in Lanzhou University.

Zhang, B., and S. Li. 2014. The performance evaluation of ecological animal husbandry cooperatives. *Journal of Qinghai Nationalities University* 1: 108–113.

Zhao, X., L. Zhao, Q. Li, et al. 2018. Using balance of seasonal herbage supply and demand to inform sustainable grassland management on the Qinghai-Tibetan plateau. *Frontiers of Agricultural Science and Engineering* 5 (1): 1–8.

Index

A

Above-ground biomass (AGB), 292
Absorbed photosynthetically active radiation (APAR), 47
Adaptation
　adaptive management, 291
　cross scale interaction, 292
　decision making, 291
　definition, 291
　livelihood, 292
　local knowledge, 292
　options, 291, 292
　typologies, 291
AGB storage
　agro-ecological settings, 308
　EO, 307
　fuelwood collection, 308
　global change, 308
　human impacts, 308
　quantitative assessment, 309
　standard forestry observations, 307
Agriculture, forestry and other land use (AFOLU), 35
Agro-forest systems, 10
Agropastoral activities
　animal husbandry, 273
　carbon input/output activities, 275
　crop planting and harvesting, 274
　Gatlang VDC, 273
　parma, 274
　women contribution, 277
　work calender, 274
Alpine grassland ecosystem, 152, 164, 175, 177
　animal husbandry, 166
　CO_2/O_2 balance, 165
　degradation, 166
　ecological function, 165
　function, 165
　life function, 166
　livelihood function, 166
Alpine grasslands, 187
Alpine grazing system, *see* Grazing exclusion; Optimized grazing management; Qinghai-Tibetan Plateau (QTP)
Alternative Energy Promotion Center (AEPC), 79
Animal distribution, 112
Animal husbandry, 80
Anthropogenic emissions, 146
Anti-poverty programs, 318
Artificial grassland
　alpine meadow and alpine grassland ecosystem, 216
　carbon cycle research, 216
　carbon sequestration, 216
　ecological environment, 215
　economic benefits, 322, 323
　global carbon cycle, 216
　legume grass pasture and gramineous pasture, 215
　local plant seeds, 321
　perennial and annual, 215
　soil carbon content, 322
　Tibetan plateau, 215
　unicast, 215
Asian Water Tower, 28
Australian TERN program, 18

B

Bali roadmap, 4
Bhutanese Himalayas
  fuel use pattern
    biogas, 82
    biomass, 82
  land use pattern and carbon management
    CBFs, 81
    environmental protection, 80
    forest resources, 80
    rangeland, grassland, pasture and pastoralism, 81
    shifting cultivation, 80, 81
  physiography and indigenous people, 79
Biochar
  alpine grasslands, 188
  application, 191, 192
  biomass, 189
  domestic yaks, 189
  economic and ecological services, 187
  environmental and socio-economic factors, 188
  grasslands ecosystem, 186
  harsh environment, 187
  mountain grasslands, 192
  porous structure, 191
  production, 190
  properties, 192
  pyrolysis units, 193
  QTP, 187, 189
  soil properties, 189
Biochar application, 193
Biodiversity, 29
Biogas Support Program Nepal (BSP-N), 79
Biomass energy, 70, 71, 77
Biome Parameter Look-up Table (BPLUT), 51

C

Carbon and livelihood balance, 202–203
Carbon compensation
  accurate and coordinative- evaluation systems, 11, 12
  carbon sinking benefit, 16, 17
  climate governance, 17
  ecological service and carbon benefit accounting, 14, 15
  field work design, 13
  innovative action, 12, 13
  livelihood benefits, 15
  livelihood improvement, 17
  social-economy strategy, 17
  sustainable development, 16
Carbon cycle, 19
  HKH region (see Hindu Kush-Himalayan (HKH))
Carbon dynamics
  beneficial ecosystem services, 47
  data analysis
    DEM, 52
    MODIS GPP, 50, 51
  ecosystem change, 47
  glaciers, 46
  GPP estimation, 58, 59
  HKH region, 48, 50
  methodology, 47
  MODIS GPP, 48, 57, 58
  physiographic landscapes, 46
  sustainable development, 47
  terrestrial GPP, 47
  variation (see GPP variation)
  vegetation productivity, 47
Carbon emission, 117, 120
  atmosphere, 120
  CH_4 emission, 117
  industrialization, 120
  livestock production emission, 119
  pastoral system, 119
Carbon fixation effect
  ecosystem plants, 219
  factors, 219
  grazing and enclosure, 219
  human management, 219
  non-degraded grassland, 219
Carbon-fixation-livelihood, 10
Carbon-livelihood equivalent model, 15
Carbon-livelihood strategies, 316
Carbon management
  agropastoral activities, 273, 274
  collaborative network and feedback mechanism, 18, 19
  decision-making, 268
  household decision making, 277
  indigenous knowledge system, 267
  labor contribution, 276
    emission, 277
    households, 276
    input, 275
    livestock dung, 276
    output, 275
  long-time monitoring and evaluating plots system, 18
  methodology
    Gatlang VDC, 269
    purposive sampling method, 270
  rangelands, 268
  repeatable evaluation techniques, 18

Index

transhumance pastoralism, 268, 270
women's role, 271, 273
Carbon management campaign, 316, 318, 330, 332
Carbon payment schemes, 309
Carbon pool, 32
   afforestation activity, 7
   carbon budget, 5
   forest ecosystem, 5
   Garhwal Himalaya, 7
   land types, 7
   land utilization, 7
   Quercus semecarpifolia, 6
   shrubland and grassland ecoregions, 6
   topographical change, 6
   vegetation coverage, 7
Carbon sequestration, 66, 282
   assessments, 305
   community-based land management projects, 305
   community-oriented, 304
   correlation, 218
   ecosystem services, 309
   grassland management, 218
   grass planting and grassland vegetation, 217
   land management practices, 309
   Medicago sativa, 217
   plant's growth and the carbon transport, 217
   pro-poor carbon storage, 310
   recovery effect, 217
   soil carbon source and sink function, 218
   storage, 305
   trade-off, 305
Carbon sinks, 138
   afforestation reward, 37
   CH_4 fluxes, 139
   ecological industry, 36
   education, 38
   emission reduction and fixed numerical criteria, 36
   global warming, 138
   infrastructure, 37, 38
   Paris Agreement, 35, 36
   public-private reasonable configuration, 37
   remote areas, 138
   resources, 30
   technical assistance, 38
   terrestrial ecosystems, 138
   wetland sediments, 137
Carbon stock changes
   carbon sink, 32
   carbon source, 32, 33

Carbon storage
   AGB, 307
   below ground carbon, 306
   climate change mitigation, 306
   SOC, 306
Cation exchange capacity (CEC), 192
CBM-CFS3 model, 15
Central Bureau of Statistic (CBS), 269, 276
China Council for International Cooperation on the Environment and Development (CCICED), 256
China's Grassland Law, 201
China's National Climate Change Program (CNCCP), 255
Classical Mountain Nomadism, 111
Climate change
   adaptation policy, 289, 291
   glaciers, 26
   natural and human factors, 26
   pastoralist, 285
   physical stimuli, 292
   third pole, 286
   uncertainties, 291
   yak husbandry, 287
Climate Change Strategy and Action Plan (CCSAP), 253
Climate governance, 4, 26
Climate variability, 283, 285
Climate warming, 286, 306
Combined Mountain Agriculture, 111
Common-pool theory, 304
Community-based forest management (CBFM), 70
Community based forests (CBFs)
   Bhutanese Himalayas, 81
   Indian Himalayas, 69, 70
   Nepalese Himalayas, 76, 77
Community based management (CBM), 304
Community cooperation programs, 326
Community forest program (CFP), 77
Community forestry, 81
   carbon storage/tree density, 325
   definition, 325
   evaluation, 325
   food security, 326
   pathways linking, 326, 327
   REDD+ program, 326
Community forestry management (CFM)
   Nepal
      benefits, 235–236
      CFUGs, 233, 234
      community plan, 234
      conservation and protection, 234
      development strategies, 234

Community forestry management
  (CFM) (*cont.*)
  environmental quality, 235
  forest products, 234
  legal and policy framework, 235
  outcomes, 235
  policies, 235
  private sector, 236–237
  protection and management, 233
Community forest user groups (CFUGs),
  76, 234
Community pasture
  community cooperation programs, 326
  cooperation model, 327
  culture and industries, 328
  ecological compensation, 327
  ELSBs, 328, 329
  natural resource management
    system, 328
Conservation islands, 330
Conservation Reserve Program (CRP), 66
Cooperation model, 327
Cordyceps
  business, 287, 288
  cultural tourism, 288
  growing grassland, 287
  harvesting, 288
  soft gold, 287
Cordyceps sinensis, 283, 284
Cropping-forage rotation, 324–325
Culture-cognitive elements, 100

D
Daily milk yield, 155
Data envelopment analysis model, 15
Decomposition, 137
Deenbandhu, 72
Deforestations, 10, 27, 35, 37, 67, 71
Degraded grassland reclaim
  artificial-grassland, 321
  ecological function, 321
  land degradation, 321
Detached Mountain Pastoralism, 111
Digital elevation model (DEM), 52
Dry matter (DM), 116
Dung, 154
  production, 158

E
Earth observation (EO), 307
Ecological compensation, 12, 14, 16, 18, 323, 327
  carbon benefits, 323
  effective practice, 316
  marketing standard, 327
  periodic joint scientific surveys, 329
Ecological, livelihood, and social benefits
  (ELSBs), 328
*Ecological Resettlement, Turning Pastureland
  into Grassland, and Nomadic
  Settlement*, 119
Ecological services function, 165, 167–170, 179
Ecological settlement, 290
Ecological supermarkets, 126
Economic prosperity
  water and poverty, 249–250
  water exploitation and use, 250–251
Ecosystem services, 29, 30
  challenges, 309
  cost-benefit analyses, 309
  optimization, 309
  regions sub-sampling, 309

F
Farms carrying capacity, 174
Feed biomass, 113
Forest and Nature Conservation Rules
  (FNCR), 81
Forest deforestation, China Himalayas, 8
Forest Reference Emission Levels (FRELs), 232
Functional subarea model, 178
Function of ecology, production and livelihood
  (FEPL)
  alpine grassland, 168
  alpine grassland ecosystem, 177
  animal husbandry population, 175
  calculation method, 172
  carrying capacity, 172
  Chinese government, 171
  climate change, 175, 176, 180
  ecological and production functions, 170
  ecological function, 168
  ecological service function, 167
  ecological services value, 178
  grassland area, 172
  grassland ecosystem, 164
  human and ecological environments, 167
  interaction mechanism, 170, 171
  livestock carrying capacity, 174
  production function, 167
  proportional relationship, 173, 174
  proportional structure model, 171
  proportions, 170
  structure variation, 169
  sustainable development, 176, 178–179
  theoretical carrying capacity, 173
  theory, 177
  types, 171

G

Glaciations, 302
Glaciers
    catchment, 301, 302
    climate warming, 301
    glaciation, 302
    global climate models, 301
    hydrological impact, 302
    retreating, 302
    third pole, 301
Global livestock production systems, 110
Global Modelling and Assimilation Office (GMAO), 51
Global warming, 138, 164
Global warming potential (GWP), 71
God grass, 283
Goths of *Kharkas*, 270
GPP variation
    Afghanistan, 55
    Bangladesh, 56
    Bhutan, 56
    India, 56
    Myanmar, 56
    Nepal, 56
    Pakistan, 56
    QTP in China, 52, 55
    spatio-temporal variation, 52, 55, 59
Grassland contract system, 92
Grassland degradation, 188
    climatic changes and overgrazing, 201–202
    herdsmen's work incentives and increased productivity, 201
    HRS, 201
Grassland ecosystem, 30, 176
Grassland ecosystem management, 10
Grassland management patterns, 91, 92, 94, 98
Grassland resources, 178
Grasslands, 186
Grazing exclusion
    simple and effective, 202
    soil carbon pools and carbon fluxes, 202
    temperature and moisture, 202
Greenhouse gas (GHG) emissions, 66, 137
Gross primary productivity (GPP), 47

H

Harmonizing conservation and development along the Silk Road, 329
Herder livelihood, 220–221
Herder perception implications
    adaptive strategies, 293
    AGB, 292
    dwindling inventories, 294
    land fragmentation, 294
    livestock, 293
    pastoralism, 295
    resettlement and re-location, 294
    yak husbandry, 295
Himalaya Hindu Kush (HKH) region
    AGB storage, 307–308
    carbon sequestration, 305
    carbon storage, 306–307
    glaciers, 301–302
    LULCCs, 302–305
    water towers, 300
Himalayan Climate Change Adaptation Program (HICAP)
    Afghanistan
        CCAP, 253
    CCSAP
        Bangladesh, 254
    China
        CCICED, 256
        CNCCP, 255
        NCCCC, 256
        NDRC, 255
    India
        the National Water Mission, 257
        NEP, 257
        PMCCC, 256
    NAPA
        Afghanistan, 253
        Bangladesh, 254
        Bhutan, 255
        Myanmar, 257
    NCCP
        Nepal, 258
        Pakistan, 259
    Nepal
        LAPA, 258
        NAP, 258
        NAPA, 258
Himalayan region
    alpine forest landscape, Nepal, 6
    river basins, 28, 50
Himalayan wetlands, 132
Hindu Kush-Himalayan (HKH), 4, 5, 127, 146
    above-ground and under-ground biomass, 212
    advantage and disadvantage, 40, 41
    anti-poverty, 39
    area and population, countries of, 49
    artificial grassland, 213 (*see also* Artificial grassland)
    degradation, 222–223
    development direction, 223–225
    development status, 214–216

Hindu Kush-Himalayan (HKH) (*cont.*)
  herdsmen's ideas and land management patterns, 221–222
  people's livelihood, 220–221
 biodiversity, 29, 41
 carbon compensation, 40 (*see also* Carbon compensation)
 carbon emission, 7, 8
 carbon fixation effect (*see* Carbon fixation effect)
 carbon management, 4 (*see also* Carbon management)
 carbon pool and economy, 32 (*see also* Carbon pool)
 carbon sequestration (*see* Carbon sequestration)
 carbon sinks (*see* Carbon sinks)
 carbon stock (*see* Carbon stock changes)
 carbon trade, 39, 40
 China's grassland ecosystem, 212
 climate change, 41
 climate change problem, 9
 cultural diversity, 41
 driving force for change
  agriculture, 35
  industry, 34
  land usage, 33
 ecosystem services, 29, 30
 elevations, range of, 48
 emission indicators, 34
 global carbon cycle, 212
 global climate governance, 19
 global industrialization process, 4
 grassland husbandry and structural schema, 214
 land use types, 49
 livelihood and ecological problem, 9, 10
 livelihoods, 27
 natural environment, 212
 natural resources, 27, 28
 people and resources, 31
 people's livelihood, 212
 poverty reduction, 262
 regional thinking, 262
 social economy and resource, 30, 31
 socio-economy, 247
 territorial emissions in Mt CO2, 33
 trans-boundary issues, 252
 water-energy-food sectors, 251
 water resources, 247–249
HKH mountain system, 29
Household Responsibility System (HRS), 201
Human appropriation of net primary production (HANPP), 116
Human-grass-animal-ecology, 166
Hydropower, 30

I
Important plant areas (IPAs), 29
Improved cooking stoves (ICS), 78
Indian Himalayan region (IHR), 65
Indian Himalayas
 fuel use pattern
  biogas, 71, 72
  biomass, 70, 71
 land use pattern and carbon management
  agricultural activities, 66
  CBFs, 69, 70
  grassland, rangeland, pasture and pastoralism, 68, 69
  IHR, 66
  shifting cultivation, 66, 67
 physiography and indigenous people, 65
Indigenous knowledge system, 268
 description, 267
 Himalayan ecosystem, 268
 rangeland management, 273
 rotational grazing, 272
Indigenous people
 age-old livelihood styles, 64
 Bhutan, 79
 Indian Himalayas, 65
 Nepal, 72
Indus River System (IRS), 259
Industrial feedlot yak production, 120
Industrialization, 119
Intended Nationally Determined Contribution (INDC), 35
Intergovernment Panel on Climate Change (IPCC), 112
International Centre for Integrated Mountain Development (ICIMOD), 18
International Vegetation Classification (IVC), 115
Irrigation, 301

J
Janata, 72
Jhum cultivation, 66, 67
Joint Forest Management (JFM), 70

K
Khadi Village Industries Commission (KVIC), 72
Kharka, 74, 75
Kyoto Protocol, 4

Index

L
Lactation period, 157
Land degradation, 35, 302, 303
Land use/land cover (LULC), 47
Land use/land cover changes (LULCCs)
    carbon sequestration, 304
    CBM, 304
    common-pool theory, 304
    interlocking systems, 302, 303
    knowledge and evidence gaps, 304
    land degradation, 303
    land tenure systems, 304
    SLM, 303
Late-Holocene period, 112
Leaf area index (LAI), 47
Length of growing period (LGP), 110–111
Light use efficiency (LUE), 50
Liquefied petroleum gas (LPG), 71, 276, 277
Livelihood benefits
    community forestry, 325–326
    community pastures, 326–329
    degraded grassland reclaim, 321–323
    livestock-chicken grazing system, 319–321
    Tibetan plateau, 323–325
    transboundary protection, 329–330
Livelihood development
    carbon sinking benefits, 316
    C management program, 318
    gender equality, 317
    grassland-livestock systems, 328
    multi-system approach, 324
    traditional customs, 318
    transboundary protection, 329
Livelihoods
    activities and survival strategies, 285
    cash-based, 286
    coping strategies, 283
    C. sinensis, 283
    production, 292
    structural adjustments, 285
    Tibetan herder communities, 283
    yak husbandry, 285, 288
Livelihood security, 64
Livestock carbon management
    archaeological and genetic evidence, 112
    biogeochemical processes, 110
    carbon cycle, 120
    carbon emissions, 117
    CH_4 emission, 117, 118
    environmental and social outcomes, 121
    FAO sub-national livestock databases, 113
    feed biomass, 113
    grasslands, 116
    HANPP, 116
    highland Asia, 110
    HKH region, 111
    IVC, 115
    LG system, 111
    material flow, 110
    NPP, 116
    pastoralism, 112
    production ratios, 112
    programs, 111
    species, 113
    terrestrial surface, 110
Livestock-chicken grazing system
    carbon balance, 319
    ecological theory, 319
    pastures, 319
    Qilian mountain, 319
Local Adaptation Plans for Action (LAPA), 258
Local ecological knowledge (LEK), 282–283, 292
LUEmax, 58

M
Mauri model, 93
Methane, 138
Millennium Ecosystem Assessment (MEA), 131
Mixed rainfed systems (MR), 119
MOD17 algorithm, 58
Moderate resolution imaging spectroradiometer (MODIS), 307
MODIS GPP, 48, 50, 51, 57–59
Monitoring, Reporting and Verification (MRV), 231–232
Monteith's theory, 48
Multi-household grazing management pattern (MMP), 92, 98

N
NASA Earth Observing System (EOS), 48
National Adaptation Plan (NAP), 258
National Adaptation Program of Action (NAPA), 253
National Climate Change Policy (NCCP), 258
National Coordination Committee on Climate Change (NCCCC), 256
National Development and Reform Commission (NDRC), 255
The National Environment Policy (NEP), 257
National Geomatics Center of China, 52
National Sanjianyuan Nature Reserve Project, 148
National Water Plan (NWP), 260
Natural resources, 27, 28

Nepalese Himalayas
 fuel use pattern
  biogas, 79
  biomass, 77, 78
 land use pattern and carbon management
  CBFs, 76, 77
  rangeland, grassland, pasture and pastoralism, 74, 75
  shifting cultivation, 73, 74
 physiography and indigenous people, 72
Net primary productivity (NPP), 47

O
Optimized grazing management
 experiment, 204, 207
 intensity and livestock weight gain, 204
 livestock production, 204
 liveweight gain and grazing intensity, 203
 productivity and plant growth ratio, 205
 stability and resilience, 205
 turnover and supplementary feeding, 205
 warm-season pasture and cold-season pasture, 205, 207
Overgrazing, 188

P
Paris Agreement, 4, 9, 27, 35, 36
Paris Agreement's formwork, 330
Parma, 274
Pastoralism, 64, 68
Pastoral transhumance system, 80
Patans, 74
Periodic joint scientific surveys, 329
Photosynthetically active radiation (PAR), 48
Policies, carbon management
 anti-poverty programs, 318
 carbon trade system, 317
 livelihood development, 317
 local governments, 317
 traditional and alternative knowledge, 318
Poverty
 alleviation mechanism, 309
 ecosystem services, 309
 land management practice, 305
 land use, 304
 resource degradation, 300
Poverty reduction
 economic prosperity, 249–251
 socio-economy, 247
Prime Minister's Council on Climate Change (PMCCC), 256

Private sector
 hurdles, 237
 opportunities of employment, 236
 Plantec Coffee Estate Pvt. Ltd, 236
 profit maximization, 236
Production function, 167
Pro-poor carbon storage, 310

Q
Qinghai-Tibetan Plateau (QTP), 32, 51, 52, 55, 91, 146, 152, 186
 adaptation and mitigation, 291–292
 China's grassland regions, 286, 287
 climate change, 282
  plant diversity, 286
  warming, 286
 daily sustenance, 198
 definition, 282
 grassland contract policy, 92
 grassland property rights and ownership, 200–201
 grazing systems, 200
 herder perception implications, 292–295
 husbandry development and ecological protection, 208
 Kunlun Mountain and Himalayas, 197
 livelihoods, 282
 livestock grazing, 201
 livestock species, 199–200
 Ningxia Hui Autonomous Region, 94
 precipitation change, 287
 sustainable development, 103, 104
 TAR, 283
 Three-River-Source region, 205–208
 yak husbandry, 285–286
Qinghai-Tibet Plateau grassland ecosystem, 177

R
Ramsar classification system, 133
Rangelands
 carbon sequestration, 268
 fuelwood, 275, 276
 Gatlang village, 271
 Nepal Himalaya, 268
 pastoralists, 273
Reducing Emissions from Deforestation and Degradation (REDD+), 77
 biodiversity considerations and livelihood goals, 239
 boost livelihood, 239
 carbon sinking, 316

Index

cases and problem solutions, 316
climate change effects, 325
communities and government, 237
community forestry (*see* Community forestry management (CFM))
conflicts, 238
mitigation strategy, 239
negative side, 238
pathways, 326
payments, 238
planning and implementation, 238
policy framework, 237
positive and negative effects, 237
positive side, 238
results-based payment, 231, 232
Remote sensing (RS), 47
Resettlement Project in High Pastures, 111
Results-based payment
conditions, 231
emission reduction, 231
identifying and measuring, 232
integrating and safeguarding, 232
Nepal, 232–233
Paris Agreement, 231
three-phased approach, 231
timing and sequencing, 232
win-win approach, 231
Rural rapid appraisals (PRA), 283

S

Sanjiangyuan Nature Reserve, 40
Sedentarization, 147, 148, 285, 290
Single-household grazing management pattern (SMP), 92, 98
Social-carbon-metrology system, 316
Social-carbon model, 17
Social-ecological system
MMP and SMP
grazing reasons, 101, 102
institutional reasons, 100, 101
social-economic system, 93, 94
soil system (*see* Soil system)
vegetation system, 94, 95
Soft gold, 287
Soil organic carbon (SOC), 67, 95, 98, 188, 300
Soil organic matter (SOM), 190, 306
Soil pH, 95
Soil quality improvement, 120
Soil respiration, 67
Soil system
animal husbandry, 96
environmental factors, 96

grassland contract policy, 96
grazing management pattern, 95
Nagchu County, 96
QTP, 95
sampling strategy, 96
seasonal grasslands, 96
soil carbon loss, SMP, 98–100
soil properties, MMP and SMP, 97
Soil total nitrogen (STN), 95, 98
Soil total phosphorus (STP), 95, 98
Solar-induced chlorophyll fluorescence (SIF), 58
Southern Himalayas
apathy and counter-productive government policies, 64
Bhutanese context (*see* Bhutanese Himalayas)
categories
fossil fuel usage, 64
land use and management, 63–64
climate change, 63
climatic characteristics, 64
Indian context (*see* Indian Himalayas)
Indo-Gangetic Plain, 63
and livelihood, 65
Nepalese context (*see* Nepalese Himalayas)
shifting cultivation, 64
Stone Age, 149
Stratification-model, 12, 13
Subsistence-level crop, 80
Sundarbans mangroves, 139
Sustainable development, 9
cold-season grazing, 103
ecological compensation, 104
livelihood adaption strategies, 103
livestock, 104
non-market method, 104
overgrazing, 103
privatization, 103
regional and global biogeochemical cycles, 104
social-economic framework, 104
warm-season grazing, 103
Sustainable land management (SLM), 300, 303

T

Tamang ethnic group, 270
Third pole, 286, 301
Three-River Headwaters region (TRHR), 57
Three zones coupled system, 324
Thriving community forest, eastern Nepal, 76

Tibetan Autonomous Region (TAR), 283
Tibetan ecosystems, 193
Tibetan grasslands, 188
Tibetan pastures, 187
Tibetan plateau, 64
  alpine grassland ecosystem, 164, 165, 170
  animal husbandry, 163, 166
  cropping-forage rotation, 325
  ecological compensation, 324
  ecological deterioration—economic poverty, 164
  environmental stresses, 175
  FEPL, 170, 178–179
  integrated model, 324
  Lhasa valley and northern Tibet, 324, 325
  livestocks, 323
  marketing and commerce mechanism, 323
  multiple system coupling, 323
Timber products, 11
Training and demonstration, 327
Transboundary protection
  conservation islands, 330
  harmonizing conservation and development along the Silk Road, 329
  HKH region, 329–331
  networks, 330
  periodic joint scientific surveys, 329
  transboundary-protecting-alliance, 329
Transhumance pastoralism
  chauri and sheep, 271, 272
  definition, 268
  Gatlang village settlement, 271
  goth, 270
  indigenous knowledge system, 272
  livestock production, 273
  Tamang ethnic groups, 270
Transhumant pastoralism, 75

U
Understandings
  AGB, 307
  carbon storage, 304, 306
  cost-benefit analyses, 309
  institutional multiplexity, 304
  land use and management systems, 305
The United Nations convention on climate change, 4
USGS Earth Resources Observation and Science (EROS), 52

V
Village Development Committee (VDC), 269
Volatile organic compounds (VOCs), 193

W
Water institution and policy, 261, 262
Water Resources Strategy (WRS), 260
Water security
  basin/sub-basin policies, 260
  flood risk management, 259
  ICIMOD, 262
  isolated projects, 260
  IWRM, 261, 262
  legal and institutional arrangements, 262
  The National Water Policy, 260
  policies and programs, 261
  Water Law, 260
Water tower of plateau, 96
Wetlands
  abrasive and threatening, 126
  aerobic processes, 138
  biogeochemical processes, 137
  and carbon, 137
  characteristics, 128
  CH_4 emission, 139
  CH_4 fluctuations, 138
  CO_2 gas, 136
  convention, 126
  decomposition and movement, 137
  energy flow, 135
  energy-rich organic matter, 136
  estuarine/coastal, 128
  factors, 135
  fast flowing streams, 136
  freshwater and saltwater, 128
  function, 131, 138
  GHGs, 138
  glacial lakes, 131
  Hindu Kush-Himalaya, 128–131, 138
  human activities and climate change, 134
  integrated ecosystem, 126
  methane emission, 140–141
  photosynthesis, 137
  population, 135
  prolific habitats, 126
  seas and lakes, 126
  types, 132, 136
  value and services, 134

Index

Y
Yak dung, 189
Yak husbandry
   Cordyceps, 287–289
   ecological settlement, 290
   gender roles, 289, 290
   livelihoods, 285
   livestock species, 285
   pastoralism, 285
   sedentarization, 285, 290
Yaks
adaptation, 149
bacteria with probiotic activity, 154
biomass carbon stock, 146
cattle, 151
Chinese government, 147
CO_2 concentration, 146
communal grazing lands, 147
curds are dried, 155
entrance, Tibetan house, 150
factors, 146
family members, 148
fatty acids, 156
government, 148
importance, 149
land carbon, 146
livelihoods, 149
livestock, 147
local herds, 148
milk, 151
milk products, 154
mobility, 147
national natural preservation zones, 148
peak biomass, 152
population, 148
skimmed milk, 155
stacking dry yak dung, 153
statue, 150
summer and autumn, 152
woman milking, 153